AF388920

Primary Immunodeficiencies: Pathogenetic Advances, Diagnostic and Management Challenges

Primary Immunodeficiencies: Pathogenetic Advances, Diagnostic and Management Challenges

Editors

Rita Consolini
Giorgio Costagliola

MDPI • Basel • Beijing • Wuhan • Barcelona • Belgrade • Manchester • Tokyo • Cluj • Tianjin

Editors
Rita Consolini
University of Pisa
Pisa, Italy

Giorgio Costagliola
University of Pisa
Pisa, Italy

Editorial Office
MDPI
St. Alban-Anlage 66
4052 Basel, Switzerland

This is a reprint of articles from the Special Issue published online in the open access journal *Journal of Clinical Medicine* (ISSN 2077-0383) (available at: https://www.mdpi.com/journal/jcm/special_issues/Primary_Immunodeficiencies).

For citation purposes, cite each article independently as indicated on the article page online and as indicated below:

LastName, A.A.; LastName, B.B.; LastName, C.C. Article Title. *Journal Name* **Year**, *Volume Number*, Page Range.

ISBN 978-3-0365-8308-2 (Hbk)
ISBN 978-3-0365-8309-9 (PDF)

Contents

Journal of
Clinical Medicine

Editorial

Primary Immunodeficiencies: Pathogenetic Advances, Diagnostic and Management Challenges

Giorgio Costagliola and Rita Consolini *

Section of Clinical and Laboratory Immunology, Division of Pediatrics, Department of Clinical and Experimental Medicine, University of Pisa, 56126 Pisa, Italy; giorgio.costagliola@hotmail.com
* Correspondence: rita.consolini@unipi.it; Tel.: +39-050-992100

The field of immunology is rapidly progressing, with new monogenic disorders being discovered every year. The heterogeneity of clinical manifestations and the genetic background of immunodeficiencies brought about the new definition of inborn errors of immunity (IEI), which was adopted by the International Union of Immunological Societies (IUIS) in 2019. This term reflects a considerable change in the viewpoint of immunologists, with a deeper recognition of the non-infectious manifestations of IEI and their atypical presentations. In this current intriguing context, this Special Issue offers an overview of some of the most updated concepts in immunology, ranging from the discussion of some peculiar aspects of well-known entities to the presentation of recently discovered diseases. The Special Issue includes six original research papers and seven review papers submitted from different countries.

As the first contribution to this Special Issue, we provided a review on the autoimmune manifestations of IEI, with a specific focus on the various molecular mechanisms involved in autoimmunity and potential targeted therapeutic strategies [1]. This work analyzes the most common autoimmune manifestations in patients with antibody deficiencies, combined immunodeficiencies, and immune dysregulation disorders, and also introduces some specific monogenic entities that are used as a paradigm of "druggable" IEI.

Following this, other interesting elements can be derived from the review paper by Pieniawska-Śmiech et al., in which some of the most relevant non-infectious presentations of IEI are discussed [2]. Specifically, the authors focus on the role of allergic manifestations, autoimmunity, lymphoproliferation, and malignancies as the first sign of IEI, deeply discussing the role of immune dysregulation.

The atypical presentation of IEI is also the main focus of the paper by Morawska et al., which explores the spectrum of atopic manifestations in patients with selective IgA deficiency (sIgAD). This review gives specific attention to the epidemiological and clinical features of the atopic diseases associated with sIgAD and discusses the most relevant diagnostic aspects [3]. Similarly, selective IgE deficiency is comprehensively discussed in the work by Picado et al. [4], which presents the spectrum of infectious, allergic, autoimmune, and neoplastic manifestations in a large cohort of patients diagnosed with this largely unknown condition.

The other papers in this Special Issue have a major focus on the molecular mechanisms responsible for IEI. Concerning this, the review paper by Romano et al. discusses the role of epigenetic alterations associated with IEI [5]. This paper offers an overview of the epigenetic mechanisms implicated in the regulation of the immune response and the most relevant known epigenetic alterations in IEI.

The review paper by Mertowska et al. [6] deeply discusses the molecular structure and function of the Foxp3 transcription factor and its role in the immune response and in the development of IEI. The work has a specific focus on the pathogenesis of IPEX syndrome, but also presents the intriguing role of FOXP3 in common variable immunodeficiency (CVID).

Citation: Costagliola, G.; Consolini, R. Primary Immunodeficiencies: Pathogenetic Advances, Diagnostic and Management Challenges. *J. Clin. Med.* **2023**, *12*, 4651. https://doi.org/10.3390/jcm12144651

Received: 5 July 2023
Accepted: 10 July 2023
Published: 13 July 2023

The paper by Votto et al. [7] deals with the role of gastrointestinal eosinophilic manifestations in patients with IEI. As these manifestations are still underdiagnosed, the article offers an interesting view on when to suspect gastrointestinal eosinophilic manifestations in individuals diagnosed with IEI, as well as when to suspect an IEI in those presenting with isolated gastrointestinal eosinophilic involvement.

Another paper analyzing a rare and poorly recognized entity is the cohort study by Alberio et al. [8], in which the clinical and laboratory features of patients with the DiGeorge-like clinical phenotype in the absence of the classical 22q11.2 deletion are described. This study evidences some new copy number variants associated with the Di George-like phenotype, strongly suggesting the use of array CGH in patients presenting with this phenotype to better identify the genotype-to-phenotype correlations.

Finally, the wide spectrum of antibody deficiencies is the main focus of four research papers and a review paper. The study by Więsik-Szewczyk et al. [9] analyzes a cohort of adult patients with CVID associated with autoimmunity or isolated infectious manifestations, reporting some peculiarities in the immunophenotype of those affected by autoimmune diseases. Indeed, the authors describe a tendency for lymphopenia, reduced NK cells, and low levels of regulatory T cells and Th17 cells in patients with autoimmunity, thus contributing to the elucidation of the immunological heterogeneity of the disease. Two papers published in this Special Issue were by the research group of Diaz Alberola et al. [10,11]. In the first paper [10], the epidemiologic, clinical, and immunological features of CVID-associated giardiasis are reviewed, evidencing that patients with giardiasis more commonly have reduced IgA levels and lower levels of switched memory B cells. The second paper published by this group [11] is original research describing a patient with CVID associated with a de novo IKZF1 variant, in which a reduced humoral response against the SARS-CoV-2 vaccine was demonstrated in the presence of an adequate T-cell response against the pathogen. Another study investigating the relationship between IEI and SARS-CoV-2 is the original research by Pieniawska-Śmiech et al. [12]. In this study, the incidence of COVID-19, its clinical course, and the anti-SARS-CoV-2 serologic response in a cohort of patients with IEI are analyzed, demonstrating a low rate of severe infections in the study cohort.

Finally, the paper by Sgrulletti et al. [13] focuses on the clinical evolution of pediatric patients with unclassified primary antibody deficiencies, highlighting the need for an appropriate follow-up to promptly identify those who will progress to definite IEI.

To conclude, the present Special Issue represents an overview of the current immunological scenario, and deals with different innovative concepts and clinical and research approaches. Indeed, the expanding availability of immunological and genetic testing offers the opportunity to identify new disease entities and elucidate the function of new genes involved in the development and regulation of the immune response. In this continuously evolving field, both researchers and clinicians need to be constantly updated on the most relevant innovations, and with this Special Issue we hope to have contributed to this extremely relevant topic.

Funding: This research received no external funding.

Conflicts of Interest: The authors declare no conflict of interest.

References

1. Costagliola, G.; Cappelli, S.; Consolini, R. Autoimmunity in Primary Immunodeficiency Disorders: An Updated Review on Pathogenic and Clinical Implications. *J. Clin. Med.* **2021**, *10*, 4729. [CrossRef] [PubMed]
2. Pieniawska-Śmiech, K.; Pasternak, G.; Lewandowicz-Uszyńska, A.; Jutel, M. Diagnostic Challenges in Patients with Inborn Errors of Immunity with Different Manifestations of Immune Dysregulation. *J. Clin. Med.* **2022**, *11*, 4220. [CrossRef] [PubMed]
3. Morawska, I.; Kurkowska, S.; Bębnowska, D.; Hrynkiewicz, R.; Becht, R.; Michalski, A.; Piwowarska-Bilska, H.; Birkenfeld, B.; Załuska-Ogryzek, K.; Grywalska, E.; et al. The Epidemiology and Clinical Presentations of Atopic Diseases in Selective IgA Deficiency. *J. Clin. Med.* **2021**, *10*, 3809. [CrossRef] [PubMed]

4. Picado, C.; Ortiz de Landazuri, I.; Vlagea, A.; Bobolea, I.; Arismendi, E.; Amaro, R.; Sellarés, J.; Bartra, J.; Sanmarti, R.; Hernandez-Rodriguez, J.; et al. Spectrum of Disease Manifestations in Patients with Selective Immunoglobulin E Deficiency. *J. Clin. Med.* **2021**, *10*, 4160. [CrossRef] [PubMed]
5. Romano, R.; Cillo, F.; Moracas, C.; Pignata, L.; Nannola, C.; Toriello, E.; De Rosa, A.; Cirillo, E.; Coppola, E.; Giardino, G.; et al. Epigenetic Alterations in Inborn Errors of Immunity. *J. Clin. Med.* **2022**, *11*, 1261. [CrossRef] [PubMed]
6. Mertowska, P.; Mertowski, S.; Podgajna, M.; Grywalska, E. The Importance of the Transcription Factor Foxp3 in the Development of Primary Immunodeficiencies. *J. Clin. Med.* **2022**, *11*, 947. [CrossRef] [PubMed]
7. Votto, M.; Naso, M.; Brambilla, I.; Caimmi, S.; De Filippo, M.; Licari, A.; Marseglia, G.L.; Castagnoli, R. Eosinophilic Gastrointestinal Diseases in Inborn Errors of Immunity. *J. Clin. Med.* **2023**, *12*, 514. [CrossRef] [PubMed]
8. Alberio, A.M.Q.; Legitimo, A.; Bertini, V.; Baroncelli, G.I.; Costagliola, G.; Valetto, A.; Consolini, R. Clinical, Immunological, and Genetic Findings in a Cohort of Patients with the DiGeorge Phenotype without 22q11.2 Deletion. *J. Clin. Med.* **2022**, *11*, 2025. [CrossRef] [PubMed]
9. Więsik-Szewczyk, E.; Rutkowska, E.; Kwiecień, I.; Korzeniowska, M.; Sołdacki, D.; Jahnz-Różyk, K. Patients with Common Variable Immunodeficiency Complicated by Autoimmune Phenomena Have Lymphopenia and Reduced Treg, Th17, and NK Cells. *J. Clin. Med.* **2021**, *10*, 3356. [CrossRef] [PubMed]
10. Díaz-Alberola, I.; Gutiérrez-Bautista, J.F.; Espuch-Oliver, A.; García-Aznar, J.M.; Anderson, P.; Jiménez, P.; Hidalgo-Tenorio, C.; López-Nevot, M.Á. Incidence, Management Experience and Characteristics of Patients with Giardiasis and Common Variable Immunodeficiency. *J. Clin. Med.* **2022**, *11*, 7007. [CrossRef] [PubMed]
11. Díaz-Alberola, I.; Espuch-Oliver, A.; García-Aznar, J.M.; Ganoza-Gallardo, C.; Aguilera-Franco, M.; Sampedro, A.; Jiménez, P.; López-Nevot, M.Á. Common Variable Immunodeficiency Associated with a De Novo IKZF1 Variant and a Low Humoral Immune Response to the SARS-CoV-2 Vaccine. *J. Clin. Med.* **2022**, *11*, 2303. [CrossRef] [PubMed]
12. Pieniawska-Śmiech, K.; Kuraszewicz, A.; Sado, J.; Śmiech, K.; Lewandowicz-Uszyńska, A. Assessment of COVID-19 Incidence and the Ability to Synthesise Anti-SARS-CoV-2 Antibodies of Paediatric Patients with Primary Immunodeficiency. *J. Clin. Med.* **2021**, *10*, 5111. [CrossRef] [PubMed]
13. Sgrulletti, M.; Costagliola, G.; Giardino, G.; Graziani, S.; Del Duca, E.; Di Cesare, S.; Di Matteo, G.; Consolini, R.; Pignata, C.; Moschese, V. The Evolutionary Scenario of Pediatric Unclassified Primary Antibody Deficiency to Adulthood. *J. Clin. Med.* **2023**, *12*, 4206. [CrossRef]

Journal of
Clinical Medicine

Review

Autoimmunity in Primary Immunodeficiency Disorders: An Updated Review on Pathogenic and Clinical Implications

Giorgio Costagliola [†], Susanna Cappelli [†] and Rita Consolini *

Section of Clinical and Laboratory Immunology, Division of Pediatrics, Department of Clinical and Experimental Medicine, University of Pisa, 56126 Pisa, Italy; giorgio.costagliola@hotmail.com (G.C.); susicappelli82@yahoo.it (S.C.)
* Correspondence: rita.consolini@med.unipi.it; Tel.: +39-050-993-889
† Equally contributed to this paper.

Abstract: During the last years, studies investigating the intriguing association between immunodeficiency and autoimmunity led to the discovery of new monogenic disorders, the improvement in the knowledge of the pathogenesis of autoimmunity, and the introduction of targeted treatments. Autoimmunity is observed with particular frequency in patients with primary antibody deficiencies, such as common variable immunodeficiency (CVID) and selective IgA deficiency, but combined immunodeficiency disorders (CIDs) and disorders of innate immunity have also been associated with autoimmunity. Among CIDs, the highest incidence of autoimmunity is described in patients with autoimmune polyendocrine syndrome 1, LRBA, and CTLA-4 deficiency, and in patients with STAT-related disorders. The pathogenesis of autoimmunity in patients with immunodeficiency is far to be fully elucidated. However, altered germ center reactions, impaired central and peripheral lymphocyte negative selection, uncontrolled lymphocyte proliferation, ineffective cytoskeletal function, innate immune defects, and defective clearance of the infectious agents play an important role. In this paper, we review the main immunodeficiencies associated with autoimmunity, focusing on the pathogenic mechanisms responsible for autoimmunity in each condition and on the therapeutic strategies. Moreover, we provide a diagnostic algorithm for the diagnosis of PIDs in patients with autoimmunity.

Keywords: 22q11.2 deletion syndrome; activated phosphoinositide 3-kinase d syndrome; common variable immunodeficiency; complement deficiency; CTLA-4; Immune dysregulation; LRBA; selective IgA deficiency; severe combined immunodeficiency; X-linked agammaglobulinemia

Citation: Costagliola, G.; Cappelli, S.; Consolini, R. Autoimmunity in Primary Immunodeficiency Disorders: An Updated Review on Pathogenic and Clinical Implications. *J. Clin. Med.* **2021**, *10*, 4729. https://doi.org/10.3390/jcm10204729

Academic Editor: Burkhard Möller

Received: 17 September 2021
Accepted: 13 October 2021
Published: 15 October 2021

1. Introduction

In recent years, the association between primary immunodeficiency disorders (PIDs) and autoimmunity has been extensively studied. Patients with PIDs can develop an immune dysregulation of variable degree, which is responsible for a clinical picture featured by infectious complications and autoimmunity [1,2]. Autoimmune manifestations are observed with considerable frequency in patients with primary antibody deficiencies, including common variable immunodeficiency (CVID) and selective IgA deficiency (sIgAD), but can also be evidenced in patients with combined immunodeficiency disorders (CID) [3]. Notably, autoimmunity can represent the presentation sign of PIDs in a significant number of patients [1]. The molecular mechanisms responsible for the immune dysregulation in patients with PIDs are multiple and not completely elucidated; impaired B cell differentiation and germ-center reactions, altered T cell central or peripheral tolerance, uncontrolled lymphocyte proliferation and differentiation, dysfunctional complement, and innate immune activation can participate in the complex pathogenic process leading to autoimmunity. In patients with PIDs, the association with autoimmunity leads to a significant impact on the quality of life, higher medicalization, and increased mortality [2]. Furthermore, the increasing use of new sequencing techniques allowed the identification of different monogenic causes of PID, the better understanding of genotype-phenotype correlations, and the improvement of the therapeutic strategies targeting the immune dysregulation in PIDs [4,5].

In this paper, we review the main autoimmune manifestations observed in patients with PIDs, focusing on the molecular mechanisms implicated in the pathogenesis of the immune imbalance in each condition. Furthermore, we provide some key issues for the diagnostic and therapeutic approach to autoimmunity in patients suffering from PIDs.

2. Autoimmunity in Primary Antibody Deficiency Disorders

Among primary antibody deficiencies, autoimmunity has been described with considerable frequency in patients with CVID, sIgAD, and hyper-IgM syndrome (HIGM), with the clinical phenotype being significantly associated with the variability of the genetic background (Table 1).

Table 1. Autoimmune/inflammatory manifestations in primary antibody deficiencies.

Disease	% of Patients with Autoimmunity	Autoimmune/Inflammatory Manifestations
CVID	20–30%	Autoimmune cytopenias (ITP, AIHA, neutropenia), organ specific autoimmune diseases (e.g., thyroiditis, T1D, ILD, IBD), systemic autoimmune diseases (RA, SLE), lymphoproliferation, lymphoma
Specific genetic associations		
TACI defect		Variable autoimmune manifestations
BAFF-R defect		Variable autoimmune manifestations
ICOS deficiency		Autoimmune cytopenias, enteropathy, RA, SLE
NF-kB1 deficiency		Autoimmune cytopenias, enteropathy, lymphoproliferation, lymphoma
NF-kB2 deficiency		Autoimmunity affecting skin, hair and nails, pituitary hormone deficiencies, autoimmune cytopenias
sIgAD	5–30%	Celiac disease, autoimmune cytopenias (ITP, AIHA), hypothyroidism, Graves' disease, T1D, RA, SLE.
Hyper IgM syndromes XHIM	10–20%	AIHA, ITP, autoimmune hepatitis, T1D, Chron's disease and uveitis, seronegative arthritis, hypothyroidism, SLE, sclerosing cholangitis
AID deficiency	21%	AIHA, ITP, autoimmune hepatitis, T1D, Chron's disease and uveitis, lymphoproliferation
NEMO		IBD, arthritis, AIHA
X-linked agammaglobulinemia	15%	Arthritis, DM, IBD, AIHA, scleroderma, alopecia, T1D, glomerulonephritis

ITP, Immune thrombocytopenic purpura; AIHA, autoimmune hemolytic anemia; T1D, Type 1 Diabetes; ILD, interstitial lung disease; IBD, inflammatory bowel disease; RA, rheumatoid arthritis; SLE, systemic lupus erythematosus; NF-kB1, Nuclear factor kappa-light chain enhancer; LRBA, LPS, responsive beige-like anchor protein; CTLA-4, cytotoxic lymphocyte antigen 4; PI3Kδ, Phosphoinositide 3-kinase δ; STAT3, signal transducer and activator of transcription; sIgA, selective immunoglobulin A deficiency; XHIM, X linked variant of hyper-IgM syndrome; AID, activation induce cytidine deaminase; DM: dermatomyositis.

2.1. Common Variable Immunodeficiency

CVID is characterized by a reduction in serum immunoglobulin (Ig) G and IgA, by at least two standard deviations below age-appropriate reference levels, with or without low

Ig M levels, accompanied by poor antibody response to vaccines or low switched-memory B cells. Clinical diagnosis of CVID requires at least one between increased susceptibility to infections, autoimmune manifestations, granulomatous disease, lymphoproliferation, or an affected family member with antibody deficiency [6]. A total of 30–50% of CVID patients have non-infectious manifestations, including autoimmune, gastrointestinal, pulmonary, lymphoproliferative, and malignant complications [7–9], which strongly contribute to morbidity and mortality [7,10]. Autoimmune diseases occur in 20–30% of CVID patients. Autoimmune cytopenias, and in particular immune thrombocytopenic purpura (ITP), autoimmune hemolytic anemia (AIHA), and Evans syndrome [9,11], are the most commonly reported, but organ-specific and systemic autoimmune diseases are also described. Interestingly, cytopenia may be the first manifestation of the immune defect, in patients without a typical history of infections [12]. Concerning systemic autoimmune diseases, in a recent study of 870 CVID patients from the United States Immunodeficiency Network (USIDNET) registry, 5% were found to have a rheumatologic disease [13]. The most common rheumatologic manifestation reported in CVID is inflammatory arthritis, occurring in about 3% of patients, but systemic lupus erythematosus (SLE), Sjögren's disease, Behçet disease, and psoriasis have also been described [10,13–15]. Among organ-specific autoimmune diseases, in a European Society of Immune Deficiencies (ESID) registry of 2700 CVID patients, hypothyroidism was the most prevalent at 3.5%, followed by alopecia areata, vitiligo, and type I diabetes (T1D) [9]. Among autoimmune manifestations, only cytopenias have been associated with decreased survival and CVID-associated noninfectious complications, including lymphoproliferation, granulomatous disease, lymphoma, hepatic disease, pulmonary involvement (granulomatous lymphocytic interstitial lung disease (GLILD)), and enteropathy [7].

Different studies suggested that both genetic background and immunological abnormalities play a significant role in explaining the link between CVID and autoimmunity. CVID patients can show deregulated immune responses at different levels, involving altered germ center reactions, class switch, and B cell proliferation [11]. Additionally, an impaired suppressive function of Bregs on activated T cells leading to excessive T cell activation has been described [16]. T cell dysfunction is a contributing factor in the development of autoimmunity in CVID [17]. Hyperactivated T cell phenotype [18], reduced number and function of regulatory T cells (Tregs) [19], and an increase in T helper type 1 (TH1), type 17 (TH17), and T helper follicular cells have been observed in CVID patients with autoimmunity [11,20,21]. Moreover, autoimmunity may be caused not only by a break in tolerance to self-antigens but also by the inability of CVID patients to completely eradicate microbial antigens, resulting in compensatory, often exaggerated, and chronic inflammatory responses [2].

Even if many CVID patients may have a polygenic disease, in about 15–30% of CVID cases, a monogenic cause has been found [22]. CVID patients with *TNFRSF13B* mutation (encoding for the B cell-activating factor (BAFF) and APRIL receptor, TACI), especially if heterozygous, have a propensity to autoimmune manifestations and lymphoid hyperplasia potentially due to lack of normal mechanisms required to establish tolerance [23]. *BAFF-R* mutations, which may impair B-cell maturation, have also been described in association with autoimmunity [24,25]. Autoimmunity and other clinical manifestations (including lymphoproliferation) have been associated with the deficiency of *NF-kB1* and *NF-kB2*, which are transcription factors that are crucial for B-cell maturation, survival, differentiation, class switching, and self-tolerance. Additionally, it is described in patients with mutations affecting the inducible T-cell co-stimulator (*ICOS*), a T cell surface receptor that is closely related to NF-kB activation and is essential for terminal B cell differentiation and immune tolerance [25]. Finally, autoimmunity has been described in patients with mutations in other genes implicated in B cell activation and proliferation, including PLCγ2, which is responsible for the PLCγ2-associated antibody deficiency and immune dysregulation (PLAID) [25,26].

2.2. Selective IgA Deficiency

sIgAD is defined, according to ESID and the International Union of Immunological Societies (IUIS), as serum levels of <7 mg/dL in individuals older than 4 years in the presence of normal levels of both IgG and IgM, normal IgG antibody response to vaccinations and exclusion of other causes of hypogammaglobulinemia and T-cell defects [6]. Although most of the patients with sIgAD are asymptomatic, some patients develop various clinical manifestations, such as minor recurrent sinopulmonary infections, allergies, and autoimmune manifestations [27]. A variety of autoimmune diseases may be overrepresented in patients with sIgAD than the normal population and sometimes autoimmunity could be the only clinical manifestation in these patients [27]. The prevalence of autoimmune disorders in patients with sIgAD varies from 5 to 30% [28–80], with celiac disease, ITP, AIHA, autoimmune thyroiditis, T1D, RA, and SLE being the most frequently observed manifestations [28,29,31,32].

Several mechanisms have been suggested in the development of autoimmunity in sIgAD [32], including the association with specific HLA haplotypes (particularly, the haplotype 8.1) [33], T and B cells or cytokine abnormalities, shared genetic susceptibility, or ineffective antigen clearance with molecular mimicry. Concerning immune dysfunction, Tregs deficiency is observed in 64% of the patients [34], and a lower number of CD4 + lymphocytes and switched memory B cells have been described in patients with sIgA [35]. Additionally, it has been observed that sIgAD patients with a lower number of switched memory B cells are more prone to infections and autoimmunity [30]. The monogenic hypothesis suggests that certain monogenic mutations predispose both to the development of sIgAD and autoimmune diseases. Interestingly, similar variants of CTLA4-ICOS have been found in celiac disease, sIgAD, and CVID [36]. Functionally, as IgA protect mucosal barriers from the entry of foreign antigens, in patients with sIgAD, pathogens can easily penetrate the mucosa and through a mechanism of molecular mimicry and cross-reaction with self-antigens might cause the formation of self-reactive antibodies [29,37]. Additionally, the lack of IgA may cause defective removal of immune complexes, thus propagating a state of persistent local and systemic inflammation, which may predispose to the sensitization of immune cells to self-antigen s [29]. Finally, IgA interact with cell receptors (as FcαRI) to downregulate immune pathways and protect against autoimmunity, and this function is impaired in patients with sIgAD [29].

2.3. Hyper IgM Syndrome

The HIGM syndromes are a group of primary immunodeficiency disorders in which defective Ig class switch recombination, with or without defects of somatic hypermutation, leads to deficiency of IgG, IgA, and IgE with preserved or elevated levels of IgM [38]. Among HIGM syndromes, there is a genetic heterogeneity supported by the existence of X-linked, autosomal recessive, and autosomal dominant inheritance. Among X-linked HIGM (XHIM), the most common form is caused by mutations in the gene encoding CD40 ligand (CD40L) a molecule transiently expressed on the surface of activated T cells [39]. Mutations of NEMO/IKKγ genes are implicated in the X-linked Anhidrotic Ectodermal Dysplasia with Immunodeficiency (EDA-ID), a syndrome associated with HIGM. Mutations of NEMO lead to an abnormal expression of multiple enzymes required for antibody switching, such as activation-induced cytidine deaminase (AIG) and uracil DNA glycosylase (UNG), while mutations in IKKγ gene result in blockage of NF-kB release into the nucleus interfering with NF-kB and downstream CD40 signaling [40]. Concerning autosomal recessive HIGM, mutations in the AID and UNG genes result in HIGM syndrome with pure humoral immunodeficiency associated with lymphoid hypertrophy [41,42]. The natural receptor of CD40L is CD40, which is expressed on antigen-presenting cells, including B cells (APCs), dendritic cells, and macrophages. Additionally, mutations in the CD40 gene have been described in patients with HIGM, who present a very similar clinical picture to boys with XHIM [43]. In addition to susceptibility to infections, HIGM patients are prone to develop autoimmune diseases, in particular those with mutations in CD40L, CD40, AID, and NEMO [44]. The prevalence of autoimmune manifestations in X-linked HIGM has been reported to be 10–20% without considering neutropenia, whose etiology is not well understood, with seronegative arthritis, thyroiditis, and SLE being the most commonly observed manifestations [45,46]. Regarding recessive HIGM, autoimmunity is described in about 20% of the patients with AID deficiency, and the manifestations consist of AIHA, ITP, hepatitis, T1D, Chron's disease, and uveitis [47]. The mechanisms responsible for autoimmunity in HIGM are heterogeneous and depend on the genetic background. In the XHIM form, the defective CD40-CD40L-mediated interaction results in the failure of elimination of self-reactive B cells, reduction of Tregs [48], and altered cytokine secretion [49], while in the recessive form, AID deficiency could result in defective regulation of self-reactive B cells [50].

2.4. X-Linked Agammaglobulinemia

X-linked agammaglobulinemia (XLA) is caused by a B lymphocyte differentiation arrest caused by mutations in the BTK gene, and is predominantly featured by recurrent infections (especially with encapsulated bacterial pathogens) caused by antibody deficiency with nearly undetectable levels of peripheral B cells [51]. XLA patients appear to be at an increased risk of developing autoimmune diseases, which can be found in up to 15% of patients [52]. Arthritis is the most frequent autoimmune presentation of XLA patients [52] but dermatomyositis, inflammatory bowel diseases (IBD), AIHA, scleroderma, alopecia, T1D, and glomerulonephritis have also been described [53–55]. Chronic inflammation due to subclinical infections significantly contributes to immune dysregulation in XLA patients. Evidence supports the notion that BKT-dependent, but antibody-independent, mechanisms may be involved in the pathophysiology of autoimmunity in XLA [56]. Excessive stimulation by pathogen molecules of Toll-like receptors (TLRs) may contribute to inducing autoimmunity. Indeed, by certain mutations of BTK and recurrent infections, in XLA patients, overstimulation of TLR9 and its secondary messengers, NF-kB, may occur [57], thus causing enhanced production of autoantibodies from innate B-1 cells [58].

2.5. Therapeutic Approach to Autoimmunity in Primary Antibody Deficiencies

The treatment strategies of autoimmune manifestations in patients with PADS are generally the same as in immune-competent patients and include the use of high dose intravenous immunoglobulins (IVIg) and immunosuppressive agents, such as corticosteroids, methotrexate, and azathioprine (resulting in an increased risk of infections). As a second-line therapy, rituximab appears to be highly effective and relatively safe for the management of severe immune cytopenias [59]. On the other hand, splenectomy is reserved as a last resource in patients who have failed all other therapies [60] and is generally disfavored because of the risk of subsequent infections.

3. Autoimmunity in Severe Combined Immunodeficiency and Related Disorders

The genetic and clinical variability among severe combined immunodeficiency (SCID) is remarkable and in some patients, the phenotype of early-onset severe infections can be associated with autoimmunity [61]. The case of Omenn syndrome (OS) is of particular interest, since in this condition the immune impairment coexists with a marked tendency towards lymphoproliferation and autoimmunity [62]. Children with OS develop severe invasive infections in the first months of life, and commonly show hepatosplenomegaly, diffuse lymphadenopathy, severe eczema, and alopecia [63]. Laboratory testing evidence peripheral eosinophilia, lymphopenia, and reduced serum immunoglobulin levels, associated with the peripheral expansion of self-reactive T cells, which represents the most peculiar aspect of OS [64].

The molecular basis underlying this clinical phenotype relies in most of the patients on a mutation of the recombinase activating genes (RAG) 1 and 2, which are central in the V(D)J recombination during T and B cell development. However, defects in other proteins (such as IL-7Ra, ZAP70, ARTEMIS, AK2, JAK3, and others) can be responsible for OS [61]. The pathogenesis of the autoimmune phenotype is not completely elucidated, but defects in central negative selection secondary to reduced AIRE expression and altered peripheral tolerance are implicated [65]. The prognosis of children with OS is severe [64], with death occurring in the first years of life unless they receive definitive treatment with HSCT [66]. The clinical expression of RAG mutations is not limited to OS. Mutations causing a partial loss of function of RAG cause an extremely variable clinical phenotype, with a wide spectrum of severity and clinical features of combined immunodeficiency, immune dysregulation with autoimmunity (mainly autoimmune cytopenia), and lymphoproliferation [62,67,68].

Among the non-OS phenotypic variants of SCID, patients with ARTEMIS deficiency show an ineffective DNA repair with genomic instability, with a consequent clinical picture of SCID associated with radiosensitivity and immune dysregulation with autoimmunity. Autoimmunity has also been described in a few cases of SCID carrying other molecular defects (i.e., IL-7Ra, ZAP70, ADA, and PNP deficiency) as the result of an altered central negative selection of T-cells [61]. Finally, mutations in the ORAI1 and STIM1 genes, encoding for calcium channels implicated in multiple cell functions (including B-cell receptor (BCR) and T-cell receptor (TCR) signaling), are responsible for a clinical phenotype featured by SCID-like manifestations, autoimmunity, hypotonia, and ectodermal dysplasia [65]. In all these conditions, HSCT represents the only curative therapeutic strategy.

4. Autoimmunity in Disorders of T-Cell Central Tolerance

The immunologic tolerance for T cells is reached through the process of central negative selection of self-reactive T cell progenitors, the peripheral induction of anergy of cells escaping central tolerance, and the action of Tregs [69]. Autoimmune polyendocrine syndrome (APS-1) and 22q11.2 deletion syndrome (22q11.2DS) represent two paradigmatic examples of autoimmunity caused by impaired central negative selection of T cells.

4.1. Autoimmune Polyendocrine Syndrome 1

APS-1 is a rare monogenic disorder with autosomal recessive inheritance caused by mutations altering the function of the AIRE gene, expressed by thymic medullary cells and thymic dendritic cells (DCs) [70]. AIRE promotes the production of a wide range of proteins expressed in other tissues, thus causing their presentation to immature thymocytes and driving the process of central negative selection. Moreover, AIRE induces the production of Tregs, contributing to the elimination of self-reactive T cells [71]. As a consequence, defective AIRE function is associated with an expansion of self-reactive lymphocytes and the production of different specificities of autoantibodies, with high variability among APS-1 patients. Although there are no specific autoantibodies allowing the diagnosis of APS-1, a significant percentage of patients exhibit antibodies directed against cytokines implicated in the immune and inflammatory response, such as interferon (IFN), IL-17, and IL-22, contributing to the immune impairment and dysregulation observed in this disease [71]. The main clinical features of APS-1 are represented by chronic mucocutaneous candidiasis, Addison's disease, and primary hypoparathyroidism, but the phenotypic spectrum comprehends autoimmune enteropathy, hepatitis, pancreatitis, nephritis, and other clinical manifestations [70,71]. Treatment strategies are not uniformed, since they are strongly influenced by the prominent clinical manifestations observed in the individual patient [72]. Indeed, while endocrine complications are treated with hormonal replacement therapy, autoimmune organ involvement often requires the use of steroids, immunosuppressive agents (such as azathioprine), or rituximab [70].

4.2. 22q11.2 Deletion Syndrome

The wide clinical spectrum of 22q11.2DS, also known as DiGeorge syndrome (DGS), comprehends both congenital abnormalities (cardiac malformation, velo/palatal dysfunction, parathyroid insufficiency) and immunological alterations [73]. Children with 22q11.2DS show variable severity of immune impairment, ranging from complete athymia to different degrees of combined immunodeficiency with reduced thymic function (low levels of naïve T cells and recent emigrants T (RTE) cells) and increased risk of autoimmunity, which is observed in about 10% of the patients [74,75]. The autoimmune manifestations more commonly evidenced in 22q11.2DS are ITP, juvenile idiopathic arthritis (JIA), and thyroiditis. Additionally, enteropathy and cutaneous autoimmunity (alopecia, psoriasis, vitiligo) have been described [76].

In 22q11.2DS, the pathogenesis of autoimmunity involves multiple mechanisms. The abnormal thymic environment [77] is associated with reduced expression of AIRE (Figure 1), thus impairing T cell negative selection, and reduced generation of Tregs [65,74,75,78]. Moreover, the ineffective immune response, with consequent persistence of microbial antigens, could lead to the phenomenon of molecular mimicry [2]. Recently, alterations in DCs subpopulations have also been described in 22q11.2DS, with reduced circulating numbers of both myeloid DCs (mDCs) and plasmacytoid DCs [79]. This could contribute to the development of autoimmunity, since pDCs have an important role in maintaining peripheral immune tolerance [79]. Interestingly, while the infectious phenotype is prevalent during early childhood, autoimmunity is commonly observed at a higher age. This partly reflects an evolution of the immunological phenotype of 22q11.2DS patients, with progressive reduction of Tregs and expansion of self-reactive T cells [79,80].

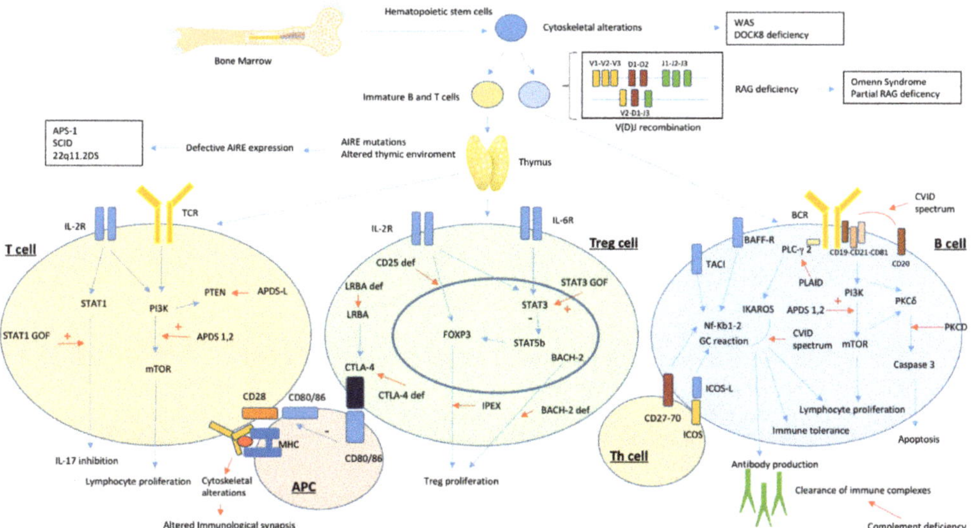

Figure 1. Pathogenesis of autoimmunity in immunodeficiency disorders. Figure legend: 22q11.2DS: chromosome 22q11.2 deletion syndrome; APC: antigen-presenting cells; APDS: Activated phosphoinositide 3-kinase d syndrome; APS-1: Autoimmune polyendocrine syndrome 1; CTLA-4: Cytotoxic lymphocyte antigen 4; CVID: Common variable immunodeficiency; LRBA: LPS-responsive beige-like anchor protein; PI3K: Phosphoinositide 3-kinase; PKCD: protein kinase C δ deficiency; PKCδ: protein kinase C δ; RAG: Recombinase activating genes; SCID: severe combined immunodeficiency; STAT: Signal Transducers and Activator of Transcription; WAS: Wiskott-Aldrich syndrome; XLA: X-linked agammaglobulinemia.

When the immune function is preserved and patients experience a low rate of infections, the use of corticosteroids and conventional immunosuppressive agents represents the initial therapeutic strategy to treat autoimmunity, while the therapeutic approach is more blurred in patients with severe infectious complications [73].

5. Autoimmunity in Disorders of T-Cell Peripheral Tolerance

Among disorders of peripheral tolerance, conditions affecting Treg function (also called "Tregopathies") have a prominent relevance [3]. The most frequent disorder of

Tregs is the immune dysregulation, polyendocrinopathy, enteropathy, X-linked (IPEX) syndrome, which is featured by a classic clinical triad of eczema, enteropathy, and endocrine autoimmunity, without a significant increase in infectious morbidity [81]. Recent genetic advances allowed the identification of different monogenic disorders featured by an IPEX-like clinical phenotype accompanied with increased susceptibility to infections, placing them at a molecular, pathogenic, and clinical crossroad between immunodeficiency and autoimmunity [81,82].

5.1. CTLA-4 Deficiency

Cytotoxic lymphocyte antigen 4 (CTLA-4) is a molecule expressed by Treg cells that has a central role in the induction of peripheral immune tolerance. Indeed, CTLA-4 reduces the expression of CD80 and CD86 on the surface of antigen-presenting cells. As CD80 and CD86 are essential proteins for the costimulatory signal in the immunologic synapsis between APC and T cells, their depletion causes a reduced activation of T cells and differentiation in effector cells [83].

Patients with CTLA-4 deficiency show a picture of combined immunodeficiency with lymphopenia (reduced naïve T cells, Tregs, CD19 cells), hypogammaglobulinemia, and susceptibility to viral and bacterial infections accompanied by a high rate of autoimmune and lymphoproliferative manifestations [2,83]. The autoimmune spectrum observed in CTLA-4 deficiency is variable, comprehending autoimmune cytopenia, arthritis, uveitis, endocrinopathies, and enteropathy. In this condition, lymphoproliferation is observed in about 50% of the patients and presenting with lymphadenopathy, hepatosplenomegaly, and, although in a reduced percentage of patients, with pulmonary involvement, in the form of GLILD [62,84]. In patients with CTLA-4 deficiency, surveillance for the risk of lymphomas is essential, and treatment of autoimmunity and lymphoproliferation comprehends the use of sirolimus and the biologic agent abatacept, a fusion molecule containing the extracellular domain of CTLA-4 [85].

5.2. LRBA Deficiency

LPS-responsive beige-like anchor protein (LRBA) is a protein implicated in intracellular trafficking, which acts by inhibiting the lysosome degradation of CTLA-4 [86]. Therefore, it is essential for maintaining adequate expression of CTLA-4 on the cellular surface, and its deficiency shares several common features with CTLA-4 deficiency. Indeed, patients with LRBA deficiency often present with recurrent sinopulmonary infections, hepatosplenomegaly, lymphadenopathy, and autoimmune cytopenia [87,88]. Among the other autoimmune manifestations observed in this condition, there are enteropathy, endocrinopathies (thyroiditis, T1D), hepatitis, and uveitis [89]. Therefore, patients can present with both an IPEX-like and a CVID-like clinical phenotype, and the immunological assessment commonly shows hypogammaglobulinemia and lymphopenia, with a reduced absolute number of Tregs and memory B cells [87]. Among classic immunosuppressive strategies, there is interest in the role of sirolimus and hydroxychloroquine, as this drug can potentially reduce CTLA-4 degradation. Although it has shown promising results on autoimmune manifestations and immune abnormalities, there are only a few reports of patients treated with abatacept and, similarly, the experience with HSCT in this condition is still limited [85,90,91].

5.3. STAT-Related Disorders

The family of Signal Transducers and Activator of Transcription (STAT) molecules is involved in multiple signaling pathways activated by different cytokines and controls the transcription of genes implicated in the immune and inflammatory response [92]. In particular, the activation of STAT-1 is mostly mediated by IFN-α and IL-2, while STAT-3 is also influenced by IL-6. The molecular mechanism leading to STAT phosphorylation requires the presence of Janus kinase (JAK) molecules [92]. Impaired or enhanced function of the JAK/STAT-dependent molecular pathways can result in a wide spectrum of immunological and clinical alterations, with immune dysregulation and susceptibility to infections being the most relevant features.

In STAT1 gain of function (GOF), patients show reduced proliferation of TH17 cells, causing increased susceptibility to different infections, and typically present with chronic mucocutaneous candidiasis [93]. Moreover, up to a third of the patients develop autoimmune manifestations, that are mainly represented by endocrinopathies, autoimmune cytopenia, and enteropathy [81].

STAT-3 GOF is also featured by an increased risk of severe infections, deriving from a combined immune defect, associated with a high incidence of autoimmunity (cytopenia, enteropathy, endocrinopathy, arthritis). In this condition, patients also frequently display lymphoproliferation with hepato-splenomegaly [94]. The molecular defect underlying this phenotype involves reduced Tregs and Th17 proliferation [95].

Finally, STAT5b deficiency causes impaired IL-2 signaling, with consequently reduced proliferation of Tregs. The disease presents with a picture of combined immunodeficiency, growth hormone insensitivity, and IPEX-like immune dysregulation [81,96].

Although the definitive treatment for STAT-related disorders is currently represented by HSCT, the use of JAK inhibitors has demonstrated promising responses in the management of autoimmunity, infections, and lymphoproliferation in STAT1 GOF and STAT3 GOF [85]. Additionally, given the role of IL-6 in the activation of STAT-3, the anti-IL-6 antibody tocilizumab is a promising alternative for this condition [85,97] (Figure 2).

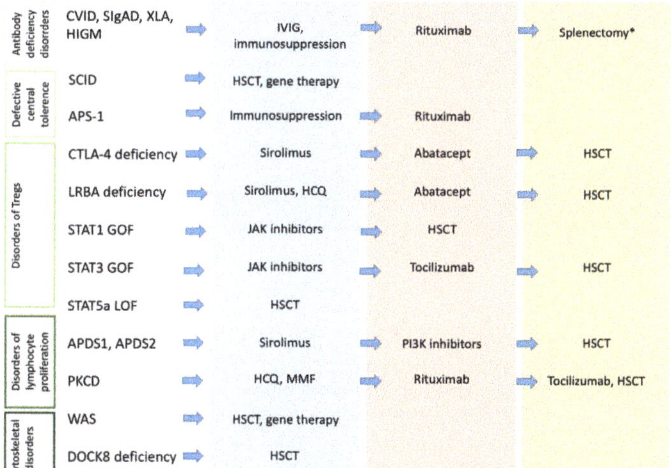

Category	Condition			
Antibody deficiency disorders	CVID, SIgAD, XLA, HIGM	IVIG, immunosuppression	Rituximab	Splenectomy*
Defective central tolerence	SCID	HSCT, gene therapy		
	APS-1	Immunosuppression	Rituximab	
Disorders of Tregs	CTLA-4 deficiency	Sirolimus	Abatacept	HSCT
	LRBA deficiency	Sirolimus, HCQ	Abatacept	HSCT
	STAT1 GOF	JAK inhibitors	HSCT	
	STAT3 GOF	JAK inhibitors	Tocilizumab	HSCT
	STAT5a LOF	HSCT		
Disorders of lymphocyte proliferation	APDS1, APDS2	Sirolimus	PI3K inhibitors	HSCT
	PKCD	HCQ, MMF	Rituximab	Tocilizumab, HSCT
Cytoskeletal disorders	WAS	HSCT, gene therapy		
	DOCK8 deficiency	HSCT		

Figure 2. Therapeutic strategies for autoimmunity in patients with PIDs. The figure shows the current therapeutic options for specific PIDs. The choice of the therapeutic strategy (immunosuppressive agents, biologic drugs, HSCT, gene therapy) depends on the clinical severity, comorbidities and also on the availability and physician's experience. * In patients with refractory autoimmune cytopenia. APDS: Activated phosphoinositide 3-kinase d syndrome; APS-1: Autoimmune polyendocrine syndrome 1; CTLA-4: Cytotoxic lymphocyte antigen 4; CVID: Common variable immunodeficiency; HCQ: Hydroxychloroquine; HIGM: Hyper-IgM syndromes; HSCT: hematopoietic stem cell transplantation; JAK: Janus kinase; LRBA: LPS-responsive beige-like anchor protein; MMF: mycophenolate mofetil; PI3K: Phosphoinositide 3-kinase; PKCD: protein kinase C δ deficiency; SCID: severe combined immunodeficiency; sIgAD: selective IgA deficiency; STAT: Signal Transducers and Activator of Transcription; Tregs: regulatory T cells; WAS: Wiskott-Aldrich syndrome; XLA: X-linked agammaglobulinemia.

5.4. Other Disorders of Regulatory T Cells

Recently, other molecular defects impairing Treg function have been described, each in a reduced number of patients. In CD25 (IL-2RA) deficiency, IL-2 signaling is significantly impaired, thus resulting in an IPEX-like clinical picture associated with infections, and lymphoproliferation [81]. The haploinsufficiency of the BACH2 transcription factor causes altered gem center reactions, reduced Tregs levels, and increased Th1 cell proliferation, finally causing hypogammaglobulinemia, sinopulmonary infections, enteropathy, and lymphoproliferation [98].

6. Autoimmunity in Disorders of Lymphocyte Differentiation and Proliferation

Altered lymphocyte proliferation and differentiation can be responsible for a heterogeneous range of clinical manifestations, ranging from severe infectious diseases to increased susceptibility to autoimmunity and lymphoproliferation. This pathogenic aspect has particular relevance in determining the clinical phenotype of the activated phosphoinositide 3-kinase d syndrome (APDS) and protein kinase C δ deficiency (PKCD).

6.1. Activated Phosphoinositide 3-Kinase d Syndrome

APDS is a combined immunodeficiency disorder associated with an increased risk of sinusitis, respiratory infections, severe herpesvirus infections, and a high rate of autoimmunity and lymphoproliferation. Indeed, autoimmune and lymphoproliferative manifestations (diffuse lymphadenopathy, hepatomegaly, splenomegaly) are the presenting sign in more than half of the patients diagnosed with APDS [99,100]. Autoimmunity occurs in about two-thirds of the patients with APDS, and the clinical expression of autoimmunity consists mostly of the finding of autoimmune cytopenia, arthritis, and enteropathy [101]. The disease is caused by mutations affecting the phosphoinositide 3-kinase (PI3K) molecular complex, which is involved in numerous signaling pathways activated after the binding of TCR and BCR with their ligands and influences cellular metabolism, proliferation, and differentiation of B and T lymphocytes [102] (Figure 1). Patients with GOF mutations in the phosphatidylinositol-4,5-bisphosphate 3-kinase Catalytic Subunit δ (PIK3CD) gene are classified as affected by APDS1, while loss of function mutation in the Phosphoinositide-3-Kinase Regulatory Subunit 1 (PIK3R1) is diagnostic for APDS2 [101]. These two molecular defects are responsible for the uncontrolled activation of the PI3K-dependent molecular pathways, including the intracellular events linked to the activation of mTOR. In APDS, a shift in cellular metabolic and proliferative activity is observed, and patients usually display a peculiar immunological phenotype, featured by high levels of senescent T lymphocytes, effector memory T cells, progressive B lymphocytopenia, and reduced absolute numbers of naïve T cells [101]. The serum levels of immunoglobulin are extremely variable, since patients can present with a CVID-like phenotype, an HIGM picture, or in some cases, hypergammaglobulinemia [99].

A correct diagnosis of APDS is mandatory since it significantly influences the therapeutic approach and the follow-up. Concerning follow-up, the surveillance against the development of lymphoid neoplasms is a central feature, since patients with APDS have an increased risk of developing lymphomas and, particularly, non-Hodgkin's lymphomas [62,101]. Treatment of APDS comprehends the immunoglobulin replacement therapy, when necessary, and the measures to control autoimmunity and lymphoproliferation. To this point, sirolimus, an mTOR inhibitor, is commonly used as a first-line strategy, while selective PI3K inhibitors (leniolisib, nemiralisib) are given in refractory cases [101]. Finally, patients with APDS could benefit from HSCT, although there is no uniform consensus on the timing of the transplantation and the conditioning regimen [103].

6.2. Protein Kinase C δ Deficiency

Mutation impairing the protein kinase C δ (PKCδ) structure or function leads to a clinical phenotype featured by an increased risk of infections with hypogammaglobulinemia, autoimmunity, and lymphoproliferation [104]. PKCδ is implicated in the activation of different transcriptional factors with a central role in the immune homeostasis of B lymphocytes [105]. It is activated in the context of a wide number of molecular pathways, including mTOR, PI3K, and mediates the transcription of STAT1 (leading to enhanced

IFN-stimulated transcription), ERK1, and other factors. PKCδ induces also the transcription of the IL-10 gene and reduces the production of IL-6. Finally, it promotes apoptosis through the interaction with caspase 3, thus limiting lymphocyte proliferation [104,106]. In PKCD, autoimmunity depends on the accumulation of self-reactive lymphocytes, the production of different autoantibodies, and defects in the process of lymphocyte negative selection [105]. The most common manifestations of autoimmunity are represented by arthritis, cytopenia, and SLE-like manifestations, such as glomerulonephritis and photosensitive rash [104]. Moreover, the ineffective control of apoptosis causes lymphoproliferation, and some patients presented with a clinical phenotype mimicking autoimmune lymphoproliferative syndrome (ALPS) [107]. As literature reports only a small number of cases of patients diagnosed with PKCD, the therapeutic approach has yet to be defined. However, available data suggest the use of conventional immunosuppressive strategies approved for SLE, including hydroxychloroquine, mycophenolate mofetil, and rituximab, and, when the lymphoproliferative aspect is prominent, the adoption of anti-mTOR drugs, such as sirolimus [104]. The administration of tocilizumab in patients with high IL-6 serum levels and the use of HSCT represent promising therapeutic strategies [104].

Interestingly, a similar clinical picture can be observed in the RAS-associated autoimmune leukoproliferative disease (RALD), a disease caused by mutations in the RAS signaling pathway (often NRAS or KRAS), which is currently classified among the phenocopies of PIDs. In RALD, uncontrolled activation of RAS-dependent molecular signaling is responsible for the development of hepatosplenomegaly, lymphadenopathy, autoimmune cytopenia, and SLE-like manifestations [108].

7. Autoimmunity in Disorders of Cytoskeletal Function

Cytoskeletal proteins are essential for multiple cellular functions, including the immunologic synapse between T cells and APCs, and the regulation of lymphocyte proliferation.

Wiskott-Aldrich syndrome (WAS) is an inherited condition caused by mutations in the WASP gene on the X chromosome, which encodes the WAS protein, an actin-nucleation promoting factor expressed in hematopoietic stem cells [109]. WAS is clinically featured by a classic triad of thrombocytopenia with small-size platelets, eczema, and lymphopenia, mainly affecting T cells [110]. The molecular defect influences multiple cellular lineages and causes complex implications on the immune function, including ineffective T cell proliferation and function, reduced Treg activity (with preserved absolute Treg values), and hyperproliferation of B cells, which show enhanced production of autoantibodies [110]. The disease severity of WAS patients is variable, and the more commonly reported autoimmune manifestations are AIHA, autoimmune neutropenia, peripheral vasculitis, and arthritis [111]. Although patients with a clinical picture dominated by thrombocytopenia could benefit from splenectomy, this intervention does not reduce the risk of autoimmunity. Consequently, the definitive treatment of patients with WAS is currently represented by HSCT or gene therapy [111].

Another disease featured by altered cytoskeletal structure and function is DOCK8 deficiency, which shares some common clinical features with WAS. In this condition, the molecular defect is responsible for an altered cytoskeletal actin regulation, which causes reduced proliferation, migration, and function of innate and adaptive immune cells, and impaired Treg activity [112].

Clinically, patients with DOCK8 deficiency show a picture of combined immunodeficiency featured by recurrent cutaneous and respiratory infections, eczema, increased risk of malignancies, and predisposition to the development of atopy (with high IgE levels), and autoimmunity (cytopenia, thyroiditis, vasculitis, uveitis). In absence of specific gene therapy, the only curative treatment for DOCK8 deficiency is HSCT [111,112].

8. Autoimmunity in Complement Deficiencies and Disorders of Innate Immunity

Genetic defects causing deficiency of components of the complement cascade are associated with an increased risk of bacterial infections caused by capsulate agents. This derives from the pivotal role of the classic and alternate complement pathways in determining bacterial lysis. However, deficiencies of specific complement proteins (C1q, C1r/s, C2, C4a, C4b) also represent a risk factor for the development of different autoimmune manifestations, such as SLE-like features, glomerulonephritis, JIA, and dermatomyositis [2,113,114]. The pathogenic mechanism linking complement deficiencies with autoimmunity is mostly dependent on a reduced clearance of apoptotic cells and immune complexes. This causes an enhanced availability of auto-antigens, which are chronically exposed to the immune system, thus triggering a self-reactive immune response [114,115]. Moreover, complement deficiency can impair B-cell negative selection, thus allowing the expansion of self-reactive B-lymphocytes, and can contribute to the development of autoimmunity [113]. Since in patients with complement deficiency the immunization against capsulated bacteria significantly improves the outcome, recognizing a patient with this condition and characterizing the specific defect is essential to improve the long-term management, including the anti-infectious prophylaxis.

Concerning disorders of innate immunity, the incidence of autoimmune manifestations is lower compared to PIDs affecting the adaptive response. However, in patients with chronic granulomatous disease (CGD), a higher incidence of SLE-like clinical features, arthritis, and other autoimmune conditions (hepatitis, nephritis) is observed [116]. Although the pathogenesis of these manifestations is not completely elucidated, it is accepted that the process involves the persistence of infectious antigens, inflammasome overactivation, altered production of neutrophil extracellular traps, and defective apoptosis [2,117]. The treatment of autoimmunity in CGD is challenging, since the need to use steroids and immunosuppressive agents should be balanced with the high infectious risk observed in this population. Currently, HSCT and gene therapy are the only curative treatments for CGD [116,117].

9. From Theory to Bedside

The above-discussed pathogenic and clinical associations carry different significant implications for the approach to children with autoimmunity, highlighting that it should represent a warning sign for the presence of a PID, particularly in pediatric age. Identifying a condition of PID in children presenting with autoimmunity offers the opportunity to provide an adequate treatment of the underlying disease (Ig replacement therapy, targeted treatments, HSCT, gene therapy), to offer supplementary immunization if needed, and optimize the management of autoimmunity itself (Figure 2). On the other hand, as different studies have demonstrated a reduced survival in the subgroup of patients with PIDs showing autoimmunity, its finding in patients with already diagnosed PID could lead to significant changes in the follow-up strategy and therapeutic approach [118,119].

9.1. Diagnosing PIDs in Children Presenting with Autoimmunity

The diagnostic approach to a child with autoimmune manifestations should include the detailed analysis of the clinical history to evidence the occurrence of infections, quantify their impact, and point out the associated clinical features suggestive for a PID (growth delay, high frequency of infections, need of hospitalization for infections, prolonged use of antibiotics, infections by unusual pathogens, and others). Additionally, a baseline immunological assessment including the determination of serum Ig levels and lymphocyte subpopulations should be performed in all the children with autoimmune manifestations. The need for other specific investigations (Figure 3) varies depending on the specific autoimmune phenotype. The approach to children with immune cytopenia is of particular interest, since it can represent the presentation sign of a wide spectrum of PIDs, but also the first manifestation of a systemic connective tissue disease, such as SLE [120]. As a consequence, the clinical and laboratory assessment of a child with suspected immune cytopenia should comprehend both the analysis of the immune response and the determination of the most relevant autoantibody subclasses, including antinucleous antibodies and antithyroid antibodies [121]. Interestingly, patients with PID have a 120-fold high risk of developing autoimmune cytopenia compared to the general population, with a higher increase in risk observed for AIHA [122]. The case of children presenting with multiple cytopenias is of significant interest. Indeed, studies on Evans syndromes showed that almost half of the children with this condition have e positive genetic testing for PID and that in this cohort of patients the incidence of systemic autoimmune diseases is also considerable [123,124]. In patients with autoimmune endocrinopathy, the suspect of PID should be posed when the disease onset is earlier than usual, when there is an association of two or more endocrine disorders, and when other signs suggestive for PID are present. In this subset of patients, the finding of eczema, elevated serum IgE levels, and peripheral eosinophilia should induce the suspect of a Treg-mediated disorder [125], while the association with chronic mucocutaneous candidiasis is observed in patients with APS-1 and STAT1 GOF [126]. Finally, in the case of SLE-like manifestations, the serum levels of the complement fractions need to be determined [127]. Specific investigations to allow a definitive diagnosis, including the analysis of the immune response to vaccines (for the clinical diagnosis of CVID), extended determination of the lymphocyte subpopulations (including memory B and T cells), functional analysis, and cytogenetic and genetic testing, should be performed based on the clinical suspect.

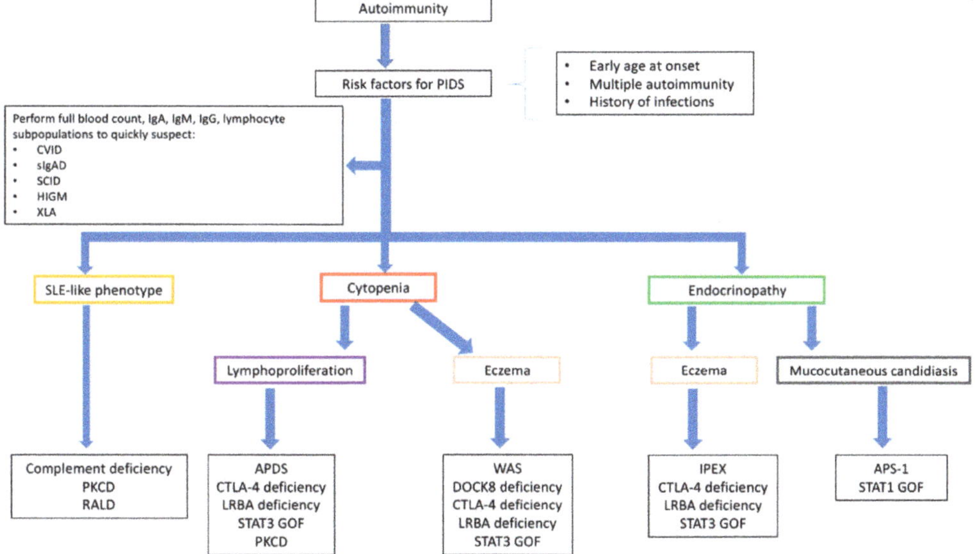

Figure 3. Diagnostic approach to autoimmunity in patients with suspect immunodeficiency. Figure legend: APDS: Activated phosphoinositide 3-kinase d syndrome; APS-1: Autoimmune polyendocrine syndrome 1; CTLA-4: Cytotoxic lymphocyte antigen 4; CVID: Common variable immunodeficiency; HIGM: Hyper-IgM syndrome; LRBA: LPS-responsive beige-like anchor protein; PKCD: protein kinase C δ deficiency; RALD: Ras-associated leukoproliferative disorder; SCID: severe combined immunodeficiency; sIgAD: Selective IgA deficiency; SLE: systemic lupus erythematosus; STAT: Signal Transducers and Activator of Transcription; WAS: Wiskott-Aldrich syndrome; XLA: X-linked agammaglobulinemia.

9.2. Diagnosing Autoimmunity in Children with PIDs

Periodic surveillance for autoimmune manifestations is mandatory in all patients with PIDs. This is particularly relevant for children carrying genetic mutations with a well-defined association with autoimmunity (i.e., CTLA-4, LRBA, PI3K mutations). Moreover, there is increasing interest in the identification of potential immunological predictors of autoimmunity in patients with PIDs. Although there are no specific immunological markers with high predictive value for the development of autoimmunity, literature data from patients with primary antibody deficiencies and CID have identified several potential candidates.

A large cohort study on CVID patients with immune cytopenia demonstrated higher levels of serum immunoglobulin, CD19hi B cells, and T CD4 effector T cells, accompanied by reduced naïve T cells [128]. Absence or reduced switched memory B cells have been associated with autoimmune cytopenias, systemic autoimmune diseases, splenomegaly, granulomatous diseases, and lymphadenopathy [129,130]. An expansion of $CD21^{low}$ B cells has been described in association with reduced Tregs in CVID patients with autoimmunity [131]. It has been found that $CD21^{low}$ B cells produce significantly more IgM than naïve B cells after stimulation with CD40L, IL-2, and IL-10 and that CVID patients with autoimmunity have higher levels of IgM compared with non-autoimmune phenotypes [15,132], thus suggesting that increased IgM levels may be a marker of autoimmunity and they may have a pathogenic role. Additionally, lower naïve CD4+ and CD8+ T cells and increased differentiated T cells have been described in CVID patients with autoimmunity [18].

Concerning combined immunodeficiencies, a recent study by Montin et al. in patients with 22q11.2DS highlighted that some immunological features, including a reduced number of naïve T cells, reduced RTE, and elevation of naïve B cells are associated with the development of hematologic autoimmunity and can be evidenced significantly before the onset of autoimmunity [133]. Moreover, the degree of T-cell lymphopenia has been suggested as a contributing factor for the development of autoimmunity in this disease [75].

10. Conclusions

It is well recognized that autoimmune manifestations are observed in a significant percentage of patients with PIDs, often representing the first sign of these conditions. Patients with early-onset autoimmunity, an association between two or more autoimmune manifestations, or increased susceptibility to infections should be promptly screened for PIDs. Although the intriguing mechanisms underlying the development of autoimmunity in patients with PIDs are far to be completely elucidated, the rapidly evolving knowledge in the genetic background of PIDs will hopefully help to characterize the defects linking immunodeficiency and autoimmunity, thus providing interesting diagnostic and therapeutic implications.

Author Contributions: G.C. and S.C. wrote the initial draft of this paper, which was critically revised by R.C. All the authors contributed to conceptualizing this work. All authors have read and agreed to the published version of the manuscript.

Funding: This research received no funding.

Institutional Review Board Statement: Not applicable.

Informed Consent Statement: Not applicable.

Data Availability Statement: Not applicable.

Conflicts of Interest: The authors declare no conflict of interest.

References

1. Walter, J.E.; Ayala, I.A.; Milojevic, D. Autoimmunity as a continuum in primary immunodeficiency. *Curr. Opin. Pediatr.* **2019**, *31*, 851–862. [CrossRef]
2. Amaya-Uribe, L.; Rojas, M.; Azizi, G.; Anaya, J.M.; Gershwin, M.E. Primary immunodeficiency and autoimmunity: A comprehensive review. *J. Autoimmun.* **2019**, *99*, 52–72. [CrossRef]
3. Allenspach, E.; Torgerson, T.R. Autoimmunity and Primary Immunodeficiency Disorders. *J. Clin. Immunol.* **2016**, *36* (Suppl 1), 57–67. [CrossRef]
4. Notarangelo, L.D.; Uzel, G.; Rao, V.K. Primary immunodeficiencies: Novel genes and unusual presentations. *Hematol. Am. Soc. Hematol. Educ. Program* **2019**, *2019*, 443–448. [CrossRef]
5. Bousfiha, A.; Jeddane, L.; Picard, C.; Ailal, F.; Bobby Gaspar, H.; Al-Herz, W. The 2017 IUIS Phenotypic Classification for Primary Immunodeficiencies. *J. Clin. Immunol.* **2018**, *38*, 129–143. [CrossRef]

6. Seidel, M.G.; Kindle, G.; Gathmann, B.; Quinti, I.; Buckland, M.; van Montfrans, J. The European Society for Immunodeficiencies (ESID) Registry Working Definitions for the Clinical Diagnosis of Inborn Errors of Immunity. *J. Allergy Clin. Immunol. Pract.* **2019**, *7*, 1763–1770. [CrossRef]

7. Chapel, H.; Lucas, M.; Lee, M.; Bjorkander, J.; Webster, D.; Grimbacher, B. Common variable immunodeficiency disorders: Division into distinct clinical phenotypes. *Blood* **2008**, *112*, 277–286. [CrossRef]

8. Gathmann, B.; Mahlaoui, N.; Gérard, L.; Oksenhendler, E.; Warnatz, K.; Schulze, I. Clinical picture and treatment of 2212 patients with common variable immunodeficiency. *J. Allergy Clin. Immunol.* **2014**, *134*, 116–126. [CrossRef]

9. Odnoletkova, I.; Kindle, G.; Quinti, I.; Grimbacher, B.; Knerr, V.; Gathmann, B. The burden of common variable immunodeficiency disorders: A retrospective analysis of the European Society for Immunodeficiency (ESID) registry data. *Orphanet J. Rare Dis.* **2018**, *13*, 201. [CrossRef]

10. Resnick, E.S.; Moshier, E.L.; Godbold, J.H.; Cunningham-Rundles, C. Morbidity and mortality in common variable immune deficiency over 4 decades. *Blood* **2012**, *119*, 1650–1657. [CrossRef]

11. Cunningham-Rundles, C. Autoimmunity in primary immune deficiency: Taking lessons from our patients. *Clin. Exp. Immunol.* **2011**, *164* (Suppl. 2), 6–11. [CrossRef]

12. Chapel, H.; Lucas, M.; Patel, S.; Lee, M.; Cunningham-Rundles, C.; Resnick, E. Confirmation and improvement of criteria for clinical phenotyping in common variable immunodeficiency disorders in replicate cohorts. *J. Allergy Clin. Immunol.* **2012**, *130*, 1197–1198.e9. [CrossRef] [PubMed]

13. Gutierrez, M.J.; Sullivan, K.E.; Fuleihan, R.; Bingham, C.O. Phenotypic characterization of patients with rheumatologic manifestations of common variable immunodeficiency. *Semin. Arthritis Rheum.* **2018**, *48*, 318–326. [CrossRef]

14. Maglione, P.J. Autoimmune and Lymphoproliferative Complications of Common Variable Immunodeficiency. *Curr. Allergy Asthma. Rep.* **2016**, *16*, 19. [CrossRef]

15. Abolhassani, H.; Amirkashani, D.; Parvaneh, N.; Mohammadinejad, P.; Gharib, B.; Shahinpour, S. Autoimmune phenotype in patients with common variable immunodeficiency. *J. Investig. Allergol. Clin. Immunol.* **2013**, *23*, 323–329.

16. Vlkova, M.; Ticha, O.; Nechvatalova, J.; Kalina, T.; Litzman, J.; Mauri, C. Regulatory B cells in CVID patients fail to suppress multifunctional IFN-γ+ TNF-α+ CD4+ T cells differentiation. *Clin. Immunol.* **2015**, *160*, 292–300. [CrossRef]

17. Azizi, G.; Rezaei, N.; Kiaee, F.; Tavakolinia, N.; Yazdani, R.; Mirshafiey, A. T-Cell Abnormalities in Common Variable Immunodeficiency. *J. Investig. Allergol. Clin. Immunol.* **2016**, *26*, 233–243. [CrossRef]

18. Warnatz, K.; Voll, R.E. Pathogenesis of autoimmunity in common variable immunodeficiency. *Front Immunol.* **2012**, *3*, 210. [CrossRef]

19. Genre, J.; Errante, P.R.; Kokron, C.M.; Toledo-Barros, M.; Câmara, N.O.; Rizzo, L.V. Reduced frequency of CD4(+)CD25(HIGH)FOXP3(+) cells and diminished FOXP3 expression in patients with Common Variable Immunodeficiency: A link to autoimmunity? *Clin. Immunol.* **2009**, *132*, 215–221. [CrossRef]

20. Kutukculer, N.; Azarsiz, E.; Aksu, G.; Karaca, N.E. CD4+CD25+Foxp3+ T regulatory cells, Th1 (CCR5, IL-2, IFN-γ) and Th2 (CCR4, IL-4, Il-13) type chemokine receptors and intracellular cytokines in children with common variable immunodeficiency. *Int. J. Immunopathol. Pharmacol.* **2016**, *29*, 241–251. [CrossRef]

21. Crotty, S. T Follicular Helper Cell Biology: A Decade of Discovery and Diseases. *Immunity* **2019**, *50*, 1132–1148. [CrossRef]

22. De Valles-Ibáñez, G.; Esteve-Solé, A.; Piquer, M.; González-Navarro, E.A.; Hernandez-Rodriguez, J.; Laayouni, H. Evaluating the Genetics of Common Variable Immunodeficiency: Monogenetic Model and Beyond. *Front Immunol.* **2018**, *9*, 636. [CrossRef] [PubMed]

23. Salzer, U.; Bacchelli, C.; Buckridge, S.; Pan-Hammarström, Q.; Jennings, S.; Lougaris, V. Relevance of biallelic versus monoallelic TNFRSF13B mutations in distinguishing disease-causing from risk-increasing TNFRSF13B variants in antibody deficiency syndromes. *Blood* **2009**, *113*, 1967–1976. [CrossRef] [PubMed]

24. Gereige, J.D.; Maglione, P.J. Current Understanding and Recent Developments in Common Variable Immunodeficiency Associated Autoimmunity. *Front Immunol.* **2019**, *10*, 2753. [CrossRef] [PubMed]

25. Bogaert, D.J.; Dullaers, M.; Lambrecht, B.N.; Vermaelen, K.Y.; De Baere, E.; Haerynck, F. Genes associated with common variable immunodeficiency: One diagnosis to rule them all? *J. Med. Genet.* **2016**, *53*, 575–590. [CrossRef] [PubMed]

26. Milner, J.D. PLAID: A Syndrome of Complex Patterns of Disease and Unique Phenotypes. *J. Clin. Immunol.* **2015**, *35*, 527–530. [CrossRef]

27. Yazdani, R.; Azizi, G.; Abolhassani, H.; Aghamohammadi, A. Selective IgA Deficiency: Epidemiology, Pathogenesis, Clinical Phenotype, Diagnosis, Prognosis and Management. *Scand. J. Immunol.* **2017**, *85*, 3–12. [CrossRef]

28. Edwards, E.; Razvi, S.; Cunningham-Rundles, C. IgA deficiency: Clinical correlates and responses to pneumococcal vaccine. *Clin. Immunol.* **2004**, *111*, 93–97. [CrossRef]

29. Jacob, C.M.; Pastorino, A.C.; Fahl, K.; Carneiro-Sampaio, M.; Monteiro, R.C. Autoimmunity in IgA deficiency: Revisiting the role of IgA as a silent housekeeper. *J. Clin. Immunol.* **2008**, *28* (Suppl. 1), S56–S61. [CrossRef]

30. Aghamohammadi, A.; Abolhassani, H.; Biglari, M.; Abolmaali, S.; Moazzami, K.; Tabatabaeiyan, M.; Asgarian-Omran, H.; Parvaneh, N.; Mirahmadian, M.; Rezaei, N. Analysis of switched memory B cells in patients with IgA deficiency. *Int. Arch. Allergy Immunol.* **2011**, *156*, 462–468. [CrossRef]

31. Aytekin, C.; Tuygun, N.; Gokce, S.; Dogu, F.; Ikinciogullari, A. Selective IgA deficiency: Clinical and laboratory features of 118 children in Turkey. *J. Clin. Immunol.* **2012**, *32*, 961–966.

32. Todoric, K.; Koontz, J.B.; Mattox, D.; Tarrant, T.K. Autoimmunity in immunodeficiency. *Curr. Allergy Asthma Rep.* **2013**, *13*, 361–370.

33. Price, P.; Witt, C.; Allcock, R.; Sayer, D.; Garlepp, M.; Kok, C.C.; French, M.; Mallal, S.; Christiansen, F. The genetic basis for the association of the 8.1 ancestral haplotype (A1, B8, DR3) with multiple immunopathological diseases. *Immunol. Rev.* **1999**, *167*, 257–274.

34. Azizi, G.; Tavakol, M.; Rafiemanesh, H.; Kiaee, F.; Yazdani, R.; Heydari, A.; Abouhamzeh, K.; Anvari, P.; Mohammadikhajehdehi, S.; Sharifia, L.; et al. Autoimmunity in a cohort of 471 patients with primary antibody deficiencies. *Expert Rev. Clin. Immunol.* **2017**, *13*, 1099–1106. [PubMed]

35. Nechvatalova, J.; Pikulova, Z.; Stikarovska, D.; Pesak, S.; Vlkova, M.; Litzman, J. B-lymphocyte subpopulations in patients with selective IgA deficiency. *J. Clin. Immunol.* **2012**, *32*, 441–448. [PubMed]

36. Haimila, K.; Einarsdottir, E.; de Kauwe, A.; Koskinen, L.L.; Pan-Hammarström, Q.; Kaartinen, T.; Kurppa, K.; Ziberna, F.; Not, T.; Vatta, S.; et al. The shared CTLA4-ICOS risk locus in celiac disease, IgA deficiency and common variable immunodeficiency. *Genes Immun.* **2009**, *10*, 151–161. [CrossRef] [PubMed]

37. Odineal, D.D.; Gershwin, M.E. The Epidemiology and Clinical Manifestations of Autoimmunity in Selective IgA Deficiency. *Clin. Rev. Allergy Immunol.* **2020**, *58*, 107–133. [PubMed]

38. Davies, E.G.; Thrasher, A.J. Update on the hyper immunoglobulin M syndromes. *Br. J. Haematol.* **2010**, *149*, 167–180.

39. Notarangelo, L.D.; Hayward, A.R. X-linked immunodeficiency with hyper-IgM (XHIM). *Clin. Exp. Immunol.* **2000**, *120*, 399–405. [PubMed]

40. Jesus, A.A.; Duarte, A.J.; Oliveira, J.B. Autoimmunity in hyper-IgM syndrome. *J. Clin. Immunol.* **2008**, *28* (Suppl 1), S62–S66. [CrossRef]

41. Revy, P.; Muto, T.; Levy, Y.; Geissmann, F.; Plebani, A.; Sanal, O.; Catalan, N.; Forveille, M.; Dufourcq-Lagelouse, R.; Gennery, A.; et al. Activation-induced cytidine deaminase (AID) deficiency causes the autosomal recessive form of the Hyper-IgM syndrome (HIGM2). *Cell* **2000**, *102*, 565–575. [CrossRef]

42. Imai, K.; Slupphaug, G.; Lee, W.I.; Revy, P.; Nonoyama, S.; Catalan, N.; Yel, L.; Forveille, M.; Kavli, B.; Krokan, H.E.; et al. Human uracil-DNA glycosylase deficiency associated with profoundly impaired immunoglobulin class-switch recombination. *Nat. Immunol.* **2003**, *4*, 1023–1028. [CrossRef]

43. Ferrari, S.; Giliani, S.; Insalaco, A.; Al-Ghonaium, A.; Soresina, A.R.; Loubser, M.; Avanzini, M.A.; Marconi, M.; Badolato, R.; Ugazio, A.G.; et al. Mutations of CD40 gene cause an autosomal recessive form of immunodeficiency with hyper IgM. *Proc. Natl. Acad. Sci. USA* **2001**, *98*, 12614–12619. [CrossRef]

44. Lehman, H.K. Autoimmunity and Immune Dysregulation in Primary Immune Deficiency Disorders. *Curr. Allergy Asthma. Rep.* **2015**, *15*, 53. [CrossRef]

45. Winkelstein, J.A.; Marino, M.C.; Ochs, H.; Fuleihan, R.; Scholl, P.R.; Geha, R.; Stiehm, E.R.; Conley, M.E. The X-linked hyper-IgM syndrome: Clinical and immunologic features of 79 patients. *Medicine* **2003**, *82*, 373–384. [CrossRef] [PubMed]

46. Levy, J.; Espanol-Boren, T.; Thomas, C.; Fischer, A.; Tovo, P.; Bordigoni, P.; Resnick, L.; Fasth, A.; Baer, M.; Gomez, L.; et al. Clinical spectrum of X-linked hyper-IgM syndrome. *J. Pediatr.* **1997**, *131*, 47–54. [CrossRef]

47. Quartier, P.; Bustamante, J.; Sanal, O.; Plebani, A.; Debré, M.; Deville, A.; Litzmane, J.; Levy, J.; Fermand, J.-P.; Lane, P.; et al. Clinical, immunologic and genetic analysis of 29 patients with autosomal recessive hyper-IgM syndrome due to Activation-Induced Cytidine Deaminase deficiency. *Clin. Immunol.* **2004**, *110*, 22–29. [CrossRef] [PubMed]

48. Lacroix-Desmazes, S.; Resnick, I.; Stahl, D.; Mouthon, L.; Espanol, T.; Levy, J.; Kaveri, S.V.; Notarangelo, L.; Eibl, M.; Fischer, A.; et al. Defective self-reactive antibody repertoire of serum IgM in patients with hyper-IgM syndrome. *J. Immunol.* **1999**, *162*, 5601–5608. [PubMed]

49. Lougaris, V.; Badolato, R.; Ferrari, S.; Plebani, A. Hyper immunoglobulin M syndrome due to CD40 deficiency: Clinical, molecular, and immunological features. *Immunol. Rev.* **2005**, *203*, 48–66. [CrossRef]

50. Fleischer, D.M.; Sicherer, S.; Greenhawt, M.; Campbell, D.; Chan, E.; Muraro, A.; Halken, S.; Katz, Y.; Ebisawa, M.; Eichenfield, L.; et al. Consensus Communication on Early Peanut Introduction and Prevention of Peanut Allergy in High-Risk Infants. *Pediatr. Dermatol.* **2016**, *33*, 103–106. [CrossRef]

51. Shillitoe, B.; Gennery, A. X-Linked Agammaglobulinaemia: Outcomes in the modern era. *Clin. Immunol.* **2017**, *183*, 54–62. [CrossRef]

52. Pessach, I.M. The relationship of x-linked primary immune deficiencies and autoimmunity. *Curr. Allergy Asthma. Rep.* **2010**, *10*, 311–319. [CrossRef] [PubMed]

53. Winkelstein, J.A.; Marino, M.C.; Lederman, H.M.; Jones, S.M.; Sullivan, K.; Burks, A.W.; Conley, M.E.; Cunningham-Rundles, C.; Ochs, H.D. X-linked agammaglobulinemia: Report on a United States registry of 201 patients. *Medicine* **2006**, *85*, 193–202. [CrossRef] [PubMed]

54. Ochs, H.D.; Notarangelo, L.D. X-linked immunodeficiencies. *Curr. Allergy Asthma. Rep.* **2004**, *4*, 339–348. [CrossRef]

55. Howard, V.; Greene, J.M.; Pahwa, S.; Winkelstein, J.A.; Boyle, J.M.; Kocak, M.; Conley, M.E. The health status and quality of life of adults with X-linked agammaglobulinemia. *Clin. Immunol.* **2006**, *118*, 201–208. [CrossRef] [PubMed]

56. Knight, A.K.; Cunningham-Rundles, C. Inflammatory and autoimmune complications of common variable immune deficiency. *Autoimmun. Rev.* **2006**, *5*, 156–159. [CrossRef]

57. Lee, K.G.; Xu, S.; Wong, E.T.; Tergaonkar, V.; Lam, K.P. Bruton's tyrosine kinase separately regulates NFkappaB p65RelA activation and cytokine interleukin (IL)-10/IL-12 production in TLR9-stimulated B Cells. *J. Biol. Chem.* **2008**, *283*, 11189–11198. [CrossRef]
58. Kubo, T.; Uchida, Y.; Watanabe, Y.; Abe, M.; Nakamura, A.; Ono, M.; Akira, S.; Takai, T. Augmented TLR9-induced Btk activation in PIR-B-deficient B-1 cells provokes excessive autoantibody production and autoimmunity. *J. Exp. Med.* **2009**, *206*, 1971–1982. [CrossRef]
59. Gobert, D.; Bussel, J.B.; Cunningham-Rundles, C.; Galicier, L.; Dechartres, A.; Berezne, A.; Bonnotte, B.; DeRevel, T.; Auzary, C.; Jaussaud, R.; et al. Efficacy and safety of rituximab in common variable immunodeficiency-associated immune cytopenias: A retrospective multicentre study on 33 patients. *Br. J. Haematol.* **2011**, *155*, 498–508. [CrossRef]
60. Wang, J.; Cunningham-Rundles, C. Treatment and outcome of autoimmune hematologic disease in common variable immunodeficiency (CVID). *J. Autoimmun.* **2005**, *25*, 57–62. [CrossRef]
61. Cirillo, E.; Giardino, G.; Gallo, V.; D'Assante, R.; Grasso, F.; Romano, R.; Lillo, C.D.; Galasso, G.; Pignata, C. Severe combined immunodeficiency—An update. *Ann. N. Y. Acad. Sci.* **2015**, *1356*, 90–106. [CrossRef] [PubMed]
62. Costagliola, G.; Consolini, R. Lymphadenopathy at the crossroad between immunodeficiency and autoinflammation: An intriguing challenge. *Clin. Exp. Immunol.* **2021**. Online ahead of print.
63. Kato, M.; Kimura, H.; Seki, M.; Shimada, A.; Hayashi, Y.; Morio, T.; Kumaki, S.; Ishida, Y.; Kamachi, Y.; Yachie, A. Omenn syndrome–review of several phenotypes of Omenn syndrome and RAG1/RAG2 mutations in Japan. *Allergol. Int.* **2006**, *55*, 115–119. [CrossRef] [PubMed]
64. Aleman, K.; Noordzij, J.G.; de Groot, R.; van Dongen, J.J.; Hartwig, N.G. Reviewing Omenn syndrome. *Eur. J. Pediatr.* **2001**, *160*, 718–725. [CrossRef] [PubMed]
65. Giardino, G.; Gallo, V.; Prencipe, R.; Gaudino, G.; Romano, R.; De Cataldis, M. Unbalanced Immune System: Immunodeficiencies and Autoimmunity. *Front. Pediatr.* **2016**, *4*, 107. [CrossRef] [PubMed]
66. Castagnoli, R.; Delmonte, O.M.; Calzoni, E.; Notarangelo, L.D. Hematopoietic Stem Cell Transplantation in Primary Immunodeficiency Diseases: Current Status and Future Perspectives. *Front. Pediatr.* **2019**, *7*, 295. [CrossRef]
67. Villa, A.; Notarangelo, L.D. RAG gene defects at the verge of immunodeficiency and immune dysregulation. *Immunol. Rev.* **2019**, *287*, 73–90. [CrossRef]
68. Notarangelo, L.D.; Kim, M.S.; Walter, J.E.; Lee, Y.N. Human RAG mutations: Biochemistry and clinical implications. *Nat. Rev. Immunol.* **2016**, *16*, 234–246. [CrossRef]
69. Klein, L.; Robey, E.A.; Hsieh, C.S. Central CD4(+) T cell tolerance: Deletion versus regulatory T cell differentiation. *Nat. Rev. Immunol.* **2019**, *19*, 7–18. [CrossRef]
70. Husebye, E.S.; Anderson, M.S.; Kämpe, O. Autoimmune Polyendocrine Syndromes. *N. Engl. J. Med.* **2018**, *378*, 1132–1141. [CrossRef]
71. Guo, C.J.; Leung, P.S.C.; Zhang, W.; Ma, X.; Gershwin, M.E. The immunobiology and clinical features of type 1 autoimmune polyglandular syndrome (APS-1). *Autoimmun. Rev.* **2018**, *17*, 78–85. [CrossRef]
72. Husebye, E.S.; Perheentupa, J.; Rautemaa, R.; Kämpe, O. Clinical manifestations and management of patients with autoimmune polyendocrine syndrome type I. *J. Intern. Med.* **2009**, *265*, 514–529. [CrossRef] [PubMed]
73. McDonald-McGinn, D.M.; Sullivan, K.E.; Marino, B.; Philip, N.; Swillen, A.; Vorstman, J.A.; Zackai, E.H.; Emanuel, B.S.; Vermeesch, J.R.; Morrow, B.E.; et al. 22q11.2 deletion syndrome. *Nat. Rev. Dis. Primers.* **2015**, *1*, 15071. [CrossRef]
74. Gennery, A.R. Immunological aspects of 22q11.2 deletion syndrome. *Cell Mol. Life Sci.* **2012**, *69*, 17–27. [CrossRef]
75. McLean-Tooke, A.; Spickett, G.P.; Gennery, A.R. Immunodeficiency and autoimmunity in 22q11.2 deletion syndrome. *Scand. J. Immunol.* **2007**, *66*, 1–7. [CrossRef]
76. Di Cesare, S.; Puliafito, P.; Ariganello, P.; Marcovecchio, G.E.; Mandolesi, M.; Capolino, R.; Digilio, M.C.; Aiuti, A.; Rossi, P.; Cancrini, C. Autoimmunity and regulatory T cells in 22q11.2 deletion syndrome patients. *Pediatr. Allergy Immunol.* **2015**, *26*, 591–594. [CrossRef]
77. Marcovecchio, G.E.; Bortolomai, I.; Ferrua, F.; Fontana, E.; Imberti, L.; Conforti, E.; Amodio, D.; Bergante, S.; Macchiarulo, G.; D'Oria, V.; et al. Thymic Epithelium Abnormalities in DiGeorge and Down Syndrome Patients Contribute to Dysregulation in T Cell Development. *Front Immunol.* **2019**, *10*, 447. [CrossRef]
78. Morsheimer, M.; Brown Whitehorn, T.F.; Heimall, J.; Sullivan, K.E. The immune deficiency of chromosome 22q11.2 deletion syndrome. *Am. J. Med Genet. Part A* **2017**, *173*, 2366–2372. [CrossRef]
79. Legitimo, A.; Bertini, V.; Costagliola, G.; Baroncelli, G.I.; Morganti, R.; Valetto, A.; Consolini, R. Vitamin D status and the immune assessment in 22q11.2 deletion syndrome. *Clin. Exp. Immunol.* **2020**, *200*, 272–286. [CrossRef] [PubMed]
80. McDonald-McGinn, D.M.; Sullivan, K.E. Chromosome 22q11.2 deletion syndrome (DiGeorge syndrome/velocardiofacial syndrome). *Medicine* **2011**, *90*, 1–18. [CrossRef] [PubMed]
81. Cepika, A.M.; Sato, Y.; Liu, J.M.; Uyeda, M.J.; Bacchetta, R.; Roncarolo, M.G. Tregopathies: Monogenic diseases resulting in regulatory T-cell deficiency. *J. Allergy Clin. Immunol.* **2018**, *142*, 1679–1695. [CrossRef]
82. Gambineri, E.; Ciullini Mannurita, S.; Hagin, D.; Vignoli, M.; Anover-Sombke, S.; DeBoer, S. Clinical, Immunological, and Molecular Heterogeneity of 173 Patients With the Phenotype of Immune Dysregulation, Polyendocrinopathy, Enteropathy, X-Linked (IPEX) Syndrome. *Front Immunol.* **2018**, *9*, 2411. [CrossRef] [PubMed]
83. Verma, N.; Burns, S.O.; Walker, L.S.K.; Sansom, D.M. Immune deficiency and autoimmunity in patients with CTLA-4 (CD152) mutations. *Clin. Exp. Immunol.* **2017**, *190*, 1–7. [CrossRef]

84. Schwab, C.; Gabrysch, A.; Olbrich, P.; Patiño, V.; Warnatz, K.; Wolff, D.; Hoshino, A.; Kobayashi, M.; Imai, K.; Takag, M.; et al. Phenotype, penetrance, and treatment of 133 cytotoxic T-lymphocyte antigen 4-insufficient subjects. *J. Allergy Clin. Immunol.* **2018**, *142*, 1932–1946. [CrossRef] [PubMed]

85. Delmonte, O.M.; Castagnoli, R.; Calzoni, E.; Notarangelo, L.D. Inborn Errors of Immunity With Immune Dysregulation: From Bench to Bedside. *Front Pediatr.* **2019**, *7*, 353. [CrossRef]

86. Lo, B.; Abdel-Motal, U.M. Lessons from CTLA-4 deficiency and checkpoint inhibition. *Curr. Opin. Immunol.* **2017**, *49*, 14–19. [CrossRef]

87. Gámez-Díaz, L.; August, D.; Stepensky, P.; Revel-Vilk, S.; Seidel, M.G.; Noriko, M.; Morio, T.; Worth, A.J.J.; Blessing, J.; Veerdonk, F.V.; et al. The extended phenotype of LPS-responsive beige-like anchor protein (LRBA) deficiency. *J. Allergy Clin. Immunol.* **2016**, *137*, 223–230. [CrossRef]

88. Habibi, S.; Zaki-Dizaji, M.; Rafiemanesh, H.; Lo, B.; Jamee, M.; Gámez-Díaz, L.; Salami, F.; Kamali, A.N.; Mohammadi, H.; Abolhassani, H.; et al. Clinical, Immunologic, and Molecular Spectrum of Patients with LPS-Responsive Beige-Like Anchor Protein Deficiency: A Systematic Review. *J. Allergy Clin. Immunol. Pract.* **2019**, *7*, 2379–2386.e5. [CrossRef]

89. Alkhairy, O.K.; Abolhassani, H.; Rezaei, N.; Fang, M.; Andersen, K.K.; Chavoshzadeh, Z.; Mohammadzadeh, I.; El-Rajab, M.A.; Massaad, M.; Chou, J.; et al. Spectrum of Phenotypes Associated with Mutations in LRBA. *J. Clin. Immunol.* **2016**, *36*, 33–45. [CrossRef]

90. Lo, B.; Zhang, K.; Lu, W.; Zheng, L.; Zhang, Q.; Kanellopoulou, C.; Zhang, Y.; Liu, Z.; Fritz, J.M.; Marsh, R.; et al. AUTOIMMUNE DISEASE. Patients with LRBA deficiency show CTLA4 loss and immune dysregulation responsive to abatacept therapy. *Science* **2015**, *349*, 436–440. [CrossRef] [PubMed]

91. Kiykim, A.; Ogulur, I.; Dursun, E.; Charbonnier, L.M.; Nain, E.; Cekic, S.; Dogruel, D.; Karaca, N.E.; Cogurlu, M.T.; Bilir, O.A.; et al. Abatacept as a Long-Term Targeted Therapy for LRBA Deficiency. *J. Allergy Clin. Immunol. Pract.* **2019**, *7*, 2790–2800.e15. [CrossRef] [PubMed]

92. Lorenzini, T.; Dotta, L.; Giacomelli, M.; Vairo, D.; Badolato, R. STAT mutations as program switchers: Turning primary immunodeficiencies into autoimmune diseases. *J. Leukoc. Biol.* **2017**, *101*, 29–38. [CrossRef] [PubMed]

93. Okada, S.; Asano, T.; Moriya, K.; Boisson-Dupuis, S.; Kobayashi, M.; Casanova, J.L.; Puel, A. Human STAT1 Gain-of-Function Heterozygous Mutations: Chronic Mucocutaneous Candidiasis and Type I Interferonopathy. *J. Clin. Immunol.* **2020**, *40*, 1065–1081. [CrossRef]

94. Fabre, A.; Marchal, S.; Barlogis, V.; Mari, B.; Barbry, P.; Rohrlich, P.S.; Forbes, L.R.; Vogel, T.P.; Giovannini-Chami, L. Clinical Aspects of STAT3 Gain-of-Function Germline Mutations: A Systematic Review. *J. Allergy Clin. Immunol. Pract.* **2019**, *7*, 1958–1969.e9. [CrossRef]

95. Vogel, T.P.; Milner, J.D.; Cooper, M.A. The Ying and Yang of STAT3 in Human Disease. *J. Clin. Immunol.* **2015**, *35*, 615–623. [CrossRef] [PubMed]

96. Hwa, V. STAT5B deficiency: Impacts on human growth and immunity. *Growth Horm. IGF Res.* **2016**, *28*, 16–20. [CrossRef]

97. Khoury, T.; Molho-Pessach, V.; Ramot, Y.; Ayman, A.R.; Elpeleg, O.; Berkman, N.; Zlotogorski, A.; Ilan, Y. Tocilizumab Promotes Regulatory T-cell Alleviation in STAT3 Gain-of-function-associated Multi-organ Autoimmune Syndrome. *Clin. Ther.* **2017**, *39*, 444–449. [CrossRef]

98. Afzali, B.; Grönholm, J.; Vandrovcova, J.; O'Brien, C.; Sun, H.W.; Vanderleyden, I.; Davis, F.P.; Khoder, A.; Zhang, Y.; Hegazy, A.N.; et al. BACH2 immunodeficiency illustrates an association between super-enhancers and haploinsufficiency. *Nat. Immunol.* **2017**, *18*, 813–823. [CrossRef]

99. Tessarin, G.; Rossi, S.; Baronio, M.; Gazzurelli, L.; Colpani, M.; Benvenuto, A.; Zunica, F.; Cardinale, F.; Martire, B.; Brescia, L.; et al. Activated Phosphoinositide 3-Kinase Delta Syndrome 1, Clinical and Immunological Data from an Italian Cohort of Patients. *J. Clin. Med.* **2020**, *9*, 3335. [CrossRef]

100. Coulter, T.I.; Chandra, A.; Bacon, C.M.; Babar, J.; Curtis, J.; Screaton, N.; Goodlad, J.R.; Farmer, G.; Steele, C.L.; Leahy, T.R.; et al. Clinical spectrum and features of activated phosphoinositide 3-kinase δ syndrome: A large patient cohort study. *J. Allergy Clin. Immunol.* **2017**, *139*, 597–606.e4. [CrossRef] [PubMed]

101. Nunes-Santos, C.J.; Uzel, G.; Rosenzweig, S.D. PI3K pathway defects leading to immunodeficiency and immune dysregulation. *J. Allergy Clin. Immunol.* **2019**, *143*, 1676–1687. [CrossRef]

102. Lucas, C.L.; Chandra, A.; Nejentsev, S.; Condliffe, A.M.; Okkenhaug, K. PI3Kδ and primary immunodeficiencies. *Nat. Rev. Immunol.* **2016**, *16*, 702–714. [CrossRef]

103. Dimitrova, D.; Nademi, Z.; Maccari, M.E.; Ehl, S.; Uzel, G.; Tomoda, T.; Okano, T.; Imai, K.; Carpenter, B.; Ip, W.; et al. International retrospective study of allogeneic hematopoietic cell transplantation for activated PI3K-delta syndrome. *J. Allergy Clin. Immunol.* **2021**, in press.

104. Salzer, E.; Santos-Valente, E.; Keller, B.; Warnatz, K.; Boztug, K. Protein Kinase C δ: A Gatekeeper of Immune Homeostasis. *J. Clin. Immunol.* **2016**, *36*, 631–640. [CrossRef] [PubMed]

105. Duquesnes, N.; Lezoualc'h, F.; Crozatier, B. PKC-delta and PKC-epsilon: Foes of the same family or strangers? *J. Mol. Cell Cardiol.* **2011**, *51*, 665–673. [CrossRef] [PubMed]

106. Steinberg, S.F. Distinctive activation mechanisms and functions for protein kinase Cdelta. *Biochem. J.* **2004**, *384 Pt 3*, 449–459. [CrossRef] [PubMed]

107. Kuehn, H.S.; Niemela, J.E.; Rangel-Santos, A.; Zhang, M.; Pittaluga, S.; Stoddard, J.L.; Hussey, A.A.; Evbuomwan, M.O.; Long Priel, D.A.; Kuhns, D.B.; et al. Loss-of-function of the protein kinase C δ (PKCδ) causes a B-cell lymphoproliferative syndrome in humans. *Blood* **2013**, *121*, 3117–3125. [CrossRef] [PubMed]
108. Calvo, K.R.; Price, S.; Braylan, R.C.; Oliveira, J.B.; Lenardo, M.; Fleisher, T.A.; Koneti Rao, V. JMML and RALD (Ras-associated autoimmune leukoproliferative disorder): Common genetic etiology yet clinically distinct entities. *Blood* **2015**, *125*, 2753–2758. [CrossRef] [PubMed]
109. Rivers, E.; Thrasher, A.J. Wiskott-Aldrich syndrome protein: Emerging mechanisms in immunity. *Eur. J. Immunol.* **2017**, *47*, 1857–1866. [CrossRef] [PubMed]
110. Massaad, M.J.; Ramesh, N.; Geha, R.S. Wiskott-Aldrich syndrome: A comprehensive review. *Ann. N. Y. Acad. Sci.* **2013**, *1285*, 26–43. [CrossRef]
111. Albert, M.H.; Freeman, A.F. Wiskott-Aldrich Syndrome (WAS) and Dedicator of Cytokinesis 8-(DOCK8) Deficiency. *Front Pediatr.* **2019**, *7*, 451. [CrossRef]
112. Biggs, C.M.; Keles, S.; Chatila, T.A. DOCK8 deficiency: Insights into pathophysiology, clinical features and management. *Clin. Immunol.* **2017**, *181*, 75–82. [CrossRef]
113. Schröder-Braunstein, J.; Kirschfink, M. Complement deficiencies and dysregulation: Pathophysiological consequences, modern analysis, and clinical management. *Mol. Immunol.* **2019**, *114*, 299–311. [CrossRef]
114. Conigliaro, P.; Triggianese, P.; Ballanti, E.; Perricone, C.; Perricone, R.; Chimenti, M.S. Complement, infection, and autoimmunity. *Curr. Opin. Rheumatol.* **2019**, *31*, 532–541. [CrossRef]
115. Vignesh, P.; Rawat, A.; Sharma, M.; Singh, S. Complement in autoimmune diseases. *Clin. Chim. Acta.* **2017**, *465*, 123–130. [CrossRef]
116. Henrickson, S.E.; Jongco, A.M.; Thomsen, K.F.; Garabedian, E.K.; Thomsen, I.P. Noninfectious Manifestations and Complications of Chronic Granulomatous Disease. *J. Pediatric Infect. Dis. Soc.* **2018**, *7* (Suppl. 1), S18–S24. [CrossRef] [PubMed]
117. Yu, H.H.; Yang, Y.H.; Chiang, B.L. Chronic Granulomatous Disease: A Comprehensive Review. *Clin. Rev. Allergy Immunol.* **2020**, *61*, 101–113. [CrossRef] [PubMed]
118. Azizi, G.; Ziaee, V.; Tavakol, M.; Alinia, T.; Yazdai, R.; Mohammadi, H.; Abolhassani, H.; Aghamohammadi, A. Approach to the Management of Autoimmunity in Primary Immunodeficiency. *Scand. J. Immunol.* **2017**, *85*, 13–29. [CrossRef]
119. Walter, J.E.; Farmer, J.R.; Foldvari, Z.; Torgerson, T.R.; Cooper, M.A. Mechanism-Based Strategies for the Management of Autoimmunity and Immune Dysregulation in Primary Immunodeficiencies. *J. Allergy Clin. Immunol. Pract.* **2016**, *4*, 1089–1100. [CrossRef] [PubMed]
120. Abraham, R.S. How to evaluate for immunodeficiency in patients with autoimmune cytopenias: Laboratory evaluation for the diagnosis of inborn errors of immunity associated with immune dysregulation. *Hematol. Am. Soc. Hematol. Educ. Program* **2020**, *2020*, 661–672. [CrossRef]
121. Consolini, R.; Costagliola, G.; Spatafora, D. The Centenary of Immune Thrombocytopenia-Part 2: Revising Diagnostic and Therapeutic Approach. *Front Pediatr.* **2017**, *5*, 179. [CrossRef] [PubMed]
122. Fischer, A.; Provot, J.; Jais, J.P.; Alcais, A.; Mahlaoui, N. Autoimmune and inflammatory manifestations occur frequently in patients with primary immunodeficiencies. *J. Allergy Clin. Immunol.* **2017**, *140*, 1388–1393.e8. [CrossRef]
123. Besnard, C.; Levy, E.; Aladjidi, N.; Stolzenberg, M.C.; Magerus-Chatinet, A.; Alibeu, O.; Nitschke, P.; Blanche, S.; Hermine, O.; Jeziorski, E.; et al. Pediatric-onset Evans syndrome: Heterogeneous presentation and high frequency of monogenic disorders including LRBA and CTLA4 mutations. *Clin. Immunol.* **2018**, *188*, 52–57. [CrossRef]
124. Hadjadj, J.; Aladjidi, N.; Fernandes, H.; Leverger, G.; Magérus-Chatinet, A.; Mazerolles, F.; Hussey, A.A.; Evbuomwan, M.O.; Long Priel, D.A.; Kuhns, D.B.; et al. Pediatric Evans syndrome is associated with a high frequency of potentially damaging variants in immune genes. *J. Am. Soc. Hematol.* **2019**, *134*, 9–21.
125. Costagliola, G.; Marco, S.D.; Comberiati, P.; D'Elios, S.; Petashvili, N.; Di Cicco, M.E.; Peroni, D. Practical Approach to Children Presenting with Eosinophila and Hypereosinophilia. *Curr. Pediatr. Rev.* **2020**, *16*, 81–88. [CrossRef]
126. Shamriz, O.; Tal, Y.; Talmon, A.; Nahum, A. Chronic Mucocutaneous Candidiasis in Early Life: Insights Into Immune Mechanisms and Novel Targeted Therapies. *Front Immunol.* **2020**, *11*, 593289. [CrossRef] [PubMed]
127. Leffler, J.; Bengtsson, A.A.; Blom, A.M. The complement system in systemic lupus erythematosus: An update. *Ann. Rheum. Dis.* **2014**, *73*, 1601–1606. [CrossRef] [PubMed]
128. Boileau, J.; Mouillot, G.; Gerard, L.; Carmagnat, M.; Rabian, C.; Oksenhendler, E.; Pasquali, J.-L.; Korganow, A.-S.; DEFI Study Group. Autoimmunity in common variable immunodeficiency: Correlation with lymphocyte phenotype in the French DEFI study. *J. Autoimmun.* **2011**, *36*, 25–32. [CrossRef] [PubMed]
129. Sánchez-Ramón, S.; Radigan, L.; Yu, J.E.; Bard, S.; Cunningham-Rundles, C. Memory B cells in common variable immunodeficiency: Clinical associations and sex differences. *Clin. Immunol.* **2008**, *128*, 314–321. [CrossRef] [PubMed]
130. Ahn, S.; Cunningham-Rundles, C. Role of B cells in common variable immune deficiency. *Expert Rev. Clin. Immunol.* **2009**, *5*, 557–564. [CrossRef] [PubMed]
131. Arumugakani, G.; Wood, P.M.; Carter, C.R. Frequency of Treg cells is reduced in CVID patients with autoimmunity and splenomegaly and is associated with expanded CD21lo B lymphocytes. *J. Clin. Immunol.* **2010**, *30*, 292–300. [CrossRef]

132. Picchianti Diamanti, A.; Rosado, M.M.; Scarsella, M.; Ceccarelli, S.; Laganà, B.; D'Amelio, R.; Carsetti, R. Increased serum IgM, immunodeficiency, and autoimmunity: A clinical series. *Int. J. Immunopathol. Pharmacol.* **2015**, *28*, 547–556. [CrossRef]
133. Montin, D.; Marolda, A.; Licciardi, F.; Robasto, F.; Di Cesare, S.; Ricotti, E.; Ferro, F.; Scaioli, G.; Giancotta, C.; Amodio, D.; et al. Immunophenotype Anomalies Predict the Development of Autoimmune Cytopenia in 22q11.2 Deletion Syndrome. *J. Allergy Clin. Immunol. Pract.* **2019**, *7*, 2369–2376. [CrossRef]

Journal of
Clinical Medicine

Review

Eosinophilic Gastrointestinal Diseases in Inborn Errors of Immunity

Martina Votto [1,2,†], Matteo Naso [1,2,†], Ilaria Brambilla [2,*], Silvia Caimmi [2], Maria De Filippo [1,2], Amelia Licari [1,2], Gian Luigi Marseglia [1,2] and Riccardo Castagnoli [1,2,*]

1   Pediatric Unit, Department of Clinical, Surgical, Diagnostic, and Pediatric Sciences, University of Pavia, 27100 Pavia, Italy
2   Pediatric Clinic, Fondazione IRCCS Policlinico San Matteo, 27100 Pavia, Italy
*   Correspondence: i.brambilla@smatteo.pv.it (I.B.); riccardo.castagnoli@unipv.it (R.C.)
†   These authors contributed equally to this work.

Abstract: Inborn errors of immunity (IEI) are disorders mostly caused by mutations in genes involved in host defense and immune regulation. Different degrees of gastrointestinal (GI) involvement have been described in IEI, and for some IEI the GI manifestations represent the main and characteristic clinical feature. IEI also carry an increased risk for atopic manifestations. Eosinophilic gastrointestinal diseases (EGIDs) are emerging disorders characterized by a chronic/remittent and prevalent eosinophilic inflammation affecting the GI tract from the esophagus to the anus in the absence of secondary causes of intestinal eosinophilia. Data from the U.S. Immunodeficiency Network (USIDNET) reported that EGIDs are more commonly found in patients with IEI. Considering this element, it is reasonable to highlight the importance of an accurate differential diagnosis in patients with IEI associated with mucosal eosinophilia to avoid potential misdiagnosis. For this reason, we provide a potential algorithm to suspect an EGID in patients with IEI or an IEI in individuals with a diagnosis of primary EGID. The early diagnosis and detection of suspicious symptoms of both conditions are fundamental to prevent clinically relevant complications.

Keywords: eosinophilic gastrointestinal disorders; eosinophilic esophagitis; inborn errors of immunity; immunodeficiency

Citation: Votto, M.; Naso, M.; Brambilla, I.; Caimmi, S.; De Filippo, M.; Licari, A.; Marseglia, G.L.; Castagnoli, R. Eosinophilic Gastrointestinal Diseases in Inborn Errors of Immunity. *J. Clin. Med.* **2023**, *12*, 514. https://doi.org/10.3390/jcm12020514

Academic Editors: Rita Consolini and Giorgio Costagliola

Received: 12 December 2022
Revised: 4 January 2023
Accepted: 6 January 2023
Published: 8 January 2023

1. Inborn Errors of Immunity and Gastrointestinal Manifestations

Inborn errors of immunity (IEI) are disorders mostly caused by mutations in genes involved in immune host defense and regulation [1–3]. These conditions are characterized by various combinations of increased susceptibility to infections, autoimmunity, autoinflammatory manifestations, lymphoproliferation, allergy, bone marrow failure, and/or malignancy [1]. The recently updated IEI classification from the International Union of Immunological Societies (IUIS) Expert Committee has increased the number of known genetic defects identified as causing IEI to 485 [4]. According to the IUIS classification, IEI are categorized into ten groups based on the specific clinical and immunological phenotype: combined immunodeficiencies (I); combined immunodeficiencies with syndromic features (II); predominantly antibody deficiencies (III); diseases of immune dysregulation (IV); congenital defects of phagocytes (V); defects in intrinsic and innate immunity (VI); autoinflammatory diseases (VII); complement deficiencies (VIII); bone marrow failure (IX); and phenocopies of inborn errors of immunity (X) [5]. Although IEI present with a broad spectrum of clinical features, in about one-third of them, various degrees of gastrointestinal (GI) involvement have been described, and for some IEI, the GI manifestations represent the characteristic clinical feature [6,7]. In addition, there has been an increasing understanding of which IEI carry an increased risk for specific atopic manifestations, with most studies focusing on atopic dermatitis, allergic rhinitis, asthma, and immunoglobulin E (IgE)-mediated food allergy [8]. Although eosinophilic esophagitis (EoE) is thought to co-occur with these atopic disorders following a common atopic pathophysiology, eosinophilic gastrointestinal diseases (EGIDs) and their association with IEI are relatively poorly understood.

2. Eosinophilic Gastrointestinal Diseases

EGIDs are emerging disorders characterized by chronic/remittent and prevalent eosinophilic inflammation affecting the GI tract from the esophagus to the anus in the absence of secondary causes of intestinal eosinophilia [9,10]. Based on the site of the inflammation, EGIDs have been recently classified into EoE and non-EoE EGIDs (Table 1). EoE affects approximately 1 in 1–2000 persons; however, it is currently considered one of the major causes of upper gastrointestinal morbidity [11]. EoE is found in 12–23% of patients with dysphagia and 50% in those with esophageal food impaction [12,13]. According to current guidelines, diagnosis of EoE requires (1) suggestive clinical symptoms; (2) an esophageal eosinophilic infiltrate greater than 15 eosinophils per high-powered field (HPF) (~60 eos/mm2) in endoscopically obtained biopsies; and (3) the exclusion of secondary causes of esophageal eosinophilia (gastroesophageal reflux disease [GERD], hypereosinophilic syndrome, inflammatory bowel diseases, autoimmune disorders, vasculitis, hyper-IgE syndrome, drug hypersensitivity, infections, pill esophagitis, and graft versus host disease). EoE symptoms are non-specific and vary with age. Feeding issues, failure to thrive, and recurrent vomiting generally prevail in infants and toddlers, whereas school-aged children present epigastric pain or GERD-like symptoms. Dysphagia and esophageal food impaction are typically prevalent symptoms in adolescents and adults.

Table 1. Clinical features of EGIDs.

	Symptoms	Diagnosis	Treatments

Eosinophilic esophagitis (EoE)	Symptoms mainly depend on the patient's age
	- Infants and toddlers: food refusal, feeding issues, recurrent vomiting, failure to thrive
	- Children: esophageal reflux not responding to conventional therapy, epigastric pain, vomiting
	- Adolescents and adults: dysphagia, esophageal food impaction.
	Change in eating behaviors
		(1) Suggestive clinical symptoms
		(2) ≥15 eos/HPF in esophageal biopsies
		(3) Exclusion of secondary causes of intestinal eosinophilia
			- Medical therapies
			  ○ Topical steroids Slurry budesonide Oral fluticasone Budesonide tablets (EMA approved)
			  ○ Biological therapy: dupilumab (anti-IL-4R, FDA approved)
			- Food elimination diets
			  ○ Empirical food elimination diet
			  ○ Elemental diet
			- Esophageal dilatation

Non-EoE EGIDs
• Eosinophilic Gastritis (EoG)
• Eosinophilic Enteritis (EoN)
  ○ Eosinophilic Duodenitis (EoD)
  ○ Eosinophilic Jejunitis (EoJ)
  ○ Eosinophilic Ileitis (EoI)
• Eosinophilic Colitis (EoC)
	Symptoms mainly depend on the site and the depth of intestinal inflammation
	- Mucosal form: abdominal pain, diarrhea, vomiting, weight loss, protein-losing enteropathy, GI bleeding
	- Muscle form: intestinal obstruction
	- Serosal form: eosinophilic ascites
		Stomach ≥ 30 eos/HPF
		Small intestine ≥ 52 eos/HPF
		Right colon ≥ 100 eos/HPF
		Transverse and descending colon ≥ 84 eos/HPF
		Rectosigmoid ≥ 64 eos/HPF
			- Medical therapies
			  ○ Systemic steroids (oral budesonide or prednisolone; IV corticosteroids)
			  ○ Immunosuppressants
			  ○ Biological therapies: infliximab, adalimumab (anti-TNF), mepolizumab, reslizumab and benralizumab (anti-IL-5 and anti-IL5R), dupilumab (anti-IL-4R)
			- Food elimination diets
			  ○ Empirical food elimination diet
			  ○ Elemental diet
			- Surgery

HPF: high power field; IV: intravenous.

In contrast, non-EoE EGIDs are still less understood disorders. Epidemiology of non-EoE EGIDs is limited to a few observational studies; however, in the general population, prevalence is estimated at 3–8/100,000 cases, although it was approximately 2% in patients with gastrointestinal symptoms [14]. Symptoms of non-EoE EGIDs depend on the site (stomach, intestine, or colon) and the depth (mucosal, muscular, or serosal layer) of the eosinophilic inflammation and are generally represented by abdominal pain, nausea, vomiting, and diarrhea [10]. In rare cases, patients with non-EoE EGIDs may develop GI complications, such as intestinal obstruction or eosinophilic ascites. However, they may commonly experience malnutrition or weight loss [15]. Diagnosis of the non-EoE EGIDs is challenging and often requires more endoscopies with potential misdiagnosis and diagnostic delays. The diagnostic cut-offs of tissue eosinophils vary according to the specific site of the GI tract (Table 1).

Allergic comorbidities are prevalent in patients with EGIDs. However, several non-allergic diseases have also been associated with EoE, including autism spectrum disorders, coeliac disease, esophageal malformation, and inflammatory bowel disorders [16–18]. EoE is now considered a type 2-mediated disease, developing from a genetic predisposition and impaired esophageal barrier functioning [19]. In this context, the esophageal exposure to allergens (mostly foods) elicits the local production of alarmins (interleukin [IL]-25, IL-33, and thymic stromal lymphopoietin) and the typical type 2 (Th2)-driven eosinophilic inflammation [20]. IL-4 has been characterized as one of the critical drivers of inflammation in EoE since it is upregulated in the esophageal mucosa and blood of affected patients [21]. While eosinophilic gastritis and enteritis show the same pathogenetic mechanisms of EoE, the pathogenesis of eosinophilic colitis is different from that of other non-EoE EGIDs and is mainly related to apoptosis gene expression, reduced epithelial cell proliferation, and minimal evidence of Th2 inflammation.

EGIDs are clinically heterogeneous diseases with symptoms depending on the age at onset, the site of inflammation, response to treatments, and related comorbidities (allergic and not allergic), thus, defining a spectrum of different diseases [22]. Recently, data from the USIDNET reported that EGIDs are more commonly found in patients with different IEI, such as common variable immunodeficiency (CVID) (43.2%), chronic granulomatous disease (CGD) (8.1%), hyper-IgE syndrome (6.8%), and autoimmune lymphoproliferative syndrome (6.8%) [23]. Nevertheless, more research is needed to confirm these findings and understand if patients with EGIDs and IEI may have distinct clinical features, responses to therapies, and disease endotype. Therefore, this study aims to analyze the potential relationship between these two entities, reviewing current evidence and proposing a potential diagnostic algorithm to help clinicians suspect IEI in EGID patients and vice-versa.

3. Material and Methods

The literature review was performed in November 2022, including all publication years. All studies that met the following criteria were included: (i) articles published in English in peer-reviewed journals, and (ii) participants were children and adult IEI patients diagnosed with EGIDs. Potentially eligible publications were manually screened and reviewed, and non-relevant publications were excluded.

The literature search was performed via the online database PubMed, combining the terms "eosinophilic gastrointestinal diseases AND primary immunodeficiency", "eosinophilic gastrointestinal diseases AND inborn errors of immunity", "eosinophilic esophagitis AND inborn errors of immunity", "eosinophilic esophagitis AND primary immunodeficiency", and "eosinophilic esophagitis AND immunodeficiency".

4. Results

The database search found 58 articles. Based on the title and abstract, fifteen articles met the inclusion criteria. After removing duplicates, seven articles were analysed for the review (Figure 1).

In 2016, Yamazaki et al. reported the case of a 30-year-old man with a diagnosis of X-linked agammaglobulinemia, who suffered from chronic diarrhea and persistent low serum IgG, despite the intravenous immunoglobulin replacement (Table 2) [24]. He underwent a colonoscopy with biopsies that detected eosinophilic infiltrate >20 eos/HPF, supporting the diagnosis of eosinophilic gastroenteritis. Treatment with prednisolone was started and led to a significant improvement in diarrhea.

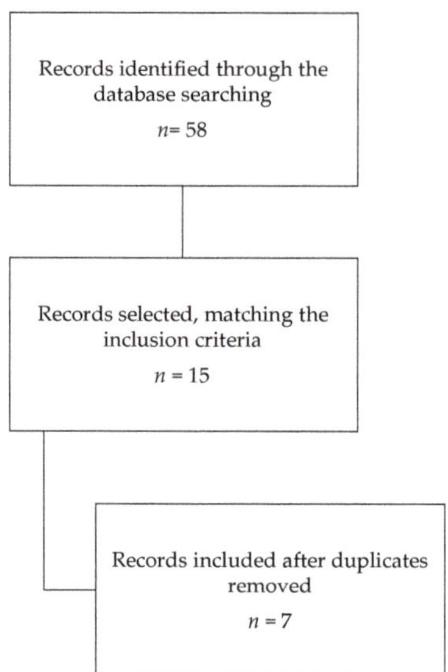

Figure 1. Search strategy.

Table 2. Summary of reviewed articles.

Author, Year [Ref]	Type of Study	IEI	EGID	Age at EGID Diagnosis	Family History	EGID Symptoms	Other Comorbidities	EGID Diagnosis	Complications	EGID Treatment
Yamakazi et al., 2016 [24]	Case report	XLA	EoC	27 years	n.a.	Chronic diarrhea, emaciation	Recurrent infections	>20 eos/HPF	n.a.	Prednisolone
Chen et al., 2016 [25]	Case report	CVID	EoE	28 years	n.a.	Dysphagia, recurrent episodes of esophageal food impaction	Recurrent sinopulmonary infections	n.a.	Esophageal stenosis	Esophageal dilatation, PPI, FED, Oral fluticasone
Hannouch et al., 2016 [26]	Case report	CVID	EoE	n.a.	n.a.	Weight loss, food impaction	Burkitt's lymphoma	n.a.	n.a.	Oral inhaled corticosteroids
Dixit et al., 2021 [27]	Case report	STAT3-HIES	EoE	n.a.	n.a.	Abdominal pain, dysphagia	Eczema, recurrent respiratory tract infections, cutaneous and retropharyngeal abscesses, and mycosis.	n.a.	n.a.	Dupilumab
Scott et al., 2022 [28]	Case report	STAT1-GOF	EoE	Late adolescence	Mother with choking episodes and CMCC; a daughter with CMCC and recurrent AOM.	Choking episodes, solid and liquid dysphagia	Vaginal candidiasis, scalp fungal infection, Candida esophagitis	22 eos/HPF	Esophageal stenosis	Balloon dilatation FED Montelukast PPI Slurry budesonide
Tang et al., 2020 [29]	Case report	XIAP-deficiency	EoC	Infancy	Mother and sister had the mutation	Abdominal distension, perianal abscess.	Anemia, respiratory tract infections, impaired growth	n.a.	n.a.	n.a.
Tran et al., 2022 [23]	Retrospective cohort study	CVID (43.2%), combined immunodeficiencies (21.6%), CGD (8.1%), HIES (6.8%), and ALPS (6.8%).	61/74 (82,5%) patients with EoE and 13/74 (17.5%) with EoG, EoN, and EoC.	n.a.	n.a.	n.a.	n.a.	n.a.	n.a.	n.a.

ALPS: autoimmune lymphoproliferative syndrome; AOM: acute otitis media; CGD: chronic granulomatous disease; CMCC: chronic mucocutaneous candidiasis; CVID: common variable immunodeficiency; EGID: eosinophilic gastrointestinal disease; EoC: eosinophilic colitis; EoE: eosinophilic esophagitis; EoG: eosinophilic gastritis; EoN: eosinophilic enteritis; FED: food elimination diet; GOF: gain of function; HIES: hyper-IgE syndromes; HPF: high power field; IEI: inborn error of immunity; N.A: not available; PPI: proton pump inhibitor; XIAP: X-linked inhibitor of apoptosis; XLA: X-linked agammaglobulinemia.

A few cases reported the association between common variable immunodeficiency (CVID) and EoE [25,26]. Chen et al. described a 34-year-old woman affected by CVID who was referred to a gastroenterologist for dysphagia, recurrent mild esophageal food impactions, and hard-textured foods that worsened in the previous 5–6 years [25]. She underwent an upper GI endoscopy that showed macro- and microscopic findings compatible with EoE. The patient partially achieved control of their symptoms with oral fluticasone. Hannouch et al. described the case of Burkitt's lymphoma development in a patient affected by CVID and EoE [26].

STAT3-hyper-IgE syndrome (HIES) has been primarily associated with GI manifestations, including gastroesophageal reflux disease, dysphagia, and abdominal pain. A recent cohort study enrolling STAT3-HIES patients investigated the GI manifestations unexpectedly observing that EoE occurred in 65% (11/17) of patients who underwent esophagogastroduodenoscopy [30]. Dixit et al. published the case of a 14-year-old boy affected by STAT3-HIES with severe atopic dermatitis and EoE, clinically characterized by dysphagia and abdominal pain. The patient was treated with dupilumab, effectively controlling skin manifestations and resolving EoE symptoms [27].

Scott et al. reported the case of a 39-year-old woman with EoE refractory on a six-food elimination diet, fluticasone, montelukast, and proton pump inhibitor, but responsive to subsequent therapy with slurry budesonide [28]. She probably developed the first GI symptoms in late adolescence, but she was not formally investigated until she was 31. The patient's family history revealed that her 70-year-old mother suffered from chronic mucocutaneous candidiasis (CMCC) and had a 50-year history of dysphagia and choking episodes, endoscopically evaluated at the age of 66 with biopsies demonstrating extensive tissue fibrosis and rare eosinophils. Even her daughter had a history compatible with CMCC but no symptoms suggestive of EoE. All three underwent a genetic evaluation, demonstrating a novel heterozygous missense variant in the N-terminal domain of STAT1 (c.194A > C; p.D65A). Through immunoblotting studies, a gain of function STAT1 phenotype was demonstrated in all family members investigated. This report first described a STAT1 gain of function mutation characterized by severe and refractory EoE as presenting clinical manifestation.

In 2020, Tang et al. reported the case of a 22-month-old boy with abdominal distension, anemia, and recurrent respiratory tract infections diagnosed with an X-linked inhibitor of apoptosis (XIAP) deficiency. He underwent a GI endoscopy that showed chronic active enteritis with different degrees of eosinophil infiltration compatible with eosinophilic colitis. XIAP deficiency is associated with inflammatory bowel diseases (IBD); however, this case report may extend the spectrum of chronic GI diseases associated with this immunodeficiency [29].

5. Discussion

Recently, Tran et al. reviewed the U.S. immunodeficiency Network (USIDNET), finding that 74 IEI patients had a concomitant diagnosis of EGID [23]. In this study, 61 patients were affected by EoE, and 27 (44.2%) had CVID. In 34.4% of patients, a specific immunodeficiency was identified, including HIES and chronic granulomatous disease (CGD). Thirteen (17.5%) patients were affected by non-EoE EGIDs (eosinophilic gastritis, enteritis, and colitis). A total of 38.4% had CVID, 46% had a combined immunodeficiency, 15.3% had CGD, and one patient had FOXP3-deficient immune dysregulation, polyendocrinopathy, and enteropathy X-linked (IPEX) syndrome. These data suggest that EGIDs may be coexisting comorbidities of patients with specific IEI and seem more common than expected. According to these results, CVID is the IEI most likely complicated by an EGID.

The potential link between IEI and EGIDs has not been elucidated yet. IEI are caused by monogenic germline mutations associated with immune function. These diseases are rare, but the prevalence is likely to be at least 1/1000–5000 [4]. Different IEI can manifest with elevated serum IgE or eosinophilia and increased Th-2 cytokine production, such as IL-5, which is an essential promoter of eosinophil differentiation, maturation, and survival [4,10].

Eosinophils are multifunctional leukocytes that play an essential role against helminth infections and are considered pro-inflammatory cells because they release pleiotropic cytokines, chemokines, lipid mediators, and cytoplasmic granule constituents [31]. Eosinophils are considered the key effector cells in EoE, since, in the absence of eosinophils, disease features (tissue remodeling, collagen accumulation, and gastric motility) are attenuated in animal models [32]. Eosinophils are also involved in the pathogenesis of allergic disorders and are implicated in EGIDs and IBD pathogenesis. Intestinal eosinophilia is not the hallmark of EGIDs, because it has been described even in IBD and celiac disease [33]. Eosinophils are also implicated in IBD pathogenesis, probably playing a significant role in the chronic inflammatory process. In recent years, a growing number of IEI manifesting with IBD have been described [7]. XIAP deficiency is considered one of the mendelian causes of inherited IBD in infancy [34]. When a XIAP deficiency patient shows recurrent and severe abdominal pain, failure to thrive, GI bleeding, and diarrhea, it is reasonable to suspect an IBD and perform a GI endoscopy. Despite this robust evidence, Tang et al. reported the case of a patient with XIAP deficiency and eosinophilic colitis, thus extending the spectrum of GI manifestations potentially related to this immunodeficiency [29]. However, the authors did not report data on long-term follow-up or the diagnostic cut-off used for EoC diagnosis [29]. Standardized international guidelines for EGID diagnosis are still lacking. Most experts agreed that a definitive diagnosis requires recurrent/chronic GI symptoms and increased intestinal eosinophilia, excluding secondary causes of EGIDs (Table 1) [10]. Considering this element, it is reasonable to highlight the importance of an accurate differential diagnosis in patients with IEI associated with mucosal eosinophilia to avoid potential misdiagnosis. We provide a potential algorithm to suspect an EGID in patients with IEI or an IEI in individuals with a diagnosis of primary EGID (Figure 2). The early diagnosis and detection of suspicious symptoms of both conditions are fundamental to prevent clinically relevant complications (severe or fatal infections, esophageal stenosis, intestinal obstruction). Of note, it is still unclear if IEI patients experience a more severe EGID phenotype than those without immunodeficiency.

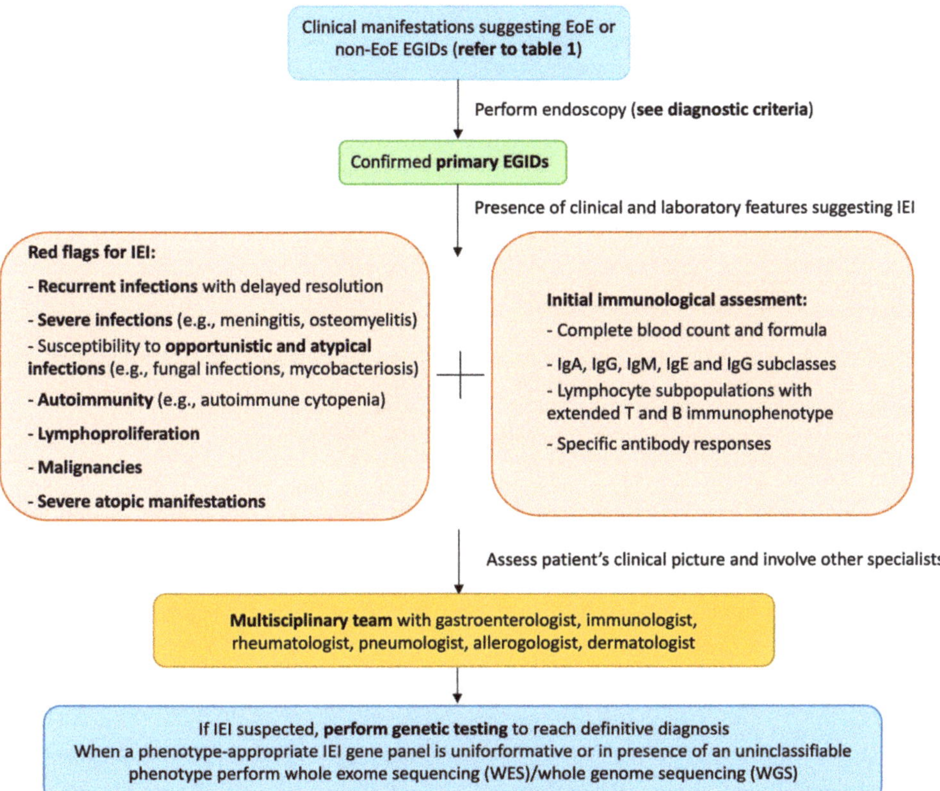

Figure 2. Proposed diagnostic algorithm. The figure can be read from the top to the bottom and vice versa. EGID: eosinophilic gastrointestinal disease; EoE: eosinophilic esophagitis; IEI: inborn errors of immunity.

6. Conclusions

This review first analyzed current evidence of a potential relationship between EGIDs and IEI. According to recent data, EGIDs seem more common in IEI patients than was already reported in the literature. It is reasonable to speculate that EGID can worsen the course of IEI, and vice versa. For this reason, early diagnosis is crucial to prevent complications and define the best personalized treatment. In this context, several unmet needs are still to be clarified. The literature data are still limited, and more research is needed to understand the pathogenetic relationship between these two chronic and invalidating clinical entities. Multicentric prospective studies should be performed to establish the real epidemiology of EGID in IEI patients, the disease-course phenotype, and the response to available treatments.

Author Contributions: Conceptualization, M.V. and R.C.; methodology, M.V. and R.C.; writing—original draft preparation, M.V. and M.N.; writing—review and editing, M.V., M.N., I.B. and R.C.; supervision, I.B., M.D.F., S.C., A.L., G.L.M. and R.C. All authors have read and agreed to the published version of the manuscript.

Funding: This research received no external funding.

Institutional Review Board Statement: Not applicable.

Informed Consent Statement: Not applicable.

Data Availability Statement: Not applicable.

Acknowledgments: The authors would like to thank Daniela Montagna and Lorenza Montagna (Pediatric Unit, Department of Clinical, Surgical, Diagnostic, and Pediatric Sciences, University of Pavia, Pavia, Italy) for their invaluable support to the research in the field of inborn errors of immunity.

Conflicts of Interest: The authors declare no conflict of interest.

References

1. Notarangelo, L.D.; Bacchetta, R.; Casanova, J.L.; Su, H.C. Human Inborn Errors of Immunity: An Expanding Universe. *Sci. Immunol.* **2020**, *5*, eabb1662. [CrossRef]
2. Castagnoli, R.; Delmonte, O.M.; Notarangelo, L.D. Congenital and Acquired Defects of Immunity: An Ever-Evolving Story. *Pediatr. Allergy Immunol.* **2022**, *33*, 61–64. [CrossRef]
3. Castagnoli, R.; Notarangelo, L.D. Updates on New Monogenic Inborn Errors of Immunity. *Pediatr. Allergy Immunol.* **2020**, *31*, 57–59. [CrossRef]
4. Tangye, S.G.; Al-Herz, W.; Bousfiha, A.; Cunningham-Rundles, C.; Franco, J.L.; Holland, S.M.; Klein, C.; Morio, T.; Oksenhendler, E.; Picard, C.; et al. Human Inborn Errors of Immunity: 2022 Update on the Classification from the International Union of Immunological Societies Expert Committee. *J. Clin. Immunol.* **2022**, *42*, 1473–1507. [CrossRef]
5. Bousfiha, A.; Moundir, A.; Tangye, S.G.; Picard, C.; Jeddane, L.; Al-Herz, W.; Rundles, C.C.; Franco, J.L.; Holland, S.M.; Klein, C.; et al. The 2022 Update of IUIS Phenotypical Classification for Human Inborn Errors of Immunity. *J. Clin. Immunol.* **2022**, *42*, 1508–1520. [CrossRef] [PubMed]
6. Castagnoli, R.; Pala, F.; Bosticardo, M.; Licari, A.; Delmonte, O.M.; Villa, A.; Marseglia, G.L.; Notarangelo, L.D. Gut Microbiota–Host Interactions in Inborn Errors of Immunity. *Int. J. Mol. Sci.* **2021**, *22*, 1416. [CrossRef]
7. Ouahed, J.D. Understanding Inborn Errors of Immunity: A Lens into the Pathophysiology of Monogenic Inflammatory Bowel Disease. *Front. Immunol.* **2022**, *13*, 1026511. [CrossRef]
8. Castagnoli, R.; Lougaris, V.; Giardino, G.; Volpi, S.; Leonardi, L.; Torre, F.L.; Federici, S.; Corrente, S.; Cinicola, B.L.; Soresina, A.; et al. Inborn Errors of Immunity with Atopic Phenotypes: A Practical Guide for Allergists. *World Allergy Organ. J.* **2021**, *14*, 100513. [CrossRef]
9. Licari, A.; Votto, M.; Scudeller, L.; De Silvestri, A.; Rebuffi, C.; Cianferoni, A.; Marseglia, G.L. Epidemiology of Nonesophageal Eosinophilic Gastrointestinal Diseases in Symptomatic Patients: A Systematic Review and Meta-Analysis. *J. Allergy Clin. Immunol. Pract.* **2020**, *8*, 1994–2003.e2. [CrossRef] [PubMed]
10. Licari, A.; Votto, M.; D'Auria, E.; Castagnoli, R.; Caimmi, S.M.E.; Marseglia, G.L. Eosinophilic Gastrointestinal Diseases in Children: A Practical Review. *Curr. Pediatr. Rev.* **2019**, *16*, 106–114. [CrossRef]
11. Dellon, E.S.; Jensen, E.T.; Martin, C.F.; Shaheen, N.J.; Kappelman, M.D. Prevalence of Eosinophilic Esophagitis in the United States. *Clin. Gastroenterol. Hepatol.* **2014**, *12*, 589–596.e1. [CrossRef]
12. Dellon, E.S.; Hirano, I. Epidemiology and Natural History of Eosinophilic Esophagitis. *Gastroenterology* **2018**, *154*, 319. [CrossRef]
13. Dellon, E.S.; Gonsalves, N.; Abonia, J.P.; Alexander, J.A.; Arva, N.C.; Atkins, D.; Attwood, S.E.; Auth, M.K.H.; Bailey, D.D.; Biederman, L.; et al. International Consensus Recommendations for Eosinophilic Gastrointestinal Disease Nomenclature. *Clin. Gastroenterol. Hepatol.* **2022**, *20*, 2474–2484.e3. [CrossRef]
14. Shaheen, N.J.; Mukkada, V.; Eichinger, C.S.; Schofield, H.; Todorova, L.; Falk, G.W. Natural History of Eosinophilic Esophagitis: A Systematic Review of Epidemiology and Disease Course. *Dis. Esophagus* **2018**, *31*, doy015. [CrossRef] [PubMed]
15. Votto, M.; De Filippo, M.; Olivero, F.; Raffaele, A.; Cereda, E.; De Amici, M.; Testa, G.; Marseglia, G.L.; Licari, A. Malnutrition in Eosinophilic Gastrointestinal Disorders. *Nutrients* **2021**, *13*, 128. [CrossRef] [PubMed]
16. Rossi, C.M.; Lenti, M.V.; Merli, S.; Licari, A.; Votto, M.; Marseglia, G.L.; Di Sabatino, A. Primary Eosinophilic Gastrointestinal Disorders and Allergy: Clinical and Therapeutic Implications. *Clin. Transl. Allergy* **2022**, *12*, e12146. [CrossRef]
17. Votto, M.; Raffaele, A.; De Filippo, M.; Caimmi, S.; Brunero, M.; Riccipetitoni, G.; Marseglia, G.L.; Licari, A. Eosinophilic Gastrointestinal Disorders in Children and Adolescents: A Single-Center Experience. *Dig. Liver Dis.* **2022**, *54*, 214–220. [CrossRef] [PubMed]
18. Capucilli, P.; Cianferoni, A.; Grundmeier, R.W.; Spergel, J.M. Comparison of Comorbid Diagnoses in Children with and without Eosinophilic Esophagitis in a Large Population. *Ann. Allergy. Asthma Immunol.* **2018**, *121*, 711–716. [CrossRef] [PubMed]
19. Votto, M.; Marseglia, G.L.; De Filippo, M.; Brambilla, I.; Caimmi, S.M.E.; Licari, A. Early Life Risk Factors in Pediatric EoE: Could We Prevent This Modern Disease? *Front. Pediatr.* **2020**, *8*, 263. [CrossRef]
20. O'Shea, K.M.; Aceves, S.S.; Dellon, E.S.; Gupta, S.K.; Spergel, J.M.; Furuta, G.T.; Rothenberg, M.E. Pathophysiology of Eosinophilic Esophagitis. *Gastroenterology* **2018**, *154*, 333. [CrossRef] [PubMed]
21. Licari, A.; Castagnoli, R.; Marseglia, A.; Olivero, F.; Votto, M.; Ciprandi, G.; Marseglia, G.L. Dupilumab to Treat Type 2 Inflammatory Diseases in Children and Adolescents. *Pediatr. Drugs* **2020**, *22*, 295–310. [CrossRef]
22. Votto, M.; Fasola, S.; Cilluffo, G.; Ferrante, G.; La Grutta, S.; Marseglia, G.L.; Licari, A. Cluster Analysis of Clinical Data Reveals Three Pediatric Eosinophilic Gastrointestinal Disorder Phenotypes. *Pediatr. Allergy Immunol.* **2022**, *33*, e13746. [CrossRef]

23. Tran, P.; Gober, L.; Garabedian, E.K.; Fuleihan, R.L.; Puck, J.M.; Sullivan, K.E.; Spergel, J.M.; Ruffner, M.A. Eosinophilic Gastrointestinal Disorders in Patients with Inborn Errors of Immunity: Data from the USIDNET Registry. *Front. Immunol.* **2022**, *13*, 987895. [CrossRef]

24. Yamazaki, S.; Ohtsuka, Y.; Yokokura, T.; Yokota, R.; Honjo, A.; Inage, E.; Baba, Y.; Mori, M.; Suzuki, R.; Iwata, T.; et al. Eosinophilic Gastroenteritis in a Patient with Bruton's Tyrosine Kinase Deficiency. *Pediatr. Int.* **2016**, *58*, 417–419. [CrossRef] [PubMed]

25. Chen, M.; Ko, H.M.; Riffle, M.E.; Andreae, D.A.; Cunningham-Rundles, C.; Chehade, M.; Maglione, P.J. Eosinophilic Esophagitis Diagnosed in a Patient with Common Variable Immunodeficiency. *J. Allergy Clin. Immunol. Pract.* **2016**, *4*, 995–997. [CrossRef] [PubMed]

26. Hannouch, K.J.; McGoey, B.A.; Hauk, M.J.; Michelis, M.A.E. Common Variable Immunodeficiency and Eosinophilic Esophagitis Complicated by Atypical Burkitt's Lymphoma: A Case Report. *J. Allergy Clin. Immunol.* **2017**, *139*, AB114. [CrossRef]

27. Dixit, C.; Thatayatikom, A.; Pappa, H.; Knutsen, A.P. Treatment of Severe Atopic Dermatitis and Eosinophilic Esophagitis with Dupilumab in a 14-Year-Old Boy with Autosomal Dominant Hyper-IgE Syndrome. *J. Allergy Clin. Immunol. Pract.* **2021**, *9*, 4167–4169. [CrossRef] [PubMed]

28. Scott, O.; Sharfe, N.; Dadi, H.; Vong, L.; Garkaby, J.; Abrego Fuentes, L.; Willett Pachul, J.; Nelles, S.; Nahum, A.; Roifman, C.M. Case Report: Eosinophilic Esophagitis in a Patient With a Novel STAT1 Gain-of-Function Pathogenic Variant. *Front. Immunol.* **2022**, *13*, 801832. [CrossRef]

29. Tang, J.; Zhou, X.; Wang, L.; Hu, G.; Zheng, B.; Wang, C.; Lu, Y.; Jin, Y.; Guo, H.; Liu, Z. Eosinophilic Colitis in a Boy with a Novel XIAP Mutation: A Case Report. *BMC Pediatr.* **2020**, *20*, 171. [CrossRef]

30. Arora, M.; Bagi, P.; Strongin, A.; Heimall, J.; Zhao, X.; Lawrence, M.G.; Trivedi, A.; Henderson, C.; Hsu, A.; Quezado, M.; et al. Gastrointestinal Manifestations of STAT3-Deficient Hyper-IgE Syndrome. *J. Clin. Immunol.* **2017**, *37*, 695–700. [CrossRef]

31. Jung, Y.; Rothenberg, M.E. Roles and Regulation of Gastrointestinal Eosinophils in Immunity and Disease. *J. Immunol.* **2014**, *193*, 999–1005. [CrossRef] [PubMed]

32. Mishra, A.; Wang, M.; Pemmaraju, V.R.; Collins, M.H.; Fulkerson, P.C.; Abonia, J.P.; Blanchard, C.; Putnam, P.E.; Rothenberg, M.E. Esophageal Remodeling Develops as a Consequence of Tissue Specific IL-5-Induced Eosinophilia. *Gastroenterology* **2008**, *134*, 204–214. [CrossRef] [PubMed]

33. Mehta, P.; Furuta, G.T. Eosinophils in Gastrointestinal Disorders: Eosinophilic Gastrointestinal Diseases, Celiac Disease, Inflammatory Bowel Diseases, and Parasitic Infections. *Immunol. Allergy Clin. N. Am.* **2015**, *35*, 413–437. [CrossRef] [PubMed]

34. Latour, S.; Aguilar, C. XIAP Deficiency Syndrome in Humans. *Semin. Cell Dev. Biol.* **2015**, *39*, 115–123. [CrossRef]

Journal of
Clinical Medicine

Article

Incidence, Management Experience and Characteristics of Patients with Giardiasis and Common Variable Immunodeficiency

Irene Díaz-Alberola [1,2,3,*], Juan Francisco Gutiérrez-Bautista [3], Andrea Espuch-Oliver [4], José María García-Aznar [5], Per Anderson [3,6], Pilar Jiménez [1,3], Carmen Hidalgo-Tenorio [3,7,†] and Miguel Ángel López-Nevot [1,3,6,†]

[1] Servicio de Análisis Clínicos e Inmunología, Hospital Universitario Virgen de las Nieves, 18014 Granada, Spain
[2] Programa de Doctorado en Biomedicina, University of Granada, 10816 Granada, Spain
[3] Instituto de Investigación Biosanitaria de Granada (ibs.GRANADA), 18012 Granada, Spain
[4] Servicio de Reproducción Asistida, Hospital Universitario de Torrecárdenas, 04009 Almería, Spain
[5] Health in Code S.L., 15008 A Coruña, Spain
[6] Departamento Bioquímica, Biología Molecular e Inmunología III, University of Granada, 18071 Granada, Spain
[7] Unidad de Enfermedades Infecciosas, Hospital Universitario Virgen de las Nieves, 18014 Granada, Spain
[*] Correspondence: idiazalberola@gmail.com
[†] These authors contributed equally to the work.

Citation: Díaz-Alberola, I.;
Gutiérrez-Bautista, J.F.;
Espuch-Oliver, A.; García-Aznar, J.M.;
Anderson, P.; Jiménez, P.;
Hidalgo-Tenorio, C.; López-Nevot,
M.Á. Incidence, Management
Experience and Characteristics of
Patients with Giardiasis and
Common Variable Immunodeficiency.
J. Clin. Med. **2022**, *11*, 7007.
https://doi.org/10.3390/
jcm11237007

Academic Editor: Ilaria Cavazzana

Received: 24 October 2022
Accepted: 25 November 2022
Published: 27 November 2022

Publisher's Note: MDPI stays neutral
with regard to jurisdictional claims in
published maps and institutional affiliations.

Abstract: Common variable immunodeficiency (CVID) is an antibody immunodeficiency with a wide variety of clinical and immunological manifestations, and whose genetic cause is found in about 25% of diagnosed cases. *Giardia lamblia* is one of the main causes of gastrointestinal infections in CVID. 5-Nitroimidazoles are the most used first-line treatment, but nitroimidazole-refractory giardiasis is increasing. Nevertheless, only a few cases of refractory giardiasis in CVID have been reported. This study aimed to determine the incidence of *Giardia* infection in our CVID cohort, shows our management experience and describes patients' phenotypic features. Clinical data collection, immunological, immunogenetics and microbiology assays were performed, and previous cases of giardiasis in CVID were reviewed. The incidence of symptomatic giardiasis was 12.9%. The main immunological features were undetectable or decreased IgA levels and reduced switched memory B cells. A probable *PTEN* pathogenic variant was detected in one. Three patients responded to metronidazole but suffered reinfections, and one was a refractory giardiasis eradicated with innovative quinacrine plus paromomycin combination. This work could contribute to the decision-making and therapeutic management of future patients with CVID and giardiasis, highlighting the importance of the early detection and treatment of infections in patients with CVID to ensure a good quality of life.

Keywords: CVID; immunodeficiency; gastrointestinal infections; *Giardia*; giardiasis; refractory; treatment; immunology; immunogenetic

1. Introduction

Common variable immunodeficiency (CVID) is the most prevalent symptomatic primary immunodeficiency (PID) in humans [1] and is included in the antibody predominant immunodeficiency category according to the International Union of Immunological Societies (IUIS) classification [2].

CVID is considered a complex group of PID due to its clinical and immunological heterogeneity, and the underlying genetic cause is mostly unknown. Genetic defects are detected in approximately 25% of the cases, involving defects in humoral and cell-mediated immunity [3,4]. Diagnostic criteria for CVID, according to the European Society for Immunodeficiencies (ESID), include a decrease in IgG (at least two standard deviations below the mean for age) and a marked decrease in at least one of the isotypes IgM or IgA, an impaired antibody production to vaccination or low percentage of switched memory B cells (<70% of age-related normal value), clinical manifestations of recurrent infections, autoimmune diseases or lymphoproliferation, the onset of clinical immunodeficiency at more than two years of age and the exclusion of other causes of hypogammaglobulinemia [5]. The defect in plasma cell differentiation causes hypogammaglobulinemia and abnormalities of circulating B cell subsets, with a normal or low absolute count of B cells [6]. Although profound T cell defects are not detected, alterations in their frequency and function can be found [7].

Severe and recurrent infections are the clinical hallmark in CVID patients. *Giardia lamblia* is the most commonly identified gastrointestinal pathogen in CVID, followed by *Campylobacter jejuni* and *Salmonella species* [8]. *Giardia lamblia* (also termed *G. duodenalis* or *G. intestinalis*) is a flagellated parasitic protozoan with a lifecycle divided into two phases: the dormant infectious cyst and the proliferating trophozoite [9] (Figure 1). Clinical manifestations of *Giardia* infection are diverse, ranging from asymptomatic cases to diarrhea, abdominal pain, nausea, anemia, malabsorption, or weight loss. Classic diagnosis is performed by microscopic detection of trophozoites or cysts in stool samples, but in recent years, rapid immunochromatographic antigen tests and more sensitive real-time polymerase chain reaction (PCR) panels have appeared [10].

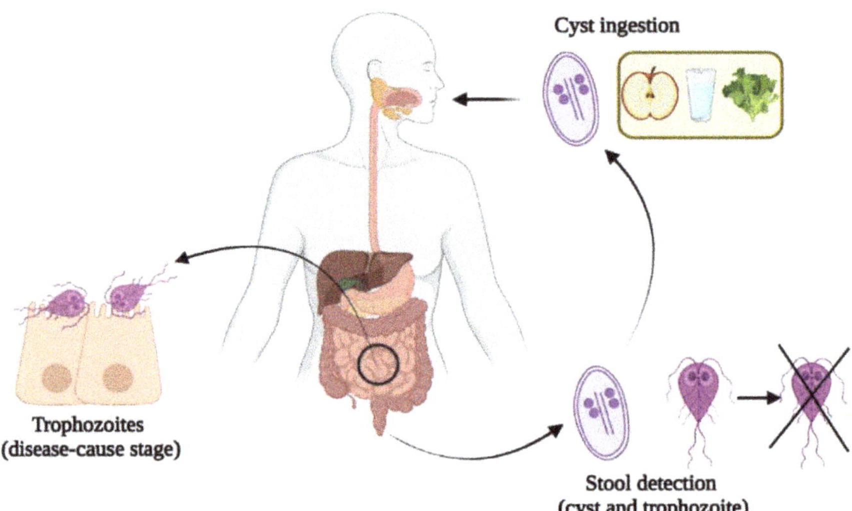

Figure 1. Lifecycle of *Giardia lamblia*. Infectious cysts are ingested via contaminated food or water, or by direct ingestion. In the human gastrointestinal tract, cysts excyst to release trophozoites, which cause disease, in part, by promoting the disruption of the intestinal epithelial barrier. Both the cysts and trophozoites can be detected in the stool, although the trophozoites released do not survive long.

5-Nitroimidazole compounds, such as tinidazole or metronidazole, are the most common first-line treatment for *Giardia* infection [11]. Nitroimidazoles are usually also effective in CVID patients. However, CVID patients have a higher risk of chronification, reinfection and relapse rate due to their immunodeficiency status or malabsorption syndrome, and often require prolonged treatment [12,13].

Nitroimidazole-refractory giardiasis is increasing in the general population, linked with parasite drug resistance and host factors [14–16]. Nevertheless, few cases of refractory giardiasis in CVID patients have been published to date [17]. Currently, resistance *Giardia* detection is not possible to perform in most laboratories, and there is no standard treatment for refractory giardiasis. Empirical treatments are currently used, highlighting the use of a nitroimidazole other than metronidazole in monotherapy or with another drug, or other agents such as quinacrine or paromomycin [18].

The aim of this study was to determine the incidence of *Giardia* infection in patients diagnosed with CVID at our hospital center, describing our management experience and their demographic, clinical, immunogenetic, and immunological characteristics. In addition, we have performed a literature review of previous reports of *Giardia* infection in CVID patients.

2. Materials and Methods

2.1. Subjects of Study

Patients diagnosed with CVID and *Giardia lamblia* infection in the University Hospital Virgen de las Nieves between 2000 and 2021 were recruited for this study. The diagnosis of CVID was established based on the ESID criteria [5], excluding patients with other types of antibody immunodeficiencies, secondary antibody deficiencies, and T-cell deficiency.

Giardia infection was determined by stool *Giardia* antigen detection test, microscopy observation, molecular technique, or a combination of these, in patients with suggestive symptoms as described below. Refractory giardiasis was considered when *Giardia* persisted after one or more strategic treatments. We collected demographic and clinical data, family and personal history, and immunoglobulin levels at CVID diagnosis. Furthermore, we performed other immunological and immunogenetics assays during subsequent follow-ups of each patient. This study was reviewed and approved by the regional ethics committee (Portal de Ética de la Investigación Biomédica de Andalucía, PEIBA, code: 1206-N-22). Patients or their legal representatives provided their written informed consent to participate.

2.2. Immunological Evaluation

Serum immunoglobulins (Ig) levels (IgG, IgA and IgM) were measured by immuno-turbidimetry using the automatic analyzer Alinity c system (Abbott Laboratories, Chicago, IL, USA). For cellular evaluation, EDTA whole blood samples were collected. Lymphocyte subpopulations (CD4+ T, CD8+ T, B and NK cells) were performed using BD Trucount tubes and the BD Multitest 6 Color BTNK kit (BD Biosciences, San Diego, CA, USA), which included the following mixtures of fluorophore-conjugated monoclonal antibodies (mAb): anti-CD45-PerCP-Cy5.5, anti-CD3-FITC, anti-CD8-APC-Cy7, anti-CD4-PE-Cy7, anti-CD19-APC, and anti-CD16+CD56-PE. B cell phenotype was performed with an eight-color panel of the following mAb: anti-CD45-APC-H7, anti-CD19-V500, anti-CD10-V450, anti-CD38-PE-Cy7, anti-CD21-PE, anti-CD27-PerCP-Cy5, anti-IgD-FITC, and anti-IgM-APC (BD Biosciences, San Diego, CA, USA), following EURO-Class classification. Cells were acquired on a BD FACSCanto II Flow Cytometer (BD Biosciences, San Diego, CA, USA), and the InfinicytTM22.0 software was employed for multiparametric analysis (Cytognos SL, Salamanca, Spain).

2.3. Immunogenetics

High-resolution genotyping of Human Leukocyte Antigen (HLA) class I (A, B and C) and II (DRB1 and DQB1) loci was performed using the LABType sequence-specific oligonucleotide typing test (One Lambda, Canoga Park, CA, USA). DNA whole blood isolation was carried out with the QIAMP DNA Blood Mini Kit, following the manufacturer's instructions (Qiagen, Hilden, Germany). Target DNA was amplified by PCR using sequence-specific primers, followed by hybridization with allele-specific oligodeoxynucleotides coupled with fluorescent phycoerythrin-labelled microspheres. Fluorescence intensity was determined using a LABScan 100 system (Luminex xMAP, Austin, TX, USA). HLA alleles were assigned using the HLA-Fusion software (One Lambda, Canoga Park, CA, USA).

We also performed a clinical exome analysis based on Next-Generation Sequencing (NGS) that covers the coding regions of 4490 genes with clinical significance (SOPHiA Clinical Exome Solution, Lausanne, Switzerland), and analyzed 237 genes associated with primary immunodeficiencies (Supplementary Material S1). Sequencing was carried out on the NextSeq 1000 platform (Illumina, San Diego, CA, USA), and the results were analyzed with DDM v.5.8.0.3 program of Sophia Genetics and the IGV informatic application (Integrative Genomics Viewer). The reference genome sequence used in the alignment phase corresponds to the GRCh37/hg19 (UCSC) version. Bioinformatic predictors (Mutation-Taster and CADD) were used to evaluate the pathogenicity of the variants found. Genetic variants found were confirmed by Sanger sequencing.

2.4. Microbiology Giardia Infection Diagnosis

Giardia infection was diagnosed using antigen detection by immunochromatography (Rida Quick Cryptosporidium/Giardia/Entamoeba, R-Biopharm AG, Darmstadt, Germany), by microscopic observation of cysts in stool samples or by molecular diagnosis (FilmArray Gastrointestinal Panel, bioMérieux, Marcy l'Étoile, France).

2.5. Systematic Literature Review

A search was performed on the PubMed database up to 2022. Search terms used were: "giardia" OR "giardiasis", AND "common variable immunodeficiency" (37 results). Articles available in English and Spanish were included, and articles that were not related to the subject of the study or did not provide sufficient data on treatment for giardiasis, microbiological and immunological diagnosis were excluded. Finally, 16 articles were included (Figure 2).

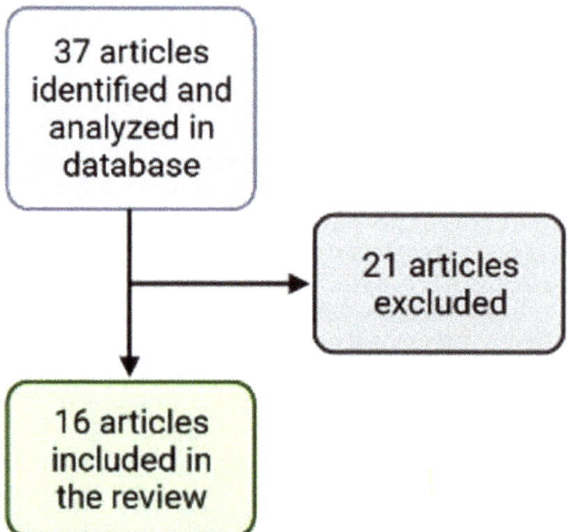

Figure 2. Flowchart of study selection for the narrative review.

3. Results

3.1. Demographic Data and Clinical Manifestations

Thirty-one patients were diagnosed with CVID and treated with immunoglobulin replacement in our hospital center between 2000 and 2021. Four CVID patients (12.9%) suffered from *Giardia* infection during their clinical course. The median age of CVID patients with *Giardia* infection was 44 years old (28–55), two were male (50%) and two were female (50%). The median age at the time of CVID diagnosis was 33 (19–49) years. *Giardia* infection was detected between the first and the fourth year after beginning intravenous immunoglobulin (IVIG) treatment. CVID patients with *Giardia* infection had variable gastrointestinal symptoms and others such as asthenia or febricula. Three of them had *Giardia* reinfections (Cases 1, 2 and 4) and one was a refractory giardiasis (Case 3). In the two females, nodular intestinal lymphoid hyperplasia (NILH) was detected by endoscopy after the first *Giardia* infection, and Case 4 developed a Crohn's disease-*like*. During the evolution of the CVID, without overlapping with the *Giardia* infection, patients also suffered from other clinical conditions, infections and comorbidities, which are summarized in Table 1.

Currently, Cases 1, 2 and 3 are clinically stable. Case 4 continues with gastrointestinal symptoms, arthritis, asthenia, anorexia and intense migraines, numbness in the face and loss of vision, probably associated with her Crohn-like disease. All CVID patients are with IVIG treatment. They receive 0,4 g/kg/day every 21 days except Case 4, which received 0,6 g/kg/day every 21 days because of her clinical condition.

3.2. Immunological Evaluation

The four CVID patients had undetectable or very low IgM and IgA levels from diagnosis. After IVIG treatment, all patients reached a normal IgG level, which is currently maintained. The analysis of lymphocyte subsets showed CD4+ T cell lymphopenia in Cases 3 and 4, a remarkable reduction of NK cells in Cases 1 and 4, and a very low B cell count in Case 1. B cell immunophenotype highlighted the reduction in switched memory B cells in all cases except in Case 3, in which they were absent. CD21*low* B cells were increased in Case 1 and transitional B cells were remarkably increased in Case 3. EURO-Class classification group and immunological data are indicated in Table 1.

3.3. Immunogenetics

HLA class I and II alleles genomic typing were performed and are shown in Supplementary Tables, along with HLA allele frequencies in the Spanish Caucasian population [19]. Case 3 had homozygosity in HLA-DQB1 alleles and Case 1 in HLA-DRB1. Case 1 had the ancestral haplotype 44.3.

The exome analysis performed was negative for Cases 1 and 3. Case 2 showed the heterozygous c.1555A>G (Lys519Glu) variant in the exon 17 of *TCF3* or *E2A* gen, which affects a region not associated with any of the major domains of the transcription factor it encodes. In silico and phylogenetic studies suggested that the affected residue is highly conserved and bioinformatic predictors did not give conclusive results on its pathogenicity. The amino acid substitution slightly modifies the physical–chemical properties of the protein. Databases consulted showed that there are asymptomatic heterozygous carriers in the general population as well as in asymptomatic carriers in families with severe agammaglobulinemia caused by biallelic variants (gnomAD frequency of 0.016%). Case 4 had the heterozygous c.1093G>A (Val365Ile) variant in exon 9 of phosphatase and tensin homolog (*PTEN*) gene, which affected the C-terminal domain of the mature protein phosphatase. In silico and phylogenetic studies suggest that the affected residue is highly conserved among vertebrate species and bioinformatic predictors showed that could be a pathogenic variant (MutationTaster score: 1, DANN score: 0.969). The amino acid substitution slightly modifies the physical–chemical properties of the protein. The database consulted points to a rare variant (gnomAD frequency < 0.01%), only present in two heterozygous carriers from the European population (Supplementary Material S2).

3.4. Giardia Infection Diagnosis and Treatment

The first *Giardia* infection was diagnosed by the *Giardia/Cryptosporidium* antigen test in Cases 1 and 2. Case 3 was diagnosed by an antigen test and multiplex PCR and Case 4 by cysts stool detection. The detection of reinfection, treatment failure and eradication of *Giardia* were performed by (a) stool cultures and/or the antigen test repetition in Case 2, (b) an antigen test and a multiplex PCR with symptom remission in Case 3, and (c) the remission of symptoms and negative antigen test in Case 4. In Case 1, it was not possible to assure *Giardia* eradication after the first infection because microbiological tests were not performed and the diagnosis was based on the remission of symptoms. In the reinfection, a stool examination was performed and *Giardia* cysts were not detected, confirming its eradication. None of the patients currently have evidence of *Giardia* infection, with antigen test negative.

Table 1. Epidemiological, clinical and immunological characteristics of patients with *Giardia* infection and CVID.

	Case 1	Case 2	Case 3	Case 4	Reference Values
Sex	Male	Male	Female	Female	
Age (years)	28	55	47	41	
Age at CVID diagnosis (years)	26	49	38	19	
Manifestation of *Giardia* infection	Diarrhea, weight loss	Diarrhea, weight loss, abdominal pain, malabsorption, rectal tenesmus, anal itching, asthenia, febricula	Diarrhea, weight loss, abdominal pain, rectal tenesmus, iron deficiency, anemia	Diarrhea, weight loss, abdominal pain, malabsorption, asthenia, iron deficiency	
Clinical and comorbid conditions	Recurrent respiratory infections, SARS-CoV-2 infection, psoriasis	Recurrent respiratory infection, otitis, *Campylobacter jejuni* infection, latent tuberculosis, verrucous cutaneous squamous cell carcinoma	Recurrent respiratory infections, NILH, ulcerative colitis, primary hypothyroidism, chronic hepatopathy, splenomegaly	Recurrent respiratory infections, bronchiectasis, HBV, SARS-CoV-2 infection, NILH, Crohn-*like* disease, arthralgias and arthritis	
Immunoglobulins (Ig) at diagnosis					
IgG (mg/dL)	164	444	461	200	540–1822
IgM (mg/dL)	<5	19	9	<5	22–240
IgA (mg/dL)	<5	10	<5	<5	70–400
Immunoglobulins (Ig) after IVIG treatment					
IgG (mg/dL)	1072	926	824	981	540–1822
IgM (mg/dL)	<5	19	<5	<5	22–240
IgA (mg/dL)	<5	<5	<5	<5	70–400
Lymphocyte subsets (cells/µL/%)					
CD3+	1441 (92)	1374 (69)	979 (60)	936 (81)	960–2600/61–84
CD3+CD4+	631 (40)	725 (36)	526 (32)	449 (39)	540–1660/32–60
CD3+CD8+	675 (43)	625 (31)	396 (24)	436 (38)	270–930/13–40
CD19+	28 (2)	356 (18)	115 (7)	127 (11)	122–632/6–27
CD3-CD56+CD16+	75 (4.8)	242 (12.1)	522 (32)	79 (6.9)	127–509/10.1–20.9
Ratio CD4/CD8	0.93	1.16	1.33	1.03	0.9–4.5
B cell subsets (%)					
CD19+ naive (IgD+CD27-)	46.2	78.3	24	75	53–86
CD19+ pre-switched memory (IgD+CD27+)	36.4	6.7	33.3	13.1	3.3–12.8
CD19+ switched memory (IgD-CD27+)	1.8	3.2	0	0.9	4–22
CD19+ CD21*low* (CD38*low*, IgM+)	10.2	2.2	0.3	1.5	0.4–4.5
CD19+ transitional (CD38*high*, IgM+)	0.4	0.7	34.5	0.6	0.9–6.3
EURO-Class classification group	smB-Tr[norm] smB-21[lo]	smB+21[norm]	smB-Tr[hi] smB-21[norm]	smB-Tr[norm] smB-21[norm]	

NILH: nodular intestinal lymphoid hyperplasia; HBV: hepatitis B virus; Ig: immunoglobulin; IVIG: intravenous immunoglobulin.

First-line *Giardia* infection treatment was metronidazole, which successfully cured infection in Cases 1, 2 and 4. Quinacrine plus paromomycin was the successful treatment used for refractory giardiasis in Case 3 (Table 2).

Table 2. *Giardia* infection treatments in each patient.

	Case 1	Case 2	Case 3	Case 4
First-line treatment	Metronidazole 250 mg/8 h for 7 days	Metronidazole 250 mg/8 h every 5 days	Metronidazole 500 mg every 8 h for 7 days	Metronidazole 250 mg every 8 h for 8 days
Other treatments	Reinfection: Metronidazole 250 mg/8 h for 20 days	Reinfection: Metronidazole 500 mg every 8 h for 10 days	1st failure: Metronidazole 500 mg every 8 h for 14 days. 2nd failure: Tinidazole 2 g spread over two consecutive days. 3rd failure: Metronidazole plus Albendazole (400 mg/12 h) for 10 days. 4th treatment: Quinacrine 100 mg plus paromomycin 400 mg every 8 h	Reinfection: Metronidazole 500 mg every 8 h for 7 days
Successful *Giardia* infection drug	Metronidazole	Metronidazole	Quinacrine + Paromomycin	Metronidazole

mg: milligrams; g: grams; h: hours.

3.5. Database Review Results

We found 17 published cases of giardiasis in CVID. The median age was 39 years old (15–62), eleven were male (64.7%) and six were female (35.3%). Six were refractory giardiasis (35.3%). The main clinical manifestations were diarrhea (14/17; 82.3%), weight loss and abdominal symptoms (9/17; 52.9%), and three patients had splenomegaly (3/17; 17.6%). All CVID patients showed undetectable or decreased IgA levels. All summarized data are collected in Table 3.

Table 3. Cases of giardiasis in CVID documented in the literature.

References	Age (Years), Gender	Giardiasis	Clinical Giardiasis Manifestations	IgA Level	Curative Treatment	Microbiological Cure	Other Characteristics
Taylor GC et al., 1987 [20]	47, Male	Refractory giardiasis	Diarrhea, abdominal cramps, weight loss	Undetectable	Metronidazole + Quinacrine	Confirmed by stool microscopy	Failure treatment: Metronidazole
Bästlein C, Burlefinger R et al., 1988 [21]	31, Male	Chronic giardiasis	Abdominal pain, splenomegaly	Decreased	Metronidazole	Not done (symptomatic relief)	NILH
Nash TE et al., 2001 [22]	46, Female	Refractory giardiasis	Nausea, diarrhea, pernicious anemia,	N.A.	Metronidazole + Quinacrine	Confirmed by *Giardia* antigen test	Failure treatments: Metronidazole
De Weerth, et al., 2002 [23]	40, Female	Giardiasis	Diarrhea, abdominal pain, weight loss	Undetectable	Metronidazole	Not done (symptomatic relief)	NILH detection, which reduces after *Giardia* eradication
Onbaşi K, Günşar F et al., 2005 [24]	39, Female	Chronic giardiasis	Diarrhea, weight loss	Decreased	Metronidazole	Not done (symptomatic relief)	
Ogershok PR, Hogan MB et al., 2006 [25]	24, Male 15, Male	Giardiasis	Diarrhea	Undetectable	N.A,	N.A.	
Ramsey NC et al., 2010 [26]	46, Male	Refractory giardiasis	Abdominal pain, diarrhea, splenomegaly and lymphadenopathy	N.A.	Tinidazole + Paromomycin + Bacitracin + Cholestyramine	Confirmed by *Giardia* antigen test (EIA)	Lymphadenopathy and splenomegaly regressed after *Giardia* eradication. Failure treatments: metronidazole, cholestyramine + tinidazole
Domínguez-López ME, González-molero I, et al., 2011 [27]	49, Male	Chronic giardiasis	Diarrhea, weight loss	N.A.	N.A.	N.A.	NILH, gastrectomy for gastric cancer
Silva GB et al., 2012 [28]	62, Female	Giardiasis	Diarrhea, weight loss, abdominal pain, and intermittent fever	Decreased	Metronidazole	Not done (symptomatic relief)	*Isospora belli* coinfection
Olmez S, Aslan M et al., 2014 [29]	38, Male	Giardiasis	Dyspeptic complaints	Decreased	N.A.	Not indicated	NILH detection
Choi JH et al., 2017 [30]	41, Female	Refractory giardiasis	Intermittent diarrhea	Undetectable	Metronidazole + Albendazole	Confirmed by stool microscopy	NILH detection, which reduces after *Giardia* eradication Failure treatments: metronidazole, albendazole, tinidazole
Paranjpe SM et al., 2017 [13]	18, Male	Chronic giardiasis	Diarrhea, weight loss, abdominal pain	Remarkably decreased	Metronidazole	Not done (symptomatic relief)	
Atalaia-Martins C, Barbeiro S et al., 2017 [31]	42, Female	Refractory giardiasis	Epigastric discomfort, postprandial fullness, diarrhea, weight loss, fatigue, anemia	Undetectable	N.A.	Not indicated	*Helicobacter pylori* coinfection, gastric dysplasia, NILH. Failure treatments: metronidazole, tinidazole, albendazole
Saurabh K, Nag VL et al., 2017 [32]	16, Male	Giardiasis	Diarrhea, nausea, vomiting, pedal edema	Decreased	Nitazoxanide + Metronidazole	Confirmed by stool microscopy	*Hymenolepis nana* coinfection. Finally, he died because of his immunological condition.
Kaya F et al., 2018 [33]	28, Male	Refractory giardiasis	Diarrhea, nausea and bloating, abdominal cramps and weight loss	Decreased	N.A.	N.A.	Failed treatments: Metronidazole, ornidazole, albendazole, nitazoxanide, trimetho-prim/sulfamethoxazole + metronidazole, nitazoxanide + paromomycin
Sousa D, 2020 [34]	33, Male	Chronic giardiasis	Diarrhea, weight loss, iron-deficiency anemia, splenomegaly	Undetectable	Metronidazole	N.A.	

N.A.: not available. EIA: enzyme immunoassay. NILH: nodular intestinal lymphoid hyperplasia.

4. Discussion

The incidence of symptomatic giardiasis in our CVID cohort was 12.9%. The main clinical manifestations of *Giardia* infection were diarrhea, weight loss and abdominal pain, both in our patients and in review patients. It is important to make a correct differential diagnosis with other entities such as celiac disease and inflammatory bowel disease (IBD), which was performed in our patients, because gastrointestinal symptoms are very common in CVID, especially transient or persistent diarrhea [35]. Gastrointestinal *Giardia* symptoms in Cases 2 and 4 produced protein loss and malabsorption, which made it difficult to maintain their IgG levels in the normal range despite IVIG treatment. Immunoglobulin replacement therapy is the basic treatment of these patients, which improve their symptoms and their quality of life, but also prophylactic and therapeutic antibiotics for their recurrent infections [8]. Antibiotics stimulate intestinal dysbiosis, producing digestive alterations and favoring chronic or refractory infections by gastrointestinal pathogens such as *Giardia*. The gut microbiome has been shown to play a key role in determining susceptibility or resistance to *Giardia* colonization by modulating immune responses, while the parasite itself can influence the immune response to the host [36]. Recent research has recognized giardiasis as an important risk factor for developing long-term postinfectious syndromes, such as IBD, chronic fatigue syndrome, and extraintestinal consequences such as arthritis or allergies, even months or years after parasite clearance [16,37], which could be one of the reasons for the Crohn's disease-*like* symptomatology in Case 4.

It should be noted that Cases 3 and 4 had nodular intestinal lymphoid hyperplasia (NILH), which is also found in CVID patients evaluated in the literature review (6/17; 35.3%). NILH is a rare condition that can occur as a form of compensation for the functionally inadequate intestinal lymphoid tissue found in patients with immunodeficiencies. However, NILH has also been associated with *Giardia lamblia* infection, and in some cases, its eradication has resulted in NILH reduction [23], or with an overregulation of the response mechanisms of the lymphoid tissue associated with the digestive tract [29]. In Cases 3 and 4, neither possibility can be ruled out, but it would be interesting to monitor its evolution because NILH could represent an intermediate stage and a risk factor for the development of lymphoma [38].

IgA, Th17 and CD4+ T cells are key in the immune response against *Giardia* [39]. Undetectable or very low levels of IgA are a characteristic in our four patients and in all available review patients, an immunoglobulin that is not replaced with IVIG treatment and that is key in the defense of the intestinal mucosa. Various studies have associated its deficiency with a greater capacity for binding and proliferation of foreign pathogens such as *Giardia* to the intestinal epithelium [12]. T CD4+ lymphopenia found in Case 3 could be another immunological factor resulting in refractory giardiasis. The remarkable reduction of B cells in Case 1 led us to suspect Bruton's disease but it was ruled out by the NGS study. A remarkable NK cell lymphopenia was detected in Cases 1 and 4. These cells are important in the antiviral and antitumor defense. However, in CVID patients, NK cells have been associated with high frequencies of severe bacterial infections and non-infectious complications, suggesting that NK cells also have a role in controlling bacterial infections [40].

The EURO-Class classification was a multicenter European effort that allowed defining different subgroups of patients with CVID based on their B cell phenotype and some clinical features [6]. In our study, all cases had a reduction in switched memory B cells. The increase in CD21*low* cells is the strongest marker associated with splenomegaly. Case 3 had splenomegaly, but CD21*low* B cells were not altered, and she had also a remarkable increase of transitional B cells, without lymphadenopathy. In one literature case with refractory giardiasis, lymphadenopathy and splenomegaly were detected, but both regressed after *Giardia* eradication [26].

Different genetic variants have been involved in the pathogenesis of CVID, but most patients do not have a specific genetic defect background [41]. The NGS analysis detected possible CVID-associated variants in two of our patients. In Case 2, a heterozygous variant in *TCF3* gene (p.Lys519Glu) was detected. This gene is located on chromosome 19p13.3 and codes for transcription factors that promote the expression of genes involved in lymphopoiesis, differentiation and maturation of B and T cells. This variant has not been previously described but its presence in the control population and in the asymptomatic carriers of families with severe agammaglobulinemia caused by biallelic variants, as well as the contradictory results of in silico predictors, point to a benign variant [42,43]. In Case 4, a heterozygous variant in *PTEN* gene (p.Val365Ile) was detected. This gene is located on chromosome 10q23.31 and codes for a phosphatase that acts as a tumor suppressor gene. This enzyme also participates in the PI3K/Akt signaling pathway and, thus, in the differentiation and homeostasis of T and B cells. It has previously been considered a variant of uncertain significance (VUS) according to human databases in patients with Cowden syndrome type 1, a condition within the group of syndromes related to the development of hamartomas (PHTS) following an autosomal dominant pattern of inheritance [44]. These patients had heterogeneous phenotypes, some of whom had PHTS with a CVID phenotype, exhibiting a decrease in switched memory B cells and a reduction in functional and mature NK cells, immunological alterations that are present in our Case 4 patient. Likewise, loss-of-function mutations in *PTEN* have also been described as being associated with activated phosphoinositide-3-kinase delta syndrome-*like* (APDS-*like*) because it acts as an antagonist in the PI3K-delta signaling pathway [45,46]. Cosegregation and functional studies will be necessary to confirm the pathogenicity of this variant in the context of CVID, which will be a future objective of our group.

The HLA class I and II genes code for cell surface molecules specialized in antigen presentation to T cells and play a key role in the immune response. Previous studies have reported a positive association between certain HLA alleles and different diseases, including susceptibility to CVID [47]. It has been described that the extended haplotype HLA-A1, -B8, -DR3 is more frequent in these patients [48] and that homozygosity in the HLA class II region, specially HLA-DQ, is associated with greater susceptibility to CVID [49]. Curiously, Case 3 patient has homozygosity in HLA class II, our giardiasis refractory case. This homozygosity could result in a lower repertoire of HLA class II molecules on the cell surface, resulting in a lower capacity to respond to foreign or pathogenic antigens and contributed to CVID susceptibility to specific environmental conditions. Moreover, Case 4 has the HLA-DRB1*13:01 allele, which has been previously associated with susceptibility to *Giardia* infection [50]. On the other hand, Case 1 patient has the C*06:02 allele, which is one of the most strongly HLA alleles associated with psoriasis susceptibility [51]. None of the four patients presented the ancestral haplotype 8.1 (HLA-A1, -B8, -DR3), although Case 1 had the ancestral haplotype 44.3 (A*29:02; B*44:03; C*16:01; DRB1*07:01; DQB1*02:02). It would be advisable to extend the study to a bigger CVID cohort to have more precise data on this evidence.

First-line giardiasis treatment is based on the use of nitroimidazoles [11]. In our cohort, 75% (3/4) responded to metronidazole, evidenced by negative microbiological tests and remission of symptoms. This is similar to the case reports included in the literature review, although reinfections occurred in all of our cases. However, one of the highlights of this work is the first refractory giardiasis case diagnosis in our hospital in a CVID patient (Case 4) and, to our knowledge, the first time that the combination of quinacrine plus paromomycin has been described and has been effective. We consider the treatment successful because stool examination has remained negative to date. So far, only six refractory giardiasis cases in CVID patients have been published, and different curative treatments have been applied [20,22,26,30,31,33]. Currently, there is no standard therapy for refractory giardiasis, and multiple drugs have been tried [15]. Tinidazole, a derivative of metronidazole, has been shown to be more effective than metronidazole [11,39] but was not effective in Case 3. The combination of metronidazole and albendazole, the latter a benzimidazole, is more effective in treating refractory disease than albendazole alone [15], but again no response was obtained as a third-line treatment. On the other hand, the efficacy and safety of the use of quinacrine in refractory giardiasis have been demonstrated in various studies, mainly in monotherapy or combined with metronidazole [22,52,53]. Quinacrine, also known as mepacrine, was the first antimalarial drug used to treat giardiasis, but its use was limited by detecting possible adverse effects of psychosis. Despite everything, it is usually well-tolerated and has a clinical efficacy of >90% [54]. In the case of paromomycin, its activity is variable against *Giardia* (55–90%) and is rarely used. It can be employed in cases of resistance or pregnancy because it has a low absorption spectrum and does not have systemic effects [14]. It is usually well-tolerated, although being an aminoglycoside, it can cause nephrotoxicity and ototoxicity [15,54]. The quinacrine and paromomycin combination therapy should be used in a larger patient cohort to confirm its efficacy and safety, and more clinical trials are needed to establish the optimal therapy for patients with refractory giardiasis.

Giardia drug resistance is currently undetectable in most microbiology laboratories. This is due, on the one hand, because the parasite culture is difficult and takes a long time, and success rates are relatively low [15] and, on the other hand, because resistance depends more on epigenetic factors and post-translational modifications than on genetic variants in the parasite genome. For these reasons, it is not yet possible to develop a routine microbiology technique to detect *Giardia* susceptibility [14]. Moreover, since there are few cases of refractory giardiasis, multicenter studies are required to define the best therapeutic alternatives. At the diagnostic level, it would be interesting to develop a strategy that combines the culture of the parasite with sequencing and comparative proteomics [55], which may allow moving from empirical treatment to targeted treatment in the future.

The early detection and treatment of infections in patients with CVID are crucial to ensure a good quality of life. Moreover, the immunological and genetic characterization of these patients is necessary to improve their clinical follow-up and to better understand the pathogenesis of CVID. One limitation of this work was that the cohort of CVID patients who suffered from symptomatic *Giardia* infection was small. Still, to our knowledge, we present a probably pathogenic *PTEN* variant associated with the CVID phenotype, and, for the first time, a case of refractory giardiasis in CVID that was successfully treated with quinacrine plus paromomycin combination. These data could contribute to the decision-making and therapeutic management of future patients with CVID and giardiasis, especially in refractory giardiasis cases.

Supplementary Materials: The following supporting information can be downloaded at https://www.mdpi.com/article/10.3390/jcm11237007/s1, Supplementary Material S1: Genes associated with primary immunodeficiency analyzed by Next-Generation sequencing; Supplementary Tables S1 and S2: Human Leukocyte Antigen (HLA) genomic typing of CVID patients with giardiasis; Supplementary Material S2: Genetic results and sanger sequencing methodology.

Author Contributions: Conceptualization, M.Á.L.-N. and C.H.-T.; formal analysis, I.D.-A., J.M.G.-A., P.J. and M.Á.L.-N.; funding acquisition, M.Á.L.-N.; investigation, I.D.-A., J.F.G.-B. and A.E.-O.; methodology, I.D.-A. and M.Á.L.-N.; supervision, M.Á.L.-N., C.H.-T. and P.A.; visualization, I.D.-A., J.F.G.-B., A.E.-O. and J.M.G.-A.; writing—original draft preparation, I.D.-A., J.F.G.-B.,A.E.-O. and P.A.; writing—review and editing, I.D.-A., J.F.G.-B., A.E.-O., M.Á.L.-N. and C.H.-T. All authors have read and agreed to the published version of the manuscript.

Funding: This study was partially financed by Palex Medical S.A.

Institutional Review Board Statement: The study was conducted in accordance with the Declaration of Helsinki, reviewed and approved by the Portal de Ética de la Investigación Biomédica de Andalucía, PEIBA (Cod. 1206-N-22).

Informed Consent Statement: Informed consent was obtained from all subjects involved in the study.

Acknowledgments: This study is part of the doctoral thesis of Irene Díaz-Alberola, within the program of Biomedicine, conducted at the University of Granada, Spain.

Conflicts of Interest: J.M.G.-A. is an employee of Health in Code S. L. All authors declare no conflict of interest.

References

1. Ameratunga, R.; Allan, C.; Woon, S.-T. Defining Common Variable Immunodeficiency Disorders in 2020. *Immunol. Allergy Clin. N. Am.* **2020**, *40*, 403–420. [CrossRef]
2. Tangye, S.G.; Al-Herz, W.; Bousfiha, A.; Cunningham-Rundles, C.; Franco, J.L.; Holland, S.M.; Klein, C.; Morio, T.; Oksenhendler, E.; Picard, C.; et al. Human Inborn Errors of Immunity: 2022 Update on the Classification from the International Union of Immunological Societies Expert Committee. *J. Clin. Immunol.* **2022**, *42*, 1473–1507. [CrossRef]
3. Ramirez, N.J.; Posadas-Cantera, S.; Caballero-Oteyza, A.; Camacho-Ordonez, N.; Grimbacher, B. There Is No Gene for CVID—Novel Monogenetic Causes for Primary Antibody Deficiency. *Curr. Opin. Immunol.* **2021**, *72*, 176–185. [CrossRef] [PubMed]
4. Ameratunga, R.; Woon, S.-T. Perspective: Evolving Concepts in the Diagnosis and Understanding of Common Variable Immunodeficiency Disorders (CVID). *Clin. Rev. Allergy Immunol.* **2020**, *59*, 109–121. [CrossRef] [PubMed]
5. Seidel, M.G.; Kindle, G.; Gathmann, B.; Quinti, I.; Buckland, M.; van Montfrans, J.; Scheible, R.; Rusch, S.; Gasteiger, L.M.; Grimbacher, B.; et al. The European Society for Immunodeficiencies (ESID) Registry Working Definitions for the Clinical Diagnosis of Inborn Errors of Immunity. *J. Allergy Clin. Immunol. Pract.* **2019**, *7*, 1763–1770. [CrossRef]
6. Wehr, C.; Kivioja, T.; Schmitt, C.; Ferry, B.; Witte, T.; Eren, E.; Vlkova, M.; Hernandez, M.; Detkova, D.; Bos, P.R.; et al. The EUROclass Trial: Defining Subgroups in Common Variable Immunodeficiency. *Blood* **2008**, *111*, 77–85. [CrossRef]
7. Azizi, G.; Rezaei, N.; Kiaee, F.; Tavakolinia, N.; Yazdani, R.; Mirshafiey, A.; Aghamohammadi, A. T-Cell Abnormalities in Common Variable Immunodeficiency. *J. Investig. Allergol. Clin. Immunol.* **2016**, *26*, 233–243. [CrossRef] [PubMed]
8. Yazdani, R.; Habibi, S.; Sharifi, L.; Azizi, G.; Abolhassani, H.; Olbrich, P.; Aghamohammadi, A. Common Variable Immunodeficiency: Epidemiology, Pathogenesis, Clinical Manifestations, Diagnosis, Classification, and Management. *J. Investig. Allergol. Clin. Immunol.* **2020**, *30*, 14–34. [CrossRef]
9. Einarsson, E.; Ma'ayeh, S.; Svärd, S.G. An Up-Date on Giardia and Giardiasis. *Curr. Opin. Microbiol.* **2016**, *34*, 47–52. [CrossRef]
10. Alharbi, A.; Toulah, F.H.; Wakid, M.H.; Azhar, E.; Farraj, S.; Mirza, A.A. Detection of Giardia Lamblia by Microscopic Examination, Rapid Chromatographic Immunoassay Test, and Molecular Technique. *Cureus* **2020**, *12*, e10287. [CrossRef] [PubMed]
11. Ordóñez-Mena, J.M.; McCarthy, N.D.; Fanshawe, T.R. Comparative Efficacy of Drugs for Treating Giardiasis: A Systematic Update of the Literature and Network Meta-Analysis of Randomized Clinical Trials. *J. Antimicrob. Chemother.* **2018**, *73*, 596–606. [CrossRef]
12. Agarwal, S.; Mayer, L. Pathogenesis and Treatment of Gastrointestinal Disease in Antibody Deficiency Syndromes. *J. Allergy Clin. Immunol.* **2009**, *124*, 658–664. [CrossRef]
13. Paranjpe, S.M. Chronic Giardiasis in a Case of Common Variable Immunodeficiency (CVID): A Case Report. *J. Clin. Diagn. Res.* **2016**, *10*, DD03-4. [CrossRef]
14. Mørch, K.; Hanevik, K. Giardiasis Treatment: An Update with a Focus on Refractory Disease. *Curr. Opin. Infect. Dis.* **2020**, *33*, 355–364. [CrossRef]
15. Carter, E.R.; Nabarro, L.E.; Hedley, L.; Chiodini, P.L. Nitroimidazole-Refractory Giardiasis: A Growing Problem Requiring Rational Solutions. *Clin. Microbiol. Infect.* **2018**, *24*, 37–42. [CrossRef] [PubMed]
16. Lalle, M.; Hanevik, K. Treatment-Refractory Giardiasis: Challenges and Solutions. *Infect. Drug Resist.* **2018**, *11*, 1921–1933. [CrossRef] [PubMed]
17. Bourque, D.L.; Neumayr, A.; Libman, M.; Chen, L.H. Treatment Strategies for Nitroimidazole-Refractory Giardiasis: A Systematic Review. *J. Travel Med.* **2022**, *29*, taab120. [CrossRef]
18. Vivancos, V.; González-Alvarez, I.; Bermejo, M.; Gonzalez-Alvarez, M. Giardiasis: Characteristics, Pathogenesis and New Insights About Treatment. *Curr. Top. Med. Chem.* **2018**, *18*, 1287–1303. [CrossRef]
19. Gonzalez-Galarza, F.F.; McCabe, A.; dos Santos, E.J.M.; Jones, J.; Takeshita, L.; Ortega-Rivera, N.D.; Cid-Pavon, G.M.D.; Ramsbottom, K.; Ghattaoraya, G.; Alfirevic, A.; et al. Allele Frequency Net Database (AFND) 2020 Update: Gold-Standard Data Classification, Open Access Genotype Data and New Query Tools. *Nucleic Acids Res.* **2020**, *48*, D783–D788. [CrossRef]
20. Taylor, G.D.; Wenman, W.M.; Tyrrell, D.L. Combined Metronidazole and Quinacrine Hydrochloride Therapy for Chronic Giardiasis. *CMAJ Can. Med. Assoc. J.* **1987**, *136*, 1179–1180.
21. Bästlein, C.; Burlefinger, R.; Holzberg, E.; Voeth, C.; Garbrecht, M.; Ottenjann, R. Common Variable Immunodeficiency Syndrome and Nodular Lymphoid Hyperplasia in the Small Intestine. *Endoscopy* **1988**, *20*, 272–275. [CrossRef]
22. Nash, T.E.; Ohl, C.A.; Thomas, E.; Subramanian, G.; Keiser, P.; Moore, T.A. Treatment of Patients with Refractory Giardiasis. *Clin. Infect. Dis.* **2001**, *33*, 22–28. [CrossRef] [PubMed]
23. De Weerth, A.; Gocht, A.; Seewald, S.; Brand, B.; van Lunzen, J.; Seitz, U.; Thonke, F.; Fritscher-Ravens, A.; Soehendra, N. Duodenal Nodular Lymphoid Hyperplasia Caused by Giardiasis Infection in a Patient Who Is Immunodeficient. *Gastrointest. Endosc.* **2002**, *55*, 605–607. [CrossRef]
24. Onbaşi, K.; Günşar, F.; Sin, A.Z.; Ardeniz, O.; Kokuludağ, A.; Sebik, F. Common Variable Immunodeficiency (CVID) Presenting with Malabsorption Due to Giardiasis. *Turk. J. Gastroenterol. Off. J. Turk. Soc. Gastroenterol.* **2005**, *16*, 111–113.
25. Ogershok, P.R.; Hogan, M.B.; Welch, J.E.; Corder, W.T.; Wilson, N.W. Spectrum of Illness in Pediatric Common Variable Immunodeficiency. *Ann. Allergy Asthma Immunol.* **2006**, *97*, 653–656. [CrossRef] [PubMed]
26. Ramsey, N.C.; Liedtke, M.D.; Greenfield, R.A.; Lockhart, S.M. Successful 4-Drug Treatment for Refractory Giardia Lamblia. *Infect. Dis. Clin. Pract.* **2010**, *18*, 275–276. [CrossRef]

27. Domínguez-López, M.E.; González-Molero, I.; Ramírez-Plaza, C.P.; Soriguer, F.; Olveira, G. Diarrea Crónica Refractaria y Malabsorción Secundaria a Hipogammaglobulinemia Común Variable, Infestación Crónica Por Giardia Lamblia y Gastrectomía Total Por Adenocarcinoma Gástrico: Un Manejo Nutricional Complejo. *Nutr. Hosp.* **2011**, *26*, 922–925.

28. Silva, G.B.E.; Fernandes, K.P.; Segundo, G.R.S. Common Variable Immunodeficiency and Isosporiasis: First Report Case. *Rev. Soc. Bras. Med. Trop.* **2012**, *45*, 768–769. [CrossRef]

29. Olmez, S.; Aslan, M.; Yavuz, A.; Bulut, G.; Dulger, A.C. Diffuse Nodular Lymphoid Hyperplasia of the Small Bowel Associated with Common Variable Immunodeficiency and Giardiasis: A Rare Case Report. *Wien. Klin. Wochenschr.* **2014**, *126*, 294–297. [CrossRef]

30. Choi, J.H.; Han, D.S.; Kim, J.; Yi, K.; Oh, Y.-H.; Kim, Y. Diffuse Nodular Lymphoid Hyperplasia of the Intestine Caused by Common Variable Immunodeficiency and Refractory Giardiasis. *Intern. Med.* **2017**, *56*, 283–287. [CrossRef]

31. Atalaia-Martins, C.; Barbeiro, S.; Marcos, P.; Cotrim, I.; Vasconcelos, H. Common Variable Immunodeficiency with Several Gastrointestinal Manifestations. *ACG Case Rep. J.* **2017**, *4*, e106. [CrossRef] [PubMed]

32. Saurabh, K.; Nag, V.L.; Khera, D.; Elhence, P. Giardiasis Mimicking Celiac Disease in a Patient of Common Variable Immunodeficiency. *Trop. Parasitol.* **2017**, *7*, 125–127.

33. Kaya, F.; İnkaya, A.Ç.; Maçin, S.; Akyön, Y.; Ergüven, S. Refractory Giardiasis in an Immunosuppressed Patient in Turkey. *J. Infect. Dev. Ctries.* **2018**, *12*, 204–207. [CrossRef] [PubMed]

34. Sousa, D.; Neto Gonçalves, T.; Marto, N.; Horta, A.B. Malabsorption Due to Chronic Giardiasis as a Presenting Symptom of Common Variable Immunodeficiency. *Cureus* **2020**, *12*, e12201. [CrossRef] [PubMed]

35. Agarwal, S.; Mayer, L. Diagnosis and Treatment of Gastrointestinal Disorders in Patients With Primary Immunodeficiency. *Clin. Gastroenterol. Hepatol.* **2013**, *11*, 1050–1063. [CrossRef]

36. Fekete, E.; Allain, T.; Siddiq, A.; Sosnowski, O.; Buret, A.G. *Giardia* spp. and the Gut Microbiota: Dangerous Liaisons. *Front. Microbiol.* **2020**, *11*, 618106. [CrossRef]

37. Leung, A.K.C.; Leung, A.A.M.; Wong, A.H.C.; Sergi, C.M.; Kam, J.K.M. Giardiasis: An Overview. *Recent Pat. Inflamm. Allergy Drug Discov.* **2019**, *13*, 134–143. [CrossRef]

38. Benedetti, I.; Hoyos, J.; Carmona, R. Hiperplasia Nodular Linfoide Asociada Con Colitis Ulcerativa: Descripción de Un Caso y Revisión de La Literatura. *Rev. Colomb. Gastroenterol.* **2017**, *31*, 288. [CrossRef]

39. Adam, R.D. Giardia Duodenalis: Biology and Pathogenesis. *Clin. Microbiol. Rev.* **2021**, *34*, e0002419. [CrossRef]

40. Ebbo, M.; Gérard, L.; Carpentier, S.; Vély, F.; Cypowyj, S.; Farnarier, C.; Vince, N.; Malphettes, M.; Fieschi, C.; Oksenhendler, E.; et al. Low Circulating Natural Killer Cell Counts Are Associated with Severe Disease in Patients with Common Variable Immunodeficiency. *EBioMedicine* **2016**, *6*, 222–230. [CrossRef]

41. Fekrvand, S.; Khanmohammadi, S.; Abolhassani, H.; Yazdani, R. B- and T-Cell Subset Abnormalities in Monogenic Common Variable Immunodeficiency. *Front. Immunol.* **2022**, *13*, 912826. [CrossRef] [PubMed]

42. Ben-Ali, M.; Yang, J.; Chan, K.W.; Ben-Mustapha, I.; Mekki, N.; Benabdesselem, C.; Mellouli, F.; Bejaoui, M.; Yang, W.; Aissaoui, L.; et al. Homozygous Transcription Factor 3 Gene (TCF3) Mutation Is Associated with Severe Hypogammaglobulinemia and B-Cell Acute Lymphoblastic Leukemia. *J. Allergy Clin. Immunol.* **2017**, *140*, 1191–1194.e4. [CrossRef]

43. Qureshi, S.; Sheikh, M.D.A.; Qamar, F.N. Autosomal Recessive Agammaglobulinemia—First Case with a Novel TCF3 Mutation from Pakistan. *Clin. Immunol.* **2019**, *198*, 100–101. [CrossRef]

44. Browning, M.J.; Chandra, A.; Carbonaro, V.; Okkenhaug, K.; Barwell, J. Cowden's Syndrome with Immunodeficiency. *J. Med. Genet.* **2015**, *52*, 856–859. [CrossRef] [PubMed]

45. Driessen, G.J.; IJspeert, H.; Wentink, M.; Yntema, H.G.; van Hagen, P.M.; van Strien, A.; Bucciol, G.; Cogulu, O.; Trip, M.; Nillesen, W.; et al. Increased PI3K/Akt Activity and Deregulated Humoral Immune Response in Human PTEN Deficiency. *J. Allergy Clin. Immunol.* **2016**, *138*, 1744–1747.e5. [CrossRef]

46. Nunes-Santos, C.J.; Uzel, G.; Rosenzweig, S.D. PI3K Pathway Defects Leading to Immunodeficiency and Immune Dysregulation. *J. Allergy Clin. Immunol.* **2019**, *143*, 1676–1687. [CrossRef] [PubMed]

47. Kralovicova, J.; Hammarström, L.; Plebani, A.; Webster, A.D.B.; Vorechovsky, I. Fine-Scale Mapping at IGAD1 and Genome-Wide Genetic Linkage Analysis Implicate HLA-DQ/DR as a Major Susceptibility Locus in Selective IgA Deficiency and Common Variable Immunodeficiency. *J. Immunol.* **2003**, *170*, 2765–2775. [CrossRef]

48. Schroeder, H.W.; Zhu, Z.-B.; March, R.E.; Campbell, R.D.; Berney, S.M.; Nedospasov, S.A.; Turetskaya, R.L.; Atkinson, T.P.; Go, R.C.P.; Cooper, M.D.; et al. Susceptibility Locus for IgA Deficiency and Common Variable Immunodeficiency in the HLA-DR3, -B8, -A1 Haplotypes. *Mol. Med.* **1998**, *4*, 72–86. [CrossRef]

49. De la Concha, E.G.; Fernandez-Arquero, M.; Martinez, A.; Vidal, F.; Vigil, P.; Conejero, L.; Garcia-Rodriguez, M.C.; Fontan, G. Hla Class II Homozygosity Confers Susceptibility to Common Variable Immunodeficiency (CVID). *Clin. Exp. Immunol.* **2001**, *116*, 516–520. [CrossRef]

50. Garzon, T.; Ortega-Tirado, D.; Lopez-Romero, G.; Alday, E.; Robles-Zepeda, R.E.; Garibay-Escobar, A.; Velazquez, C. Immunoinformatic Identification of T-Cell and B-Cell Epitopes From Giardia Lamblia Immunogenic Proteins as Candidates to Develop Peptide-Based Vaccines Against Giardiasis. *Front. Cell. Infect. Microbiol.* **2021**, *11*, 769446. [CrossRef]

51. Chen, L.; Tsai, T.-F. HLA -Cw6 and Psoriasis. *Br. J. Dermatol.* **2018**, *178*, 854–862. [CrossRef]

52. Requena-Méndez, A.; Goñi, P.; Rubio, E.; Pou, D.; Fumadó, V.; Lóbez, S.; Aldasoro, E.; Cabezos, J.; Valls, M.E.; Treviño, B.; et al. The Use of Quinacrine in Nitroimidazole-Resistant Giardia Duodenalis: An Old Drug for an Emerging Problem. *J. Infect. Dis.* **2017**, *215*, 946–953. [CrossRef] [PubMed]

53. Ydsten, K.A.; Hellgren, U.; Asgeirsson, H. Quinacrine Treatment of Nitroimidazole-Refractory Giardiasis. *J. Infect. Dis.* **2022**, *225*, 1773–1776. [CrossRef] [PubMed]

54. Riches, A.; Hart, C.J.S.; Trenholme, K.R.; Skinner-Adams, T.S. Anti- Giardia Drug Discovery: Current Status and Gut Feelings. *J. Med. Chem.* **2020**, *63*, 13330–13354. [CrossRef] [PubMed]

55. Loderstädt, U.; Frickmann, H. Antimicrobial Resistance of the Enteric Protozoon Giardia Duodenalis—A Narrative Review. *Eur. J. Microbiol. Immunol.* **2021**, *11*, 29–43. [CrossRef]

Journal of
Clinical Medicine

Review

Diagnostic Challenges in Patients with Inborn Errors of Immunity with Different Manifestations of Immune Dysregulation

Karolina Pieniawska-Śmiech [1,2,*], Gerard Pasternak [2,3], Aleksandra Lewandowicz-Uszyńska [2,3] and Marek Jutel [1,4,*]

1   Department of Clinical Immunology, Wroclaw Medical University, 50-368 Wroclaw, Poland
2   Department of Clinical Immunology and Paediatrics, Provincial Hospital J. Gromkowski, 51-149 Wroclaw, Poland;
     gerard.pasternak@umw.edu.pl (G.P.); aleksandra.lewandowicz-uszynska@umw.edu.pl (A.L.-U.)
3   3rd Department and Clinic of Paediatrics, Immunology and Rheumatology of Developmental Age,
     Wroclaw Medical University, 50-367 Wroclaw, Poland
4   ALL-MED Medical Research Institute, 53-201 Wroclaw, Poland
*   Correspondence: karolina.pieniawska-smiech@student.umw.edu.pl (K.P.-Ś.); marek.jutel@umw.edu.pl (M.J.)

Abstract: Inborn errors of immunity (IEI), formerly known as primary immunodeficiency disorders (PIDs), are inherited disorders caused by damaging germline variants in single genes, which result in increased susceptibility to infections and in allergic, autoimmune, autoinflammatory, nonmalignant lymphoproliferative, and neoplastic conditions. Along with well-known warning signs of PID, attention should be paid to signs of immune dysregulation, which seem to be equally important to susceptibility to infection in defining IEI. The modern diagnostics of IEI offer a variety of approaches but with some problems. The aim of this review is to discuss the diagnostic challenges in IEI patients in the context of an immune dysregulation background.

Keywords: allergy; autoimmunity; autoimmune lymphoproliferative syndrome; inborn errors of immunity; lymphoproliferation; malignancy; primary immunodeficiency

Citation: Pieniawska-Śmiech, K.; Pasternak, G.; Lewandowicz-Uszyńska, A.; Jutel, M. Diagnostic Challenges in Patients with Inborn Errors of Immunity with Different Manifestations of Immune Dysregulation. *J. Clin. Med.* **2022**, *11*, 4220. https://doi.org/10.3390/jcm11144220

Academic Editors: Rita Consolini and Giorgio Costagliola

Received: 24 June 2022
Accepted: 18 July 2022
Published: 20 July 2022

Publisher's Note: MDPI stays neutral with regard to jurisdictional claims in published maps and institutional affiliations.

1. Introduction

Inborn errors of immunity (IEI), formerly known as primary immunodeficiency disorders (PIDs), are inherited disorders caused by damaging germline variants in single genes, resulting not only in increased susceptibility to infections but also in allergic, autoimmune, autoinflammatory, nonmalignant lymphoproliferative, and malignant manifestations. According to the most recent report by the International Union of Immunological Societies (IUIS), the identified IEI were classified in 10 tables with subtables segregating groups of disorders into overlapping phenotypes: (1) immunodeficiencies affecting cellular and humoral immunity (combined immunodeficiencies); (2) combined immunodeficiencies with associated or syndromic features; (3) predominantly antibody deficiencies; (4) diseases of immune dysregulation; (5) congenital defects of phagocyte number or function; (6) defects in intrinsic and innate immunity; (7) autoinflammatory diseases; (8) complement deficiencies; (9) bone marrow failure disorders; and (10) phenocopies of IEI. The 55 novel monogenic gene defects positioned in the last IEI update enhanced the total number of IEI to 485 [1,2].

The COVID-19 pandemic had an impact on various fields of medicine. In the context of clinical immunology and IEI, it has uncovered several new IEI [1]. Each time, the appearance of new pathogens is a potential challenge for the general population and also healthcare systems because of the lack of significant pre-existing immune memory. Similarly, in the case of pathogens learned about so far, patients with specific germline genetic variants (causing known and unknown IEI) may be more exposed to severe disease than the general population. Research on the COVID-19 pandemic course led to the detection of genes and mechanisms necessary for anti-SARS-CoV-2 immunity. About 2–3% of cases of severe SARS-CoV2 infection resulted from germline LOF/LOE variants in the type 1 IFN signaling pathway: *TLR3, UNC93B1, TICAM1, TBK1, IRF3, IRF7, IFNAR1,* and *IFNAR2* [1]. According to Asano et al., X-linked recessive TLR7 deficiency is a highly penetrant genetic etiology of severe COVID-19 among 1.8% of males below the age of 60 years [3].

The defects of the number or the function of immune system elements determine the clinical presentation of an IEI. Family history, as well as personal and clinical data, are considered a core element of patient initial management. Extensive anamnesis and clinical evaluation are the main tools for a suspected diagnosis of IEI [4]. The early diagnosis of IEI can be life-saving but remains challenging due to the low prevalence of these pathologies. This can result in the delay of diagnosis and consequently in a worse prognosis [5].

Disease manifestation appearance (i.e., Nijmegen breakage syndrome (NBS), Shwachman-Diamond syndrome, and DiGeorge syndrome), as well as subject growth during both in utero life and later, may suggest the diagnosis of IEI and provide an important diagnostic clue [6]. Severe and/or recurrent infections, consanguinity, or an unexplained death in one's family are well-known signs of IEI; however, more attention should be paid to signs of immune dysregulation. Immune dysregulation is defined as a breakdown or malfunction of molecular control of immune system processes, and it is used to characterize an array of autoimmune and inflammatory conditions [7]. According to IUIS classification, there are 10 IEI categories based on their underlying molecular defect. One of them is called 'diseases of immune dysregulation'. Moreover, it has been established that other patients with humoral, cellular, or innate immune system deficiencies are also at risk of autoimmune or inflammatory conditions [8]. Currently, signs of immune dysregulation are of great importance in defining IEI, as well as an increased tendency to infection.

The modern diagnostics of IEI include various diagnostic measures, such as a simple blood count with particular attention paid to the total absolute lymphocyte count, the serum immunoglobulin levels, and the complete sequencing of the exome or genome [9]. However, during the clinical evaluation of a patient with suspected or confirmed IEI, we should be aware of the possible problems and finer points that may restrict diagnosis in patients with IEI. The aim of this review is to summarize these diagnostic challenges, in particular, in the context of immune dysregulation in IEI patients.

2. Allergic Disease

Allergy develops on account of disturbed function of the immune system. The immune system depends on a complex balance of activation, to defend against invasive, foreign pathogens, and control, to differentiate between self and foreign matter. Allergic reactions are exaggerated immune responses against specific allergens [10,11]. The comorbidity of IEI and allergy appears because of the impairment of the immune system, leading to infectious susceptibility; however, it is still able to trigger an allergic response [8]. The mechanisms underlying the relationship between atopy and immunodeficiency are better recognized, thanks to the discovery and characterization of genetic variants, often showing "a new face of old disorders" [8]. Several studies indicated the potential mechanisms leading to such dysregulation, which include the failure of central thymic tolerance, an imbalance between the effector and regulatory T-cell function, a failure in the production of counter regulating interferon-gamma (IFN-γ), disturbed cytokine production, and possible differences in microbial colonization and infection patterns [8,12,13].

Thanks to growing interest in the coexistence of allergy and IEI, the topic has been investigated in a number of studies. However, the results are still inconsistent. For example, in one Iranian study atopic dermatitis (AD) was present in 52% of patients with selective IgA deficiency (sIgAD) [14], while among Brazilian patients with sIgAD, AD was found in 2.3% [8,15]. In the USIDENT study, AD was most commonly reported in patients with a deficiency of the nuclear factor κB (NFkB) essential modulator (62.5%), the Wiskott–Aldrich syndrome (WAS: 41.5%), combined immunodeficiency (CID: 33.3%), selective IgM deficiency (33.3%), and autosomal-dominant hyper-IgE syndrome (AD-HIES; 25%) [8,16]. A cohort study of patients with early onset severe combined immunodeficiency due to adenosine deaminase deficiency (ADA-SCID) demonstrated that atopy was present in 56% of the patients, including mild AD in 11.1%. Severe AD was not a common feature [17]. A possible explanation of the diverse results are ethnic and geographical diversity and differences in methodological approaches.

Potential diagnostic difficulties may start even at the beginning in diagnosing IEI. An underlying, sometimes severe immune deficiency can manifest as common allergic symptoms, and IEI may masquerade allergic atopic patients [10]. In clinical practice, there are few warning signs of an underlying IEI among atopic phenotypes, and these include severe atopic disease, usually with a poor response to standard therapies, early-onset of the disease, a positive family history for IEI and/or severe familial atopy, and immunological abnormalities [11].

The standard screening tests for antibody deficiency include the measurement of immunoglobulin, IgG, IgA, and IgM levels in serum and the interpretation according to age-related reference values [18]. The routine measurement of serum IgE is not obligatory in the management of patients with suspected antibody deficiency and a history of recurrent infections. Previously, the level of total IgE was considered as a marker to catch allergic patients, but because it is nonspecific, it cannot confirm the allergy status of a patient [19,20]. Non-immunodeficient patients have variable IgE concentrations associated with atopic disease such as allergic rhinitis (AR), asthma, food allergy (FA), and AD, as well as other conditions, including parasitic disease [21]. However, in the context of PID, IgE measurement plays a role, especially in patients with concomitant eczema. Elevated IgE is common in a number of IEI, such as HIES, WAS, Netherton syndrome, immune dysregulation, polyendocrinopathy, enteropathy, X-linked (IPEX) syndrome, and Omenn syndrome [22]. One phenotype of complete DiGeorge syndrome, which is known as atypical complete DiGeorge syndrome, has oligoclonal T cell expansion with elevated IgE levels with concomitant generalized rash and lymphadenopathy [23]. The pathophysiological role of increased IgE in these disorders was not clearly characterized; however, there are few hypotheses [13]. Increased IgE production is associated not only with well-defined genetic syndromes but also with humoral, cellular, innate, and combined immunodeficiency disorders [5]. However, a high IgE (>180 IU/mL) is very rare in common variable immunodeficiency (CVID) (0.3% of patients) [21].

There are particular PIDs associated with atopy, especially eczema and elevated serum IgE, which can be confirmed by genetic tests and the identification of specific mutations. Mutations in the *WAS* gene on the X chromosome, which encodes the WAS protein (WASP), are a cause of Wiskott–Aldrich syndrome, characterized by recurrent infections, thrombocytopenia with small platelets, and eczema [8]. The mechanism for atopy in WAS is not fully described; however, impairment of regulatory T-cell (Treg) function is a possible contributor [8,24–26]. In total, 33% of patients with WAS and 20% of patients with X-linked thrombocytopenia (XLT) had positive food allergen-specific IgE (sIgE), in a study conducted by Lexmond et al. [8,27]. Food sensitization was generally detected with greater sensitivity using sIgE testing than by skin prick testing (SPT).

A dominant-negative heterozygous mutation in signal transduction and the activator of transcription 3 (STAT3) leads to autosomal-dominant hyper-IgE syndrome (AD-HIES), previously known as Job syndrome, with characteristic features such as chronic eczema, recurrent staphylococcal skin infections, pneumonia, increased serum IgE, and eosinophilia [10]. Skin findings distinguishing it from AD include a distinctive thickened texture of the facial skin, retroauricular fissures, and severe folliculitis of the axillae and groin [5]. Serum IgE levels are often >2000 IU/mL, and eosinophilia levels are often >700 cells/mL (eosinophilia does not correlate with the elevation in IgE), but patients usually do not suffer from symptomatic allergic disease such as AR, FA, or anaphylaxis [10,28]. Disturbances in the inflammatory process, and associated immune regulatory defects, are present. In clinical practice, a lower limit of 2000 IU/mL is often considered as a cutoff for AD-HIES. However, patients with HIES with lower IgE levels and STAT3 pathogenic variants have been reported [29]. Moreover, the serum IgE level does not correlate with the severity and activity of the disease, and paradoxically patients with STAT3 loss-of-function (LOF) mutations are rather protected from severe allergic reactions. A potential explanation of this protection is disturbed mast cell degranulation, as well as vascular reaction to histamine caused by the STAT3 mutation itself [8,30–32]. SPT results and clinical symptoms of allergy are consistent with the specific IgE (sIgE) results in AD-HIES. Both skin and blood test results are comparable between patients with AD-HIES and healthy controls [32]. Defective neutrophil chemotaxis has been described among AD-HIES patients, and variable specific antibody production is seen [5,33]. Patients may require immunoglobulin replacement.

At the end of 20th century, the National Institutes of Health HIES scoring system was originally presented where a score of 30 has a sensitivity of 87.5 percent and a specificity of 80.6 percent [34]. It is noteworthy that some patients (e.g., some young children), may not meet the scoring criteria. Thereupon, in cases of positive family history of HIES and some distinctive features, according to experts, molecular screening should still be performed even if the score is below 30. Other diagnostic guidelines takes into account five cardinal clinical features (recurrent pneumonia, newborn rash, pathologic bone fractures, characteristic facies, and high palate) with total IgE level and Th17 cell count [35]. Molecular genetic testing is crucial to establish the diagnosis of the AD-HIES.

Autosomal-recessive-HIES (AR-HIES) is characterized by highly elevated serum levels of IgE, eczema, recurrent staphylococcal abscesses, and hypereosinophilia. In contrast to AD-HIES, where patients are usually free from allergic manifestations, 50% to 70% of patients with AR-HIES suffer from severe allergies, i.e., eczema, anaphylaxis to food, and environmental allergies, and 30% have asthma [10,32]. Pulmonary disease is usually asthma-related as compared with AD-HIES, with pneumatocele and lung damage due to prior infections [10].

Some patients with DOCK8 or TYK2 deficiency were previously classified as AR-HIES with harmful allergic symptoms [36]. Now, we better recognize the differences in the clinical features. DOCK8 deficiency is a combined immunodeficiency characterized by allergic inflammation, severe atopy, high IgE, susceptibility towards cutaneous viral infections, and malignancy [37]. TYK2 deficiency is also a combined immunodeficiency with recurrent skin viral infections, while eczema and elevated IgE are variably found. A study conducted by Boos et al. revealed that total serum IgE levels similarly increased in STAT3-HIES, DOCK8 deficiency, and AD patients. The ratio of aeroallergen-specific IgE to total IgE was the highest in AD, whereas patients with DOCK8 deficiency showed the highest specific serum IgE against food allergens. Th2-cell numbers were significantly increased in DOCK8 deficiency and AD patients compared to STAT3-HIES patients and controls. The study showed that hyper-IgE syndromes and atopic dermatitis patients showed a different sensitization pattern of serum IgE corresponding to the allergic disease manifestations and Th-cell subset data, suggesting a key role of DOCK8 in the development of FA [32]. Moreover, according to Wilkie et al., defective Treg function may contribute to the increased skin inflammation and the eczema in DOCK8 deficient patients [38]. IEI with elevated IgE are summarized in Table 1.

Table 1. Inborn errors of immunity with elevated IgE.

Disease	IUIS Classification	Inheritance	Mutation	Characteristics	Immunological Features
Hyper IgE syndrome (HIES)	Combined immunodeficiencies with associated syndromic features	AD LOF	*STAT3*	Infectious disease and immunological manifestations (skin abscesses, recurrent sinopulmonary infections, bacterial infections, pulmonary aspergillus, Pneumocystis jirovecii, and chronic mucocutaneous candidiasis) Craniofacial, dental, musculoskeletal, neurological, and vascular abnormalities	Eosinophilia ↑ IgE ↓-specific antibody production Intermittent chemotactic defectsImpaired inflammatory cytokine production Reduced or absent Th17 cells Defective Th17 cell production of IL-17 Decreased IFN-γ production upon stimulation Decreased CD8⁺ memory T cellsDiminished delayed-type hypersensitivity and lymphoproliferative responses to antigenic stimulation
ZNF341 deficiency (phenocopy of AD-HIES)	Combined immunodeficiencies with associated syndromic features	AR	*ZNF341*	Mild facial dysmorphism Early onset eczema Recurrent bacterial infections (respiratory, skin infections) Lung abscesses and pneumatoceles Musculosceletal abnormalities Retention of primary teeth	↑ IgE- and IgG ↓-specific antibody production ↓ memory B cells excess of Th2 cells ↓ Th17 and NK cells
Loeys–Dietz syndrome (TGFBR deficiency)	Combined immunodeficiencies with associated syndromic features	AD	*TGFBR1TGFBR2*	Recurrent respiratory infections Eczema Food allergy Musculosceletal abnormalities Retention of primary teeth Vascular abnormalities	↑ IgE
PGM3 deficiency (hyperimmunoglobulin E-like syndrome with glycosylation defects)	Combined immunodeficiencies with associated syndromic features	AR	*PGM3*	Impaired immunity (recurrent respiratory tract infections, abscesses) Severe atopy, asthma, eczema, and food allergy Autoimmunity Neurocognitive impairment Skeletal dysplasia	Neutropenia T and B cell lymphopenia Eosinophilia ↑ IgE levels N/↑ IgG and IgA Progressive bone marrow failure
Comel–Netherton syndrome	Combined immunodeficiencies with associated syndromic features	AR	*SPINK5*	Congenital ichthyosis Bamboo hair Recurrent bacterial infections Atopy Failure to thrive	↑ IgE and IgA ↓ switched and non-switched B cells
CARD11 deficiency	Combined immunodeficiencies with associated syndromic features	AD LOF	*CARD11*	Severe atopic dermatitis Food allergy Molluscum contagiosum infection Recurrent respiratory infections Lymphoma Various phenotypes from SCID to combined immunodeficiency, associated with atopy and elevated IgE levels or isolated severe atopy	↑ IgE Poor specific antibody production Impaired activation of both NF-kB and mTORC1 pathways N/↓ B cell numbers Defective T-cell activation and proliferation Skewing toward Th2
ERBIN deficiency	Combined immunodeficiencies with associated syndromic features	AD	*ERBB2IP*	Recurrent respiratory infections Susceptibility to S.aureus Eczema Atopy Joint hypermobility, sometimes vascular abnormalities	↑ IgE ↑ circulating Treg

Table 1. *Cont.*

Disease	IUIS Classification	Inheritance	Mutation	Characteristics	Immunological Features
IL6R deficiency	Combined immunodeficiencies with associated syndromic features	AR	*IL6R*	Immunodeficiency (recurrent pyogenic infections, cold abscesses) Atopy Abnormal inflammatory responses	High circulating IL-6 levels Normal/$\downarrow$ serum IgM, IgG, and IgA Very $\uparrow$ IgE $\downarrow$-specific antibody productionReduced switched memory B
Interleukin 6 signal transducer (IL6ST) deficiency	Combined immunodeficiencies with associated syndromic features	AR	*IL6ST*	Recurrent infections Boils Eczema Bronchiectasis Pulmonary abscesses Skeletal abnormalities (scoliosis, bone fractures, and craniosynostosis) Retention of primary teeth	Eosinophilia $\uparrow$ IgE, Specific antibody production variably affected Impaired B cell memory and acute-phase response $\downarrow$ Th17 cells
DOCK8 deficiency	Immunodeficiencies affecting cellular and humoral immunity	AR	*DOCK8*	Recurrent viral and bacterial infections Cutaneous infections (staphylococcal, viral, and fungal) Severe atopy Often multiple severe allergies to food and environmental allergens Hepatic disorders Early-onset malignancy	Eosinophilia $\downarrow$ T cell numbers (with normal CD4/CD8 ratio) and variably decreased or normal B- and NK-cell numbers $\downarrow$ production of TNFα and IFNγ $\downarrow$ numbers of Th17 T cells $\uparrow$ Th 2 $\uparrow$ IL-4 and IL-13 Few Treg with poor function $\downarrow$ IgM levels and variable IgA and IgG levels $\uparrow$ IgE Poor antibody responses
TYK2 deficiency	Defects in intrinsic and innate immunity	AR	*TYK2*	Susceptibility to intracellular bacteria (mycobacteria, Salmonella) and viruses Eczema	Impaired cellular responses to IL-10, IL-12, and IL-23 and type I IFNs
Omenn syndrome (OS)	Immunodeficiencies affecting cellular and humoral immunity (usually a T-B-NK⁺ SCID)	AR	various	Erythroderma Alopecia Aplasia/hypoplasia of the eyebrow Desquamation of skin Dry skin Edema Chronic diarrhea Failure to thrive Hepatosplenomegaly Lymphadenopathy Pneumonia Sometimes anemia, autoimmunity, hypothyroidism, and lymphoma	Eosinophilia $\uparrow$ IgE Abnormal secretion of IL-4 and IL-5 from activated T cells Exaggerated Th2 response Absence of B cells in the circulation
Wiscott–Aldrich syndrome (WAS)	Combined immunodeficiencies with associated syndromic features	XL	*WAS*	Recurrent bacterial and viral infections Bloody diarrhea Eczema Thrombocytopenia with small platelets $\uparrow$ risk of malignancy Autoimmune diseases IgA nephropathy	Eosinophilia Often $\uparrow$ IgE and IgA $\downarrow$ IgM $\downarrow$ antibody responses to polysaccharides Progressive $\downarrow$ in T cells numbers Abnormal lymphocyte responses to anti-CD3

Table 1. *Cont.*

Disease	IUIS Classification	Inheritance	Mutation	Characteristics	Immunological Features
Atypical DiGeorge syndrome with deletion of chromosome 22q11.2	Combined immunodeficiencies with associated syndromic features	AD	Deletion typically in chromosome 22	Pharyngeal pouch defects Thymus hypoplasia/aplasia Hypoparathyroidism Congenital heart disease Eczema, erythroderma Lymphadenopathy	Eosinophilia ↑ IgE Partial T cell deficiency Oligoclonal T cells expansion T cell count is higher than typical complete DiGeorge patients
IPEX syndrome (immunodysregulation, polyendocrinopathy, and enteropathy X-linked syndrome)	Diseases of immune dysregulation	XL	*FOXP3*	Multiple endocrinopathies Severe chronic enteropathy Dermatitis Eczema Anemia Thrombocytopenia	↑ IgE and IgA Lack of (and/or impaired function of) CD4$^+$ CD25$^+$ FOXP3$^+$ regulatory T cells (Tregs)

Abbreviations: ↓—decreased; ↑—increased; γ—gamma; AD—autosomal dominant; AR—autosomal recessive; LOF—loss-of-function; N—normal; SCID – severe combined immunodeficiency; Treg—T regulatory cell; and XL—X-linked inheritance.

On the other hand, low levels of IgE interest immunologists. Selective IgE deficiency (defined as a significant decrease in the levels of IgE (<2.5 IU/mL) in patients whose other immunoglobulin levels, including IgG, IgG subclasses, and IgA levels, are normal) has not been included in international classification systems for IEI [1]. Low serum levels of IgE can be associated with some well-defined IEI: common variable immunodeficiency (CVID), IgG subclass deficiencies, sIgAD, ataxia-telangiectasia (A-T), and agammaglobulinemia [39–41]. According to studies, an undetectable serum IgE (<2 IU/mL) occurs in only 3.3% of the general population [21]. In contrast, Lawrence et al. found that an undetectable IgE occurs in 75.6% of patients with CVID [21]. Another finding was a significant correlation between serum IgE with serum IgG, suggesting that lower IgE occurs in patients with more severe hypogammaglobulinemia. Moreover, false-negative results may appear using traditional methods of sIgE measurement, and allergen sIgE was not detectable in 96.5% of patients with CVID. Many patients with CVID report symptoms of rhinitis, wheeze, or adverse reactions to antibiotics, but it is difficult to detect allergic sensitization among them, especially using SPT or serum sIgE [41–43]. In these cases, sensitization should be confirmed using different methods, for example, an oral provocation challenge, and bronchial provocation tests with allergens [41]. The interpretation of food-specific IgE values and their usefulness in predicting symptomatic food allergies in the context of IEI patients is a potential field for further studies.

While diagnostics among PID patients during Ig replacement therapy (IRT) are often challenging, in the context of allergy, studies suggest that current Ig products are not a significant source of IgE [21].

3. Autoimmunity

There is also a high degree of overlap between autoimmune diseases and IEI in the context of genetic linkages and causes [44]. The molecular mechanisms responsible for the immune dysregulation in patients with IEI still are not fully recognized [45]. The usage of genetic analysis and a better understanding of the involved immune regulatory and signaling mechanisms is revealing the complex relationships between IEI syndromes and autoimmune diseases [44]. In the past, IEI and autoimmune diseases were considered as opposites; now, we know that genetic mutations may affect multiple immune cells and molecules, and in consequence IEI does not exclude autoimmunity. Furthermore, autoimmune diseases often coexist with some IEI [46].

The potential mechanisms associated with the pathogenesis of autoimmunity include impaired B cell differentiation and germ-center reactions, altered T cell central or peripheral tolerance, uncontrolled lymphocyte proliferation and differentiation, disturbances in Treg/Th17 balance, dysfunctional complement and innate immune activation, and the defective clearance of the infectious agents [45,46].

A French national study by Fischer et al. includes all types of IEI and autoimmune manifestations. The study demonstrated that autoimmunity is a significant component of clinical presentation of all types of IEI: one or more autoimmune and inflammatory manifestations were noted in 26.2% of 2183 retrospectively screened IEI patients, with a risk of onset throughout the patient's lifetime. The risk of autoimmune cytopenia (AIC) was at least 120 times higher than in the general population; among children the risk of inflammatory bowel disease (IBD) was 80 times higher, while the risk of arthritis was 40 times higher. The risk of other autoimmune complications was approximately 10 times higher. Autoimmune manifestations occurred in patients with all types of IEI; however, patients with T-cell defects or CVID had, statistically speaking, the highest risk for autoimmunity [47].

The signs and symptoms of most rheumatic diseases are classified in international American College of Rheumatology (ACR) or European League Against Rheumatism (EULAR) criteria. The management of autoimmunity in patients with IEI is often challenging because immune dysregulation, as well as permanent inflammation, may influence the diagnostic process. Moreover, when assessing a patient with IEI for possible autoimmunity, it is important to consider a broad differential diagnosis, because infectious diseases, adverse effects of medications, and malignancies can mimic autoimmune processes. Thereupon, a complete diagnostic process is not effortless and requires a history, a complete physical examination, wide laboratory testing, imaging, and even pathological investigations [48]. Clinicians must be aware of the characteristic clinical features of autoimmune diseases among IEI patients. These include polyautoimmunity, which is defined as the presence of more than one autoimmune disease in a single patient and early onset autoimmunity (the presence of autoimmune disease at any age that is earlier than usual) [46]. Some IEI are associated with specific autoimmune diseases, and the awareness of these patterns also allows clinicians to monitor patients more effectively.

During evaluation of a patient with IEI and suspected autoimmunity, some laboratory tests are needed. This includes a complete blood count with differential, acute phase reactants, autoantibodies, serologies, flow cytometry, cytokine analysis, levels of complement, human leukocyte antigen (HLA) typing, and comprehensive endocrine and/or metabolic panels [48].

On the other hand, laboratory tests may help to catch patients with IEI among heterogenous group of patients with already diagnosed autoimmunity. Immune phenotyping and immunoglobulin (Ig) levels are indispensable. The ratio of naïve and memory T cells (CD45RA/CD45RO) may differentiate patients with late-onset or profound combined immunodeficiency disorders [49–52].

In addition, specific subsets of T and B cells have been linked to IEI with autoimmunity. These include the expansion of TCR$\alpha\beta$ CD4$^-$CD8$^-$ (double-negative) T cells in autoimmune lymphoproliferative syndrome (ALPS), CD19^{hi}21^{lo} B cells in CVID with autoimmunity, an abnormal count of Treg in Tregopathies, Th17 cells in STAT1 GOF patients, and expanding follicular helper T cells (Tfh) in CTLA4 and LRBA deficiency. Changes in these subsets may also predict the progression of autoimmune complications or a response to therapy [52,53].

Primary antibody deficiencies (PADs) are the most common inherited IEI in humans, with recurrent infections as a predominant presenting complaint. However, various types of PADs are also associated with inflammatory disorders, granulomatous lesions, lymphoproliferative diseases, and cancer. Several studies have reported that PAD patients are predisposed to autoimmune complications [47,54].

X-linked agammaglobulinemia (XLA), also known as Bruton agammaglobulinemia, is the prototype antibody deficiency [55]. Function-loss mutations in Bruton's tyrosine kinase (BTK) lead to a block in B-cell maturation, a near total absence of B cells in the periphery, and severe reductions in serum immunoglobulins. Surprisingly, most patients with XLA have a small number of B cells, or "leaky B cells", in the peripheral blood [54,56]. Patients with XLA are rather at a low risk of autoimmune or inflammatory diseases compared with other IEI patients, but several studies suggest that some XLA patients show symptoms with similar diagnostic features to rheumatoid arthritis (RA), IBD, alopecia, enteropathy, autoimmune hemolytic anemia (AIHA), immune thrombocytopenic purpura (ITP), neutropenia, and Kawasaki disease [54,57–59]. These patients are not expected to produce autoantibodies; however, surprisingly, the "leaky" production of autoantibodies and defects in B-cell central tolerance has been reported [54,60,61].

Autoimmune diseases occur in 20–30% of CVID patients. The most reportable are autoimmune cytopenias such as ITP, AIHA, and Evans syndrome; however, organ-specific and systemic autoimmune diseases are also described [45,62–64].

It is worth mentioning that it is not uncommon that autoimmune complications are the first or the only clinical manifestation of CVID during diagnostics [54,65].

A cohort study on CVID patients with immune cytopenia showed higher levels of serum immunoglobulin, CD19hi B cells, and T CD4 effector T cells, accompanied by reduced naïve T cells [45,66]. Moreover, according to several studies, Treg frequency and their functional characteristics are disturbed in CVID patients [54,67–69], which may result in elevated levels of activated T cells; autoimmunity; and chronic inflammation. Defects in Tregs are also correlated with the expansion of CD21low B cells in CVID patients with autoimmunity [70–72]. In a study by Boileau et al., the serum IgG level in CVID patients with autoimmunity (cytopenia and others) was greater than in CVID patients without autoimmunity [66]. Other studies revealed that CVID patients with autoimmunity have higher levels of IgM compared with non-autoimmune phenotypes [73,74]. On the other hand, markedly depressed serum immunoglobulin levels have been reported in patients with RA, Sjogren's syndrome (SS), and systemic lupus erythematosus (SLE), prompting suspicion of IEI [75,76].

Autoantibodies circulating in the serum and/or plasma, as well as the immune complex deposits containing autoantibodies and complement, are essential diagnostic tools in most autoimmune diseases. In patients with hypogammaglobulinemia (i.e., CVID, XLA etc.) and some types of CIDs, diagnostic tests that are based on antibodies may be not useful and provide false-negative results. For example, the diagnosis of definite autoimmune hepatitis (AIH) in CVID patients is definitely challenging. According to the European Association for the Study of the Liver (EASL), both histologic evidence of moderate to severe interface hepatitis and the positivity of the typical autoantibodies are required to make an AIH diagnosis [77,78]. It is not surprising that CVID patients generally may not have autoantibodies, even in the case of noticeable autoimmune complications.

However, in a study by Tahiat et al. among 299 IEI patients with a dominance of PAD (27.8%) and CID (26.1%), autoantibodies were found in 32.4% of all IEI patients, compared with 15.8% of healthy subjects. Anti-nuclear antibodies (ANA) (10.0%), transglutaminase antibody (TGA) (8.4%), RBC antibodies (6.7%), anti-smooth muscle antibody (ASMA) (5.4%), and ASCA (5.0%) were the most common autoantibodies. The authors have concluded that considering the association of some autoimmune diseases with certain PIDs, screening for corresponding autoantibodies would be recommended. However, due to the low positive predictive value of the autoantibodies, the results should be interpreted with caution in patients with IEI [79].

Oppositely, the production of specific antibodies may by impaired even when the level of main classes of immunoglobulins is normal in specific antibody deficiency (SAD). Consequently, most autoantibodies are not found in these patients [48,80,81]. In sIgAD, as well as in CVID with IgA deficiency, it is obvious that there is a lack of antibodies in this immunoglobulin class (for example, tissue transglutaminase IgA–tTg IgA). On the other hand, among patients during IRT, exogenous Ig may interfere with some of the special immunologic tests. That is why it is worth considering if some screening tests such as autoantibodies should be performed before the therapy is being initiated or the serum should be frozen for future testing [48].

Some IEI patients are constantly negative for disease-specific autoantibodies, and in the case of clinical suspicion of autoimmune disease, other diagnostic methods should be considered. Medical imaging is often a part of the clinical evaluation of patients with suspected autoimmune disorder. In the case of IEI patients, some difficulties may appear at this point too. In particular types of IEI there is a problem with radiosensitivity, which limits the use of medical radiation for the diagnosis of autoimmunity [82–84]. Genetic instability, defective DNA repair, and a predisposition to malignancy are associated with specific types of IEI. A-T and NBS are well-defined IEI connected with defective DNA repair [85], where patients might be sensitive to radiation. X-ray exposure should be limited to diagnostic purposes only when it is medically necessary because patients should be protected from unnecessary medical techniques that incorporate radiation. Substitution with magnetic resonance imaging (MRI) or ultrasound is desirable [48].

Histopathological examination is sometimes crucial and clinically indicated in a diagnostic process. Diagnostic challenges may occur here as well. In IEI patients, as an effect of immunoglobulins and immune cells deficiency, affected tissue can have a different histological appearance in comparison to healthy individuals [48,78,86–88].

Since autoimmune cytopenia (AIC) is a common finding in IEI patients, Westermann-Clark et al. evaluated 154 pediatric patients with AIC in the context of IEI. Splenomegaly, short stature, and recurrent or chronic infections were common clinical features among patients with AIC and IEI. IEI patients were more likely to have AIHA or Evans syndrome than AIC-only patients. Patients with both IEI and AIC more often had low CD3 and CD8 cells; low IgA and IgG levels; and a higher prevalence of autoantibodies to red blood cells, platelets, or neutrophils. AIC diagnosis preceded IEI diagnosis by 3 years on average, except among those with partial DiGeorge syndrome [89]. The early detection of patients with comorbid IEI and AIC may improve treatment outcomes.

The main molecular defects and common autoimmune complications among IEI are summarized in Table 2.

Table 2. Common autoimmune presentation in inborn errors of immunity (IEI).

IUIS Classification	Disease	Main Molecular Defect	Common Autoimmune Disease
Immunodeficiencies affecting cellular and humoral immunity	ICOS deficiency	ICOS	Arthritis, SLE, MS, and enteropathy
Combined immunodeficiencies with associated syndromic features	22q11 deletion syndrome (DiGeorge syndrome)	Large deletion typically in chromosome 22	AIC, AIT, and arthritis
	Wiskott–Aldrich syndrome	WAS	AIC, IBD, GN, arthritis, and vasculitis
Predominantly antibody deficiencies	X-linked agammaglobulinemia	Btk	RA, JIA, IBD, AIC, AIT, PND, KD, DM, T1D, SD, and alopecia
	CVID	Various	AIC (ITP, AIHA, AN), RA, JIA, SLE, IBD, AIT, PA, SS, and vitiligo
	Selective IgA deficiency	Unknown	AIC (ITP, AIHA), IBD, CD, PV, MG, SLE, RA, JIA, T1D, and AIT
	P110 delta deficiency	PIK3CD	IBD, AIC
	Hyper IgM syndrome	CD40, CD40L	AIT, IBD, RA, JIA, AIHA, and AGN
Diseases of immune dysregulation	LRBA deficiency	LRBA	AIC (AIHA, ITP, AN), IBD, RA, and JIA
	APECED	AIRE	T1D, AD, AIT, hypoparathyroidism, enteropathy, adrenal corticotropic hormone insufficiency, growth hormone insufficiency, vitiligo, alopecia, autoimmune hepatitis, and ovarian/testicular failure
	IPEX	FOXP3	IBD, AIC, AIT, vitiligo, alopecia, hepatitis, and early onset diabetes
	CTLA4 haploinsufficiency	CTLA4	IBD, AIC, SLE, and arthritis
	XIAP deficiency	XIAP	IBD, AIC, and hepatitis
	Early onset inflammatory bowel disease syndromes	various	IBD, arthritis
	STAT3 GOF	STAT3	IBD, AIC, hepatitis, and early-onset T1D
	ALPS	various	AIC, GN, endocrinopathies, and SLE
Congenital defects of phagocyte number, function, or both	Chronic granulomatous disease	CYBB	IBD, AIC, AIT, JIA, GN, SLE, APLA, and autoimmune pulmonary disease
Defects in innate immunity	STAT1 deficiency	STAT1 GOF	AIC, AIT, T1D, and SLE
Autoinflammatory disorders	Type 1 interferonopathies	various	SLE, AIC, and vasculopathy
Complement deficiencies	Complement deficiencies	various	SLE, vasculitis

Abbreviations: AD—Addison's disease; AIC—autoimmune cytopenia; AIHA—autoimmune hemolytic anemia; AIT—autoimmune thyroid disease; AN—autoimmune neutropenia; ALPS—autoimmune lymphoproliferative syndrome; APECED—autoimmune polyendocrinopathy-candidiasis-ectodermal dystrophy; APLA—antiphospholipid antibodies; CD—celiac disease; CVID—common variable immunodeficiency; GN—glomerulonephritis; GOF—gain-of-function; IBD—inflammatory bowel disease; IUIS—International Union of Immunological Societies; JIA—juvenile idiopathic arthritis; IPEX—immune dysregulation, polyendocrinopathy, enteropathy, and X-linked syndrome; ITP—immune thrombocytopenia; MS – multiple sclerosis; RA—rheumatoid arthritis; SLE—systemic lupus erythematosus; and T1D—type 1 diabetes.

4. Non-Malignant Lymphoproliferation

Ranging from reactive polyclonal hyperplasia (associated with immune disorders) to true monoclonal disease (malignant process), lymphoproliferative disorders (LPDs) constitute a heterogeneous group of diseases in clinical and genetic terms. LPDs occur when the physiological control of proliferation of both T and B cells collapses. Disturbances in this control may occur in many conditions where immunity is compromised. This creates difficulties (both in the clinical assessment of the patient and in the identification of pathogenic mechanisms) to differentiate LPDs [90]. They are observed in patients with immunodeficiency or immune dysregulation syndromes such as CVID, SCID, WAS, A-T, Chediak-Higashi syndrome (CHS), and X-linked lymphoproliferative disorders [91]. Additionally, splenomegaly and/or generalized lymphadenopathy are described in disorders such as CD27 deficiency, CD70, ITK deficiency, and XLP type 1. Autoimmune disorders, hypersensitivity reactions, and viral infections, including human immunodeficiency virus (HIV) infection, are also prone to developing lymphoproliferative disorders. Lymphoproliferation as well as lymphomas (both Hodgkin's and non-Hodgkin's lymphomas) are often associated with Epstein–Barr virus (EBV) infection. Moreover, both lymphadenopathy and splenomegaly can be caused by nonspecific infections, in CVID but also in almost any other PID, and they are not always primarily associated with immune dysregulation [92]. Transplant patients, as well as those taking immunosuppressants such as cyclosporine, sirolimus, and tacrolimus, are also at risk of developing benign LPDs [93].

Autoimmune lymphoproliferative syndrome (ALPS) is an example of a disease resulting from impaired apoptosis of lymphocytes, mostly as a consequence of abnormalities associated with programmed cell death mediated by Fas. Fas is a transmembrane receptor located on the cell surface and is one of the tumor necrosis factor receptors (TNFR). It is responsible for the induction of apoptosis, which is triggered after binding with the appropriate ligand (FasL). When the *FAS* gene is mutated, there are defects in the external pathway of programmed cell death [94]. Clinically, patients develop chronic lymphoproliferation and an increased number of T cells, which are referred to as "double negative T cells" (DNT) with $CD4^-/CD8^-$, $CD3^+$, and $TCR\alpha\beta^+$ phenotype [95].

ALPS usually presents in infancy or early childhood (the median age is 31–36 months), most often in the form of nonmalignant lymphoid expansion with lymphadenopathy, splenomegaly, and/or hepatomegaly and AIC, including hemolytic anemia and thrombocytopenia. In a minority of patients, clinical symptoms may appear later in life (18 to 35 years). In a French cohort, patients with later disease onset often presented autoimmune manifestations rather than LPD [96,97]. Patients often do not present symptoms that would suggest an infectious or neoplastic etiology. Most patients have an increased number of T and B lymphocytes, as well as polyclonal hypergammaglobulinemia. Hypogammaglobulinaemia, often not associated with increased susceptibility to infections, may occur in approximately 10% of cases. Autoimmunity is a common feature of ALPS and can be the first ALPS manifestation; however, it is not always present at the time of diagnosis. Autoantibodies are detected in up to 80% of patients, most often anticardiolipin antibodies or direct Coombs' antibodies, but only half of them actually have an autoimmune disease, usually AIHA, ITP, or autoimmune neutropenia (AIN). A pledge of hemolysis during examination of blood smears, as well as the detection of autoantibodies and a degree of reticulocytosis, are helpful in distinguishing AIC from the effects of coexistent hypersplenism. Another helpful diagnostic tip is that AIC often manifests clinically. Autoantibodies typically have high affinity and are IgG-derived, in contrast to naturally occurring autoantibodies of the same specificity that are low-affinity and IgM-derived. Autoimmune diseases that affect other systems than the haematopoietic system can also occur but are much rarer [98]. Regardless of the time since the disease onset, symptoms such as lymphadenopathy and/or splenomegaly will ultimately be seen in 100% of ALPS patients and are required for diagnosis. The areas most commonly affected by lymphadenopathy are the neck, mediastinum, armpits, groin, and pelvis, although virtually any lymph node can become enlarged. Lymphoproliferation tends to subside over time, and by the age of 20, as much as 66% of patients achieve complete remission, while the rest of the patients experience a significant improvement. Infections are sporadic but can also occur as a result of neutropenia and/or nasopharyngeal obstruction due to lymphadenopathy [99]. Moreover, patients with ALPS are characterized by an increased risk of cancer (estimated at 10–20%); the most common forms of cancer are Hodgkin's lymphoma and non-Hodgkin's lymphoma [100].

Lymphoma can develop at any age in ALPS–FAS but is rare as a presenting feature. Distinguishing a benign node from a questionable node is a diagnostic challenge because of the frequent concomitant presence of benign/typical lymphadenopathy and splenomegaly seen with ALPS. Important clues for lymphoma are classic alarm symptoms (B symptoms), including fever, night sweats, itching, and weight loss. Positron emission tomography (PET)-based imaging may be helpful for distinguishing "good" from "bad" nodes on the basis of the presumed higher metabolic activity of malignant lymphoid tissue [101]. The nonmalignant lymphadenopathy fluctuates, and PET scan results fluctuate similarly. Lymphoma nodes more often are continuously chemically active ("hot"). Lymphoma typically originates in the B cell lineage, but T cell lymphomas have also occurred.

The required criteria for the diagnosis of ALPS include chronic lymphoproliferation lasting more than 6 months with the exclusion of neoplastic and infectious lymphoproliferation. In isolated lymphadenopathy, they must involve two distinct nodal regions. The second of the required criteria includes elevated counts of double negative T cells in peripheral blood that exceed 1.5% of the total number of lymphocytes or 2.5% in the case of T lymphocytes [102]. In addition, the diagnostics include genetic, biochemical (increased concentration of vitamin B12/IL-10/IL-18/sFASL/FAS), and histopathological tests.

5. Neoplastic Manifestations

Along with a predisposition to severe and recurrent infections and autoimmunity, neoplasms form a triad that identifies the most common symptoms in a variety of IEI. Despite this, there is a lack of systematic data on the cancer risk and type of neoplasms seen in most IEI. The development of malignant neoplasms most often occurs in patients with CVID, and in patients with defects in genes regulating DNA repair, cell cycle, apoptosis, or bone marrow maturation. Available population cohort studies suggest that the increased risk of developing cancer is limited to specific and rare forms of IEI and is mainly due to an increased risk of developing lymphoma [103–106]. The highest risk of lymphomas was reported in NBS (49%), X-linked lymphoproliferative syndrome (XLP; 24–30%), A-T (15–19%), ALPS (7–15%), and the mentioned CVID (1.8–8.2%) [96,103,107,108]. Among CVID patients, there is a 7- to 10-fold increase in gastric cancer incidence, which is related to the lack of secretory IgA [109,110]. In patients with CVID, extra-nodal non-Hodgkin's B-cell lymphomas and mucosa-associated lymphomas are the most common [111]. Unlike most IEI, lymphomas in CVID are more common in people in the 4th to 7th decade of life and are usually EBV-negative [111,112]. In a study by Ludvigsson et al., individuals with IgA deficiency were at a moderately increased risk of cancer, with excess risks of gastrointestinal cancer. Children with IgA deficiency were at no increased risk of cancer, but the statistical power was limited in subanalyses [113].

Common high-grade DNA strand repair defects with chromosomal instability are seen in the A-T. Ruptures of dsDNA cause a high percentage of malignant tumors, chromosome instability, and abnormal rearrangements of V (D) J genes; a recombination of class switches and/or somatic hypermutations (the *ATM* gene in A-T, the *NBN* gene in NBS, the *DCLRE1C* gene in severe combined deficiency immunodeficiency with sensitivity to ionizing radiation and Omenn syndrome, the *LIG4* gene in the LIG4 syndrome, and the *LIG1* gene in DNA ligase 1 deficiency) cause complex immunodeficiencies and malignant neoplasms, most often lymphomas [114,115]. Patients with Bloom's syndrome (*BLM* gene) age prematurely and are susceptible to non-Hodgkin's lymphoma (NHL). Patients with Schimke syndrome (*SMARCAL1* gene) show chromosomal instability and an increased risk of malignant neoplasm, including NHL and osteosarcoma [116,117].

Malignancies associated with impaired telomere maintenance are observed in genetically heterogeneous congenital dyskeratosis and its clinically severe variant of Hoyeraal Hreidarsson syndrome, NBS and A-T. Disorders of telomerase lead to the defective function of rapidly dividing cells and increased susceptibility to hematological and solid tumors [114].

IEI, which inherently affect hematopoiesis, make it susceptible to malignant neoplasms. In Fanconi anemia, a genetically heterogeneous disorder, pancytopenia, hematologic malignancies, solid tumors, and clinical immunodeficiency phenotypes are observed. Mutations of the *WAS* gene coding for the WASP disrupt the connection between GTPases and the actin cytoskeleton, thus disrupting the regulation of signaling in hematopoietic cells. Myelodysplasia, leukemias, and lymphomas in patients with WAS are seen more frequently [107,114,118]. The deficiency of the hematologic transcription factor GATA2 leads to phenotypically variable immunodeficiency, primary alveolar proteinosis, Emberger syndrome with lymphedema and/or a predisposition to myelodysplastic syndrome, acute myeloid leukemia (AML), chronic myelomonocytic leukemia (CMML), and EBV lymphoma [119]. The risk of leukemia is increased with some severe congenital neutropenia (ELANE, HAX1, and WASP) but not increased with the *ELANE* mutation that causes cyclic neutropenia. An increased risk of leukemia has not been reported in other PIDs associated with neutropenia [120]. Mutations in the *CD40L* gene cause X-linked immunodeficiency with hyperimmunoglobulin M. In the case of CD40L and CD40 ligand deficiencies, a Cryptosporidium biliary tract infection may lead to sclerosing cholangitis, cirrhosis, and an increased risk of hepatocellular carcinoma and biliary tract cancer [121–123].

Almost 20% of all human malignancies are associated with chronic infections with such pathogens as HBV, HCV, HPV, EBV, HHV8/KSHV, HTLV-I, HIV-1, HIV-2, JCV, Merkel cell carcinoma (MCV), Helicobacter pylori, schistosomes, or hepatic flukes [124,125]. Additionally, in IEI patients, chronic infections are often associated with malignancies. They were mostly described in connection with EBV, HPV, and HHV8 infections [107,126–128]. HPV can cause cancer of the cervix, vagina, vulva, anus, and penis, as well as squamous cell carcinoma of the oral cavity. Patients with warts, hypogammaglobulinemia, infections, and myelokathexis (WHIM) syndrome are particularly prone to HPV infection, resulting in numerous warts, condylomata acuminate, and subsequent severe papillomatosis and malignant transformation of the lesions [128].

EBV in patients with IEI may cause chronic EBV viremia, hemophagocytic lymphohistiocytosis (HLH), dysgammaglobulinemia, atypical EBV-associated lymphoproliferative disorders (polymorphic B-cell hyperplasia, plasmocytic hyperplasia), and EBV-associated lymphomas [105,129,130]. In the rare heterogeneous KID syndrome (keratitis, ichthyosis, and deafness), mainly caused by mutations in the connexin 26 (*GJB2*) gene, 15% of patients develop squamous cell carcinoma, often in sun-exposed areas [131,132].

The estimated risk for developing cancer in patients with IEI ranges from 4 to 25 percent [133]. Furthermore, the diagnosis of the malignancy, both clinical and histological, can be challenging in the presence of non-malignant lymphoproliferation or bone marrow abnormalities. These states, as well as concomitant infections or complex co-morbidities, all can mimic a developing malignancy clinically, radiologically, and even histopathologically. Due to the statistically higher risk of the above-mentioned types of neoplasms, patients with IEI should undergo periodic age-appropriate screening tests, just like healthy people. However, the guidelines in this regard may differ depending on the IEI type and national or international recommendations. Patients with epidermodysplasia verruciformis (EV) should undergo regular dermatological check-ups due to an increased risk of skin cancer. Patients with A-T and their female family members with heterozygous mutant *ATM* should start the screening for breast cancer earlier than the general population, and this age depends on the type of the mutation in the *ATM* gene [134,135].

It is also worth mentioning that both NHL and Hodgkin lymphoma are diagnosed at younger ages in patients with IEI, and NHL is more common in males with IEI [136,137]. In patients with suspected lymphoma, medical management is the same as in immunocompetent patients; however, diagnostic difficulties may appear. Diagnostic tests useful in cancer screening include uric acid, lactate dehydrogenase (LDH), and erythrocyte sedimentation rate (ESR). Even histopathology, which is a gold standard of diagnosing malignancy, can be challenging in patients with IEI, particularly during the investigation of possible lymphoid malignancy.

If clinically indicated, a surgical biopsy providing sufficient material for the assessment of tissue architecture and ancillary diagnostic techniques is a better diagnostic option than needle core biopsy. Histological diagnosis may be difficult even when appropriate, high-quality material is gained [137,138]. For example, non-malignant lymphoproliferative lesions may precede, as well as co-exist with, lymphoid malignancies. Often, diagnostic boundaries between non-neoplastic and neoplastic lesions are ill-defined and difficult to apply. Lymphocyte clonality assessed by molecular techniques may help during diagnostics, but these alone cannot provide diagnostic certainty, and clonal B-cell and T-cell proliferations falling short of malignancy are not uncommon in IEI [138,139].

Patients with specific immunodeficiencies, including A-T, NBS, and CVID, should be informed about the increased risk of neoplasia associated with increased sensitivity to ionizing radiation. Before performing tests or therapy with the use of radiation, they should consult this fact with the attending immunologist. On the other hand, medical personnel should consider the benefit–risk ratio in terms of interventions with the use of ionizing radiation in the context of the underlying disease, taking into account the need to perform the examination, and the possibility of replacing the examination with radiation with alternative techniques without the use of ionizing radiation.

Advances in the diagnosis and treatment of patients with IEI contributed to a significant extension of the life of those patients who previously had no chance to live to adulthood. Patients with IEI require multidisciplinary care; therefore, physicians of various specialties should be aware of the increased tendency to develop neoplasms in these patients. Patients should be thoroughly informed about the alarm symptoms of malignant neoplasms, especially lymphoma. Cancer in a patient with IEI is more often extensive or disseminated at the time of diagnosis, which is associated with a worse prognosis. Patients with IEI are more likely to develop NHL with B-cell origin, with high histologic grades and extranodal involvement, especially in the gastrointestinal tract or central nervous system. Early diagnosis can provide better treatment options before serious organ damage occurs.

The most prevalent types of malignancies among IEI patients have been summarized in Table 3.

Table 3. Most common types of cancer among patients with IEI.

Disease	IUIS Classification	Type of Malignancy
SCID	Immunodeficiencies affecting cellular and humoral immunity (Ia)	Lymphoma
ITK deficiency	Immunodeficiencies affecting cellular and humoral immunity (Ib)	EBV-associated lymphoproliferation Lymphoma
IKAROS deficiency (CD154)	Immunodeficiencies affecting cellular and humoral immunity (Ib)	T-ALL
DOCK8 deficiency	Immunodeficiencies affecting cellular and humoral immunity (Ib)	Vulvar, facial, and anal squamous cell dysplasia and carcinomas;T cell lymphoma-leukemiaBurkitt lymphomaNHL
STK4 deficiency	Immunodeficiencies affecting cellular and humoral immunity (Ib)	Lymphoma
RHOH deficiency	Immunodeficiencies affecting cellular and humoral immunity (Ib)	Lymphoma
OX40 deficiency	Immunodeficiencies affecting cellular and humoral immunity (Ib)	Kaposi sarcoma
CD40/CD40L deficiency	Immunodeficiencies affecting cellular and humoral immunity (Ib)	Hepatocarcinoma Cholangiocarcinoma Peripheral neuroectodermal tumors of the gastrointestinal tract and the pancreas Lymphoma

Table 3. *Cont.*

Disease	IUIS Classification	Type of Malignancy
Wiskott–Aldrich syndrome	Combined immunodeficiency of T and B cell with associated or syndromic features	Lymphoma EBV-related B-cell lymphoma Leukemia Cerebellar astrocytoma Kaposi sarcoma Smooth muscle tumors
Ataxia-telangiectasia	Combined immunodeficiency of T and B cell with associated or syndromic features	Leukemia Lymphoma Breast cancer Gastrointestinal malignancies (possible)
Nijmegen breakage syndrome	Combined immunodeficiency of T and B cell with associated or syndromic features	Lymphoma Acute leukemia Solid tumors
Bloom syndrome	Combined immunodeficiency of T and B cell with associated or syndromic features	Leukemia Lymphoma
PMS2 deficiency	Combined immunodeficiency of T and B cell with associated or syndromic features	Lymphoma Colorectal carcinoma Brain tumors
MCM4 deficiency	Combined immunodeficiency of T and B cell with associated or syndromic features	B cells lymphoma
Ligase I deficiency	Combined immunodeficiency of T and B cell with associated or syndromic features	Lymphoma
Cartilage-hair hypoplasia	Combined immunodeficiency of T and B cell with associated or syndromic features	Lymphoma Leukemia Squamous cell carcinoma Basal cell carcinoma
Schimke syndrome	Combined immunodeficiency of T and B cell with associated or syndromic features	Osteosarcoma NHL
Autosomal dominant hyper-IgE syndrome (AD-HIES)	Combined immunodeficiency of T and B cell with associated or syndromic features	NHL
CID with early-onset asthma, eczema and food allergies, autoimmunity ID with atopic dermatitis (CARD11)	Combined immunodeficiency of T and B cell with associated or syndromic features	Lymphoma
X-linked agammaglobulinemia	Predominantly antibody deficiencies	Lymphoreticular malignancies Gastric and colorectal adenocarcinoma Squamous cell carcinoma of the lung
Common variable immunodeficiency (CVID)	Predominantly antibody deficiencies	Lymphoma Thymus cancer Gastric cancer
Selective IgA deficiency	Predominantly antibody deficiencies	Gastrointestinal cancer
X-linked lymphoproliferative disease (XLP1)	Diseases of immune dysregulation	Lymphoma
CD27 deficiency	Diseases of immune dysregulation	Lymphoma
RASGRP1 deficiency	Diseases of immune dysregulation	EBV-associated lymphoma
CD70 deficiency	Diseases of immune dysregulation	Hodgkin lymphoma
CTPS1 deficiency	Diseases of immune dysregulation	B-cell NH lymphoma
CD137 deficiency	Diseases of immune dysregulation	B-cell lymphoma

Table 3. *Cont.*

Disease	IUIS Classification	Type of Malignancy
XL magnesium EBV and neoplasia (XMEN)	Diseases of immune dysregulation	Lymphoma
ALPS–FAS	Diseases of immune dysregulation	Lymphoma
Severe congenital neutropenia	Congenital defects of phagocyte number, function, or both	MDS/leukemia
HAX1 deficiency	Congenital defects of phagocyte number, function, or both	MDS/leukemia
Shwachman-Diamond syndrome	Congenital defects of phagocyte number, function, or both	Leukemia
GATA2 deficiency	Congenital defects of phagocyte number, function, or both	AML/CMML
WHIM syndrome	Defects in intrinsic and innate immunity	HPV-related cancers Lymphoma
Epidermodysplasia verruciformis	Defects in intrinsic and innate immunity	Squamous cell carcinoma

Abbreviations: AML—acute myelogenous leukemia; CMML—chronic myelomonocytic leukemia; EBV—Epstein–Barr virus; HPV—human papillomavirus; MDS—myelodysplastic syndrome; NHL—non-Hodgkin lymphoma; and T-ALL—T-cell acute lymphoblastic leukemia.

6. Diseases of Immune Dysregulation

Diseases of immune dysregulation are a separate and independent category of IEI in IUIS classification [1]. This category includes i.a. familial hemophagocytic lymphohistiocytosis (FHL syndromes), FHL syndromes with hypopigmentation, regulatory T cell defects, autoimmunity with or without lymphoproliferation, immune dysregulation with colitis, ALPS, and a susceptibility to EBV and lymphoproliferative conditions. This category is often the most difficult to define clinically and to diagnose without extensive sequencing since there is a significant phenotypic overlap between different genetic causes, the evolution of features over time, and phenotypic heterogeneity. On the other hand, these diseases have improved our understanding of the pathways that drive autoimmunity in IEI.

Early-onset autoimmunity, autoimmunity that involves multiple organs, a strong family history of autoimmunity, autoimmunity in combination with susceptibility to infection, or significant lymphoproliferation all suggest an immune dysregulation defect.

Diseases of immune dysregulation, according to IUIS classification, are summarized in Table S1.

Over the years, the wide application of whole-exome sequencing/whole-genome sequencing has significantly promoted the discovery and further study of new IEI and its number has doubled from 2009 to 2019 [1,140]. It is worth mentioning that the number of cases for any particular IEI is usually few, and because of that, a large-scale study of IEI can hardly be conducted [140]. Furthermore, there are several difficulties in identifying IEI connected with immune dysregulation. There are still countries where genetic tests are not widespread and freely available, mostly because of their costs. Moreover, in some patients more than one mutation is present, which makes it even more difficult to find [140,141]. In addition, phenotypes of the same mutation vary between patients, ranging from mild or uncharacteristic symptoms to even life-threatening manifestations [140,142,143]. In conclusion, patients with immune dysregulation should be examined scrupulously, and genetic diagnostics should be conducted in cases when it is necessary and possible [140]. Early and proper diagnosis seems crucial when we consider IEI patients. In cases of IEI patients with immune dysregulation, it is even more important.

The treatment is often challenging and sometimes requires balancing between increased susceptibility to infection and the additional suppression of the immune system [144]. Not so long ago, treatment options for IEI patients remained limited. They included the intensive treatment of infections; IRT; and bone marrow transplant in some cases. IRT has been a standard, often live-saving treatment for IEI that has affected antibody production for the past four decades. Both intravenous (IVIg) and subcutaneous (SCIg) immunoglobulins are often suitable for lifelong therapy. High-dose IVIg, together with corticosteroids, is a standard therapy for ITP [144]. A significant increase in the field of clinical immunology, including molecular biology techniques, gene therapy, or the use of immune modulators, allowed the development of modern and precise therapies [145]. Equally, having better knowledge of IEI pathophysiology enables the implementation of targeted therapy. IEI is an excellent example of disease where such "precision medicine" can be applied. Precision medicine is an approach based on advances in genetic research and data analysis. It offers breakthroughs in the treatment of the disease and has the potential to overturn traditional methods of practicing medicine.

Such medicines (new or repurposed) modify intracellular pathways whose function is disturbed because of specific genetic defect [144]. Thanks to precision medicine, the treatment can selectively influence a specific cell function instead of affecting the entire immune system. Moreover, the adverse side effects that affect other tissues are possible to avoid.

Although the term "precision medicine" is relatively new, it has been part of healthcare for many years. For example, a person who needs a blood transfusion does not receive blood from a randomly selected donor; instead, the donor's blood group is matched to that of the recipient to reduce the risk of complications. Precision medicine is already used in the treatment of diabetes and cancer. It is especially useful in cases of breast, lung, skin, colon, prostate, and pancreatic cancer. Its other promising applications include cardiology, signs of aging, rare childhood diseases, cystic fibrosis, and HIV.

In the context of immunedysregulation, the usage of small molecules and biologics effectively helps with reversing the clinical manifestations of immunedysregulation and hyperinflammation. Knowledge about the genetic etiology of activated phosphoinositide 3-kinase delta (PI3Kδ) syndrome (APDS) allowed one to explore PI3Kδ inhibition as a precision medicine [146,147]. Leniolisib, a small-molecule, selective PI3Kδ inhibitor, causes the dose-dependent suppression of PI3Kδ pathway hyperactivation. Clinical trials are currently underway to establish the safety and efficacy of selective PI3K δ inhibitors as a possible therapeutic option in patients with APDS. One is related to the oral administration of leniolisib (NCT02435173), the other to the inhaled administration of nemiralisib (NCT02593539). So far, the 12-week dose escalation of leniolisib has been shown to be safe and effective in reducing lymphadenopathy, splenomegaly, and cytopenia [144,147].

7. Conclusions

IEI is a group of rare diseases that can be camouflaged or not considered because of the predominant clinical features of atopy, autoimmunity, or lymphoproliferation. Consequently, some patients will remain undiagnosed. This risk impairs their quality of life, morbidity, and mortality, especially when exposed to agents reducing the immune competence. An underlying IEI should be particularly considered, especially in severe cases of atopic disease with concomitant signs of autoimmunity and unusual, recurrent or severe infections, so appropriate treatment regimens can be initiated and inappropriate immune suppression avoided.

In terms of the scientific evidence, it is still debatable whether allergy and cancer should be considered as risk factors or rather the consequences of the underlying IEI. Autoimmunity, as well as malignancy, worsen the IEI patients' prognosis. Another important issue in IEI is their exact pathogenesis, as well as the gene–phenotype relationship. The recent advances in genetics also revolutionized the field of IEI. Until now, the increased use of new sequencing techniques allowed for the identification of different monogenic causes of IEI. They enabled the better understanding of genotype–phenotype correlations and consequently led to better therapeutic strategies targeting the immune dysregulation in IEI [45]. The unmet needs include the unified nomenclature; the pathophysiological mechanisms assessment, for example, the lymphoma' genesis in IEI patients; and better, more personalized treatment strategies [148].

Novel diagnostic approaches, as well as evidence-based treatment guidelines that consider the underlying immunodeficiency rather than using extrapolation from non-IEI settings, are necessary. The recommendations for validated screening of cohorts at risk of allergy, autoimmunity, and malignancy are of the utmost importance.

Supplementary Materials: The following supporting information can be downloaded at: https://www.mdpi.com/article/10.3390/jcm11144220/s1, Table S1: Diseases of immune dysregulation according to IUIS classification. Accessed on 8 July 2022.

Author Contributions: K.P.-Ś. and G.P. wrote the initial draft of this paper, which was critically revised by A.L.-U. and M.J. All the authors contributed to conceptualizing this work. All authors have read and agreed to the published version of the manuscript.

Funding: This research was financially supported by The Ministry of Health subvention according to number of STM.A020.20.063 from the IT Simple system of Wrocław Medical University.

Institutional Review Board Statement: Not applicable.

Informed Consent Statement: Not applicable.

Data Availability Statement: Not applicable.

Conflicts of Interest: The authors declare no conflict of interest.

References

1. Tangye, S.G.; Al-Herz, W.; Bousfiha, A.; Cunningham-Rundles, C.; Franco, J.L.; Holland, S.M.; Klein, C.; Morio, T.; Oksenhendler, E.; Picard, C. Human Inborn Errors of Immunity: 2022 Update on the Classification from the International Union of Immunological Societies Expert Committee. *J. Clin. Immunol.* **2022**, 1–35. [CrossRef] [PubMed]
2. Bousfiha, A.; Jeddane, L.; Picard, C.; Al-Herz, W.; Ailal, F.; Chatila, T.; Cunningham-Rundles, C.; Etzioni, A.; Franco, J.L.; Holland, S.M.; et al. Human Inborn Errors of Immunity: 2019 Update of the IUIS Phenotypical Classification. *J. Clin. Immunol.* **2020**, *40*, 66–81. [CrossRef] [PubMed]
3. Asano, T.; Boisson, B.; Onodi, F.; Matuozzo, D.; Moncada-Velez, M.; Maglorius Renkilaraj, M.; Zhang, P.; Meertens, L.; Bolze, A.; Materna, M.; et al. X-linked recessive TLR7 deficiency in ~1% of men under 60 years old with life-threatening COVID-19. *Sci. Immunol.* **2021**, *6*, eabl4348. [CrossRef]
4. Solé, D. Primary immunodeficiencies: A diagnostic challenge? *J. Pediatr.* **2021**, *97* (Suppl. S1), S1–S2. [CrossRef] [PubMed]
5. Ponsford, M.J.; Klocperk, A.; Pulvirenti, F.; Dalm, V.; Milota, T.; Cinetto, F.; Chovancova, Z.; Rial, M.J.; Sediva, A.; Litzman, J.; et al. Hyper-IgE in the allergy clinic—When is it primary immunodeficiency? *Allergy* **2018**, *73*, 2122–2136. [CrossRef] [PubMed]
6. Pieniawska-Śmiech, K.; Bar, K.; Babicki, M.; Śmiech, K.; Lewandowicz-Uszyńska, A. Assessment of weight and height of patients with primary immunodeficiency disorders and group of children with recurrent respiratory tract infections. *BMC Immunol.* **2020**, *21*, 42. [CrossRef]
7. Mauracher, A.A.; Gujer, E.; Bachmann, L.M.; Güsewell, S.; Schmid, J.P. Patterns of Immune Dysregulation in Primary Immunodeficiencies: A Systematic Review. *J. Allergy Clin. Immunol. Pract.* **2021**, *9*, 792–802.e10. [CrossRef]
8. Sokol, K.; Milner, J.D. The overlap between allergy and immunodeficiency. *Curr. Opin. Pediatr.* **2018**, *30*, 848–854. [CrossRef]
9. Lewandowicz-Uszyńska, A.; Pasternak, G.; Świerkot, J.; Bogunia-Kubik, K. Primary Immunodeficiencies: Diseases of Children and Adults—A Review. *Adv. Exp. Med. Biol.* **2021**, *1289*, 37–54. [CrossRef]
10. Chan, S.K.; Gelfand, E.W. Primary Immunodeficiency Masquerading as Allergic Disease. *Immunol. Allergy Clin. N. Am.* **2015**, *35*, 767–778. [CrossRef]
11. Castagnoli, R.; Lougaris, V.; Giardino, G.; Volpi, S.; Leonardi, L.; La Torre, F.; Federici, S.; Corrente, S.; Cinicola, B.L.; Soresina, A.; et al. Inborn errors of immunity with atopic phenotypes: A practical guide for allergists. *World Allergy Organ. J.* **2021**, *14*, 100513. [CrossRef]
12. Lyons, J.J.; Milner, J.D. Primary atopic disorders. *J. Exp. Med.* **2018**, *215*, 1009–1022. [CrossRef]
13. Ozcan, E.; Notarangelo, L.D.; Geha, R.S. Primary immune deficiencies with aberrant IgE production. *J. Allergy Clin. Immunol.* **2008**, *122*, 1054–1064. [CrossRef]
14. Aghamohammadi, A.; Cheraghi, T.; Gharagozlou, M.; Movahedi, M.; Rezaei, N.; Yeganeh, M.; Parvaneh, N.; Abolhassani, H.; Pourpak, Z.; Moin, M. IgA deficiency: Correlation between clinical and immunological phenotypes. *J. Clin. Immunol.* **2009**, *29*, 130–136. [CrossRef]
15. Jacob, C.M.; Pastorino, A.C.; Fahl, K.; Carneiro-Sampaio, M.; Monteiro, R.C. Autoimmunity in IgA deficiency: Revisiting the role of IgA as a silent housekeeper. *J. Clin. Immunol.* **2008**, *28* (Suppl. S1), S56–S61. [CrossRef]

16. Tuano, K.S.; Orange, J.S.; Sullivan, K.; Cunningham-Rundles, C.; Bonilla, F.A.; Davis, C.M. Food allergy in patients with primary immunodeficiency diseases: Prevalence within the US Immunodeficiency Network (USIDNET). *J. Allergy Clin. Immunol.* **2015**, *135*, 273–275. [CrossRef]

17. Lawrence, M.G.; Barber, J.S.; Sokolic, R.A.; Garabedian, E.K.; Desai, A.N.; O'Brien, M.; Jones, N.; Bali, P.; Hershfield, M.S.; Stone, K.D.; et al. Elevated IgE and atopy in patients treated for early-onset ADA-SCID. *J. Allergy Clin. Immunol.* **2013**, *132*, 1444–1446. [CrossRef]

18. Available online: https://primaryimmune.org/about-primary-immunodeficiencies-diagnosis-information/laboratory-tests (accessed on 17 February 2022).

19. Wittig, H.J.; Belloit, J.; De Fillippi, I.; Royal, G. Age-related serum immunoglobulin E levels in healthy subjects and in patients with allergic disease. *J. Allergy Clin. Immunol.* **1980**, *66*, 305–313. [CrossRef]

20. Ansotegui, I.J.; Melioli, G.; Canonica, G.W.; Caraballo, L.; Villa, E.; Ebisawa, M.; Passalacqua, G.; Savi, E.; Ebo, D.; Gómez, R.M.; et al. IgE allergy diagnostics and other relevant tests in allergy, a World Allergy Organization position paper. *World Allergy Organ. J.* **2020**, *13*, 100080. [CrossRef]

21. Lawrence, M.G.; Palacios-Kibler, T.V.; Workman, L.J.; Schuyler, A.J.; Steinke, J.W.; Payne, S.C.; McGowan, E.C.; Patrie, J.; Fuleihan, R.L.; Sullivan, K.E.; et al. Low Serum IgE Is a Sensitive and Specific Marker for Common Variable Immunodeficiency (CVID). *J. Clin. Immunol.* **2018**, *38*, 225–233. [CrossRef]

22. Mogensen, T.H. Primary Immunodeficiencies with Elevated IgE. *Int. Rev. Immunol.* **2016**, *35*, 39–56. [CrossRef] [PubMed]

23. Vu, Q.V.; Wada, T.; Toma, T.; Tajima, H.; Maeda, M.; Tanaka, R.; Oh-Ishi, T.; Yachie, A. Clinical and immunophenotypic features of atypical complete DiGeorge syndrome. *Pediatr. Int.* **2013**, *55*, 2–6. [CrossRef] [PubMed]

24. Adriani, M.; Aoki, J.; Horai, R.; Thornton, A.M.; Konno, A.; Kirby, M.; Anderson, S.M.; Siegel, R.M.; Candotti, F.; Schwartzberg, P.L. Impaired in vitro regulatory T cell function associated with Wiskott-Aldrich syndrome. *Clin. Immunol.* **2007**, *124*, 41–48. [CrossRef] [PubMed]

25. Humblet-Baron, S.; Sather, B.; Anover, S.; Becker-Herman, S.; Kasprowicz, D.J.; Khim, S.; Nguyen, T.; Hudkins-Loya, K.; Alpers, C.E.; Ziegler, S.F. Wiskott-Aldrich syndrome protein is required for regulatory T cell homeostasis. *J. Clin. Investig.* **2007**, *117*, 407–418. [CrossRef]

26. Maillard, M.H.; Cotta-de-Almeida, V.; Takeshima, F.; Nguyen, D.D.; Michetti, P.; Nagler, C.; Bhan, A.K.; Snapper, S.B. The Wiskott-Aldrich syndrome protein is required for the function of CD4$^+$CD25$^+$Foxp3$^+$ regulatory T cells. *J. Exp. Med.* **2007**, *204*, 381–391. [CrossRef]

27. Lexmond, W.S.; Goettel, J.A.; Lyons, J.J.; Jacobse, J.; Deken, M.M.; Lawrence, M.G.; DiMaggio, T.H.; Kotlarz, D.; Garabedian, E.; Sackstein, P.; et al. FOXP3fl Tregs require WASP to restrain Th2-mediated food allergy. *J. Clin. Investig.* **2016**, *126*, 4030–4044. [CrossRef]

28. Kumánovics, A.; Wittwer, C.T.; Pryor, R.J.; Augustine, N.H.; Leppert, M.F.; Carey, J.C.; Ochs, H.D.; Wedgwood, R.J.; Faville, R.J., Jr.; Quie, P.G.; et al. Rapid molecular analysis of the *STAT3* gene in Job syndrome of hyper-IgE and recurrent infectious diseases. *J. Mol. Diagn.* **2010**, *12*, 213–219. [CrossRef]

29. Yong, P.F.; Freeman, A.F.; Engelhardt, K.R.; Holland, S.; Puck, J.M.; Grimbacher, B. An update on the hyper-IgE syndromes. *Arthritis Res. Ther.* **2012**, *14*, 228. [CrossRef]

30. Siegel, A.M.; Stone, K.D.; Cruse, G.; Lawrence, M.G.; Olivera, A.; Jung, M.Y.; Barber, J.S.; Freeman, A.F.; Holland, S.M.; O'Brien, M.; et al. Diminished allergic disease in patients with STAT3 mutations reveals a role for STAT3 signaling in mast cell degranulation. *J. Allergy Clin. Immunol.* **2013**, *132*, 1388–1396. [CrossRef]

31. Hox, V.; O'Connell, M.P.; Lyons, J.J.; Sackstein, P.; Dimaggio, T.; Jones, N.; Nelson, C.; Boehm, M.; Holland, S.M.; Freeman, A.F.; et al. Diminution of signal transducer and activator of transcription 3 signaling inhibits vascular permeability and anaphylaxis. *J. Allergy Clin. Immunol.* **2016**, *138*, 187–199, Correction in *J. Allergy Clin. Immunol.* **2017**, *140*, 320. [CrossRef]

32. Boos, A.C.; Hagl, B.; Schlesinger, A.; Halm, B.E.; Ballenberger, N.; Pinarci, M.; Heinz, V.; Kreilinger, D.; Spielberger, B.D.; Schimke-Marques, L.F.; et al. Atopic dermatitis, STAT3- and DOCK8-hyper-IgE syndromes differ in IgE-based sensitization pattern. *Allergy* **2014**, *69*, 943–953. [CrossRef]

33. Hill, H.R.; Quie, P.G.; Pabst, H.F.; Ochs, H.D.; Clark, R.A.; Klebanoff, S.J.; Wedgwood, R.J. Defect in neutrophil granulocyte chemotaxis in Job's syndrome of recurrent "cold" staphylococcal abscesses. *Lancet* **1974**, *304*, 617–619. [CrossRef]

34. Grimbacher, B.; Holland, S.M.; Gallin, J.I.; Greenberg, F.; Hill, S.C.; Malech, H.L.; Miller, J.A.; O'Connell, A.C.; Puck, J.M. Hyper-IgE syndrome with recurrent infections—An autosomal dominant multisystem disorder. *N. Engl. J. Med.* **1999**, *14*, 692–702. [CrossRef]

35. Woellner, C.; Gertz, E.M.; Schäffer, A.A.; Lagos, M.; Perro, M.; Glocker, E.O.; Pietrogrande, M.C.; Cossu, F.; Franco, J.L.; Matamoros, N.; et al. Mutations in STAT3 and diagnostic guidelines for hyper-IgE syndrome. *J. Allergy Clin. Immunol.* **2010**, *125*, 424–432.e8. [CrossRef]

36. Hafsi, W.; Yarrarapu, S. Job syndrome. In *StatPearls*; StatPearls Publishing: Treasure Island, FL, USA, 2021.

37. Zhang, Q.; Davis, J.C.; Lamborn, I.T.; Freeman, A.F.; Jing, H.; Favreau, A.J.; Matthews, H.F.; Davis, J.; Turner, M.L.; Uzel, G.; et al. Combined immunodeficiency associated with DOCK8 mutations. *N. Engl. J. Med.* **2009**, *361*, 2046–2055. [CrossRef]

38. Wilkie, H.; Leyva-Castillo, J.M.; Janssen, E.; Geha, R.S. DOCK8 Deficiency Exacerbates Skin Contact Hypersensitivity: 194. *J. Allergy Clin. Immunol.* **2019**, *143*, AB64. [CrossRef]

39. Stites, D.P.; Ishizaka, K.; Fudenberg, H.H. Serum IgE concentrations in hypogammaglobulinaemia and selective IgA deficiency. Studies on patients and family members. *Clin. Exp. Immunol.* **1972**, *10*, 391–397.

40. Waldmann, T.A.; Polmar, S.H.; Balestra, S.T.; Jost, M.C.; Bruce, R.M.; Terry, W.D. Immunoglobulin E in immunologic deficiency diseases. II. Serum IgE concentration of patients with acquired hypogammaglobulinemia, thymoma and hypogammaglobulinemia, myotonic dystrophy, intestinal lymphangiectasia and Wiskott-Aldrich syndrome. *J. Immunol.* **1972**, *109*, 304–310.

41. Agondi, R.C.; Barros, M.T.; Rizzo, L.V.; Kalil, J.; Giavina-Bianchi, P. Allergic asthma in patients with common variable immunodeficiency. *Allergy* **2010**, *65*, 510–515. [CrossRef]

42. Agondi, R.C.; Barros, M.T.; Kokron, C.M.; Cohon, A.; Oliveira, A.K.; Kalil, J.; Giavina-Bianchi, P. Can patients with common variable immunodeficiency have allergic rhinitis? *Am. J. Rhinol. Allergy* **2013**, *27*, 79–83. [CrossRef]

43. Hartman, H.; Schneider, K.; Hintermeyer, M.; Bausch-Jurken, M.; Fuleihan, R.; Sullivan, K.E.; Cunningham-Rundles, C.; Bonilla, F.A.; USIDNET Consortium; Verbsky, J. Lack of Clinical Hypersensitivity to Penicillin Antibiotics in Common Variable Immunodeficiency. *J. Clin. Immunol.* **2017**, *37*, 22–24. [CrossRef]

44. Schmidt, R.E.; Grimbacher, B.; Witte, T. Autoimmunity and primary immunodeficiency: Two sides of the same coin? *Nat. Rev. Rheumatol.* **2017**, *14*, 7–18. [CrossRef]

45. Costagliola, G.; Cappelli, S.; Consolini, R. Autoimmunity in Primary Immunodeficiency Disorders: An Updated Review on Pathogenic and Clinical Implications. *J. Clin. Med.* **2021**, *10*, 4729. [CrossRef]

46. Amaya-Uribe, L.; Rojas, M.; Azizi, G.; Anaya, J.M.; Gershwin, M.E. Primary immunodeficiency and autoimmunity: A comprehensive review. *J. Autoimmun.* **2019**, *99*, 52–72. [CrossRef]

47. Fischer, A.; Provot, J.; Jais, J.P.; Alcais, A.; Mahlaoui, N.; members of the CEREDIH French PID study group. Autoimmune and Inflammatory Manifestations Occur Frequently in Patients with Primary Immunodeficiencies. *J. Allergy Clin. Immunol.* **2017**, *140*, 1388–1393.e8. [CrossRef]

48. Azizi, G.; Ziaee, V.; Tavakol, M.; Alinia, T.; Yazdai, R.; Mohammadi, H.; Abolhassani, H.; Aghamohammadi, A. Approach to the Management of Autoimmunity in Primary Immunodeficiency. *Scand. J. Immunol.* **2017**, *85*, 13–29. [CrossRef]

49. Bertinchamp, R.; Gérard, L.; Boutboul, D.; Malphettes, M.; Fieschi, C.; Oksenhendler, E.; DEFI study group. Exclusion of Patients with a Severe T-Cell Defect Improves the Definition of Common Variable Immunodeficiency. *J. Allergy Clin. Immunol. Pract.* **2016**, *4*, 1147–1157. [CrossRef] [PubMed]

50. Speckmann, C.; Doerken, S.; Aiuti, A.; Albert, M.H.; Al-Herz, W.; Allende, L.M.; Scarselli, A.; Avcin, T.; Perez-Becker, R.; Cancrini, C. A prospective study on the natural history of patients with profound combined immunodeficiency: An interim analysis. *J. Allergy Clin. Immunol.* **2017**, *139*, 1302–1310.e4. [CrossRef] [PubMed]

51. Von Spee-Mayer, C.; Koemm, V.; Wehr, C.; Goldacker, S.; Kindle, G.; Bulashevska, A.; Proietti, M.; Grimbacher, B.; Ehl, S.; Warnatz, K. Evaluating laboratory criteria for combined immunodeficiency in adult patients diagnosed with common variable immunodeficiency. *Clin. Immunol.* **2019**, *203*, 59–62. [CrossRef] [PubMed]

52. Walter, J.E.; Ayala, I.A.; Milojevic, D. Autoimmunity as a continuum in primary immunodeficiency. *Curr. Opin. Pediatr.* **2019**, *31*, 851–862. [CrossRef] [PubMed]

53. Alroqi, F.J.; Charbonnier, L.M.; Baris, S.; Kiykim, A.; Chou, J.; Platt, C.D.; Algassim, A.; Keles, S.; Al Saud, B.K.; Alkuraya, F.S.; et al. Exaggerated follicular helper T-cell responses in patients with LRBA deficiency caused by failure of CTLA4-mediated regulation. *J. Allergy Clin. Immunol.* **2018**, *141*, 1050–1059.e10. [CrossRef]

54. Azizi, G.; Ahmadi, M.; Abolhassani, H.; Yazdani, R.; Mohammadi, H.; Mirshafiey, A.; Rezaei, N.; Aghamohammadi, A. Autoimmunity in Primary Antibody Deficiencies. *Int. Arch. Allergy Immunol.* **2016**, *171*, 180–193. [CrossRef]

55. Bruton, O.C. Agammaglobulinemia. *Pediatrics* **1952**, *9*, 722–728. [CrossRef]

56. Nonoyama, S.; Tsukada, S.; Yamadori, T.; Miyawaki, T.; Jin, Y.Z.; Watanabe, C.; Morio, T.; Yata, J.; Ochs, H.D. Functional analysis of peripheral blood B cells in patients with X-linked agammaglobulinemia. *J. Immunol.* **1998**, *161*, 3925–3929.

57. Jacobs, Z.D.; Guajardo, J.R.; Anderson, K.M. XLA-associated neutropenia treatment: A case report and review of the literature. *J. Pediatr. Hematol. Oncol.* **2008**, *30*, 631–634. [CrossRef]

58. Behniafard, N.; Aghamohammadi, A.; Abolhassani, H.; Pourjabbar, S.; Sabouni, F.; Rezaei, N. Autoimmunity in X-linked agammaglobulinemia: Kawasaki disease and review of the literature. *Exp. Rev. Clin. Immunol.* **2012**, *8*, 155–159. [CrossRef]

59. Hernandez-Trujillo, V.P.; Scalchunes, C.; Cunningham-Rundles, C.; Ochs, H.D.; Bonilla, F.A.; Paris, K.; Yel, L.; Sullivan, K.E. Autoimmunity and inflammation in X-linked agammaglobulinemia. *J. Clin. Immunol.* **2014**, *34*, 627–632. [CrossRef]

60. Ng, Y.S.; Wardemann, H.; Chelnis, J.; Cunningham-Rundles, C.; Meffre, E. Bruton's tyrosine kinase is essential for human B cell tolerance. *J. Exp. Med.* **2004**, *200*, 927–934. [CrossRef]

61. Samuels, J.; Ng, Y.S.; Coupillaud, C.; Paget, D.; Meffre, E. Human B cell tolerance and its failure in rheumatoid arthritis. *Ann. N. Y. Acad. Sci.* **2005**, *1062*, 116–126. [CrossRef]

62. Odnoletkova, I.; Kindle, G.; Quinti, I.; Grimbacher, B.; Knerr, V.; Gathmann, B.; Ehl, S.; Mahlaoui, N.; Van Wilder, P.; Bogaerts, K. The burden of common variable immunodeficiency disorders: A retrospective analysis of the European Society for Immunodeficiency (ESID) registry data. *Orphanet J. Rare Dis.* **2018**, *13*, 201. [CrossRef]

63. Cunningham-Rundles, C. Autoimmunity in primary immune deficiency: Taking lessons from our patients. *Clin. Exp. Immunol.* **2011**, *164* (Suppl. S2), 6–11. [CrossRef]

64. Grześk, E.; Dąbrowska, A.; Urbanczyk, A.; Ewertowska, M.; Wysocki, M.; Kołtan, S. Common variable immunodeficiency: Different faces of the same disease. *Postepy Dermatol. Alergol.* **2021**, *38*, 873–880. [CrossRef]

65. Quinti, I.; Soresina, A.; Spadaro, G.; Martino, S.; Donnanno, S.; Agostini, C.; Claudio, P.; Franco, D.; Maria Pesce, A.; Borghese, F.; et al. Long-term follow-up and outcome of a large cohort of patients with common variable immunodeficiency. *J. Clin. Immunol.* **2007**, *27*, 308–316. [CrossRef]

66. Boileau, J.; Mouillot, G.; Gérard, L.; Carmagnat, M.; Rabian, C.; Oksenhendler, E.; Pasquali, J.L.; Korganow, A.S.; DEFI Study Group. Autoimmunity in common variable immunodeficiency: Correlation with lymphocyte phenotype in the French DEFI study. *J. Autoimmun.* **2011**, *36*, 25–32. [CrossRef]

67. Arandi, N.; Mirshafiey, A.; Jeddi-Tehrani, M.; Abolhassani, H.; Sadeghi, B.; Mirminachi, B.; Shaghaghi, M.; Aghamohammadi, A. Evaluation of CD4$^+$CD25$^+$FOXP3$^+$ regulatory T cells function in patients with common variable immunodeficiency. *Cell. Immunol.* **2013**, *281*, 129–133. [CrossRef]

68. Carter, C.R.; Aravind, G.; Smalle, N.L.; Cole, J.Y.; Savic, S.; Wood, P.M. CVID patients with autoimmunity have elevated T cell expression of granzyme B and HLA-DR and reduced levels of Treg cells. *J. Clin. Pathol.* **2013**, *66*, 146–150. [CrossRef]

69. Yu, G.P.; Chiang, D.; Song, S.J.; Hoyte, E.G.; Huang, J.; Vanishsarn, C.; Nadeau, K.C. Regulatory T cell dysfunction in subjects with common variable immunodeficiency complicated by autoimmune disease. *Clin. Immunol.* **2009**, *131*, 240–253. [CrossRef]

70. Arumugakani, G.; Wood, P.M.; Carter, C.R. Frequency of Treg cells is reduced in CVID patients with autoimmunity and splenomegaly and is associated with expanded CD21lo B lymphocytes. *J. Clin. Immunol.* **2010**, *30*, 292–300. [CrossRef] [PubMed]

71. Chew, G.Y.; Sinha, U.; Gatenby, P.A.; DeMalmanche, T.; Adelstein, S.; Garsia, R.; Hissaria, P.; French, M.A.; Wilson, A.; Whittle, B. Autoimmunity in primary antibody deficiency is associated with protein tyrosine phosphatase nonreceptor type 22 (PTPN22). *J. Allergy Clin. Immunol.* **2013**, *131*, 1130.e1–1135.e1. [CrossRef] [PubMed]

72. Brouet, J.C.; Chedeville, A.; Fermand, J.P.; Royer, B. Study of the B cell memory compartment in common variable immunodeficiency. *Eur. J. Immunol.* **2000**, *30*, 2516–2520. [CrossRef]

73. Abolhassani, H.; Amirkashani, D.; Parvaneh, N.; Mohammadinejad, P.; Gharib, B.; Shahinpour, S. Autoimmune phenotype in patients with common variable immunodeficiency. *J. Investig. Allergol. Clin. Immunol.* **2013**, *23*, 323–329.

74. Picchianti Diamanti, A.; Rosado, M.M.; Scarsella, M.; Ceccarelli, S.; Laganà, B.; D'Amelio, R.; Carsetti, R. Increased serum IgM, immunodeficiency, and autoimmunity: A clinical series. *Int. J. Immunopathol. Pharmacol.* **2015**, *28*, 547–556. [CrossRef]

75. Lin, L.H.; Tsai, C.N.; Liu, M.F.; Wang, C.R. Common variable immunodeficiency mimicking rheumatoid arthritis with Sjogren's syndrome. *J. Microbiol. Immunol. Infect.* **2005**, *38*, 358–360.

76. Jesus, A.A.; Liphaus, B.L.; Silva, C.A.; Bando, S.Y.; Andrade, L.E.; Coutinho, A.; Carneiro-Sampaio, M. Complement and antibody primary immunodeficiency in juvenile systemic lupus erythematosus patients. *Lupus* **2011**, *20*, 1275–1284. [CrossRef]

77. European Association for the Study of the Liver. EASL clinical practice guidelines: Autoimmune hepatitis. *J. Hepatol.* **2015**, *63*, 971–1004. [CrossRef]

78. Pecoraro, A.; Crescenzi, L.; Varricchi, G.; Marone, G.; Spadaro, G. Heterogeneity of Liver Disease in Common Variable Immunodeficiency Disorders. *Front. Immunol.* **2020**, *11*, 338. [CrossRef]

79. Tahiat, A.; Yagoubi, A.; Ladj, M.S.; Belbouab, R.; Aggoune, S.; Atek, L.; Bouziane, D.; Melzi, S.; Boubidi, C.; Drali, W.; et al. Diagnostic and Predictive Contribution of Autoantibodies Screening in a Large Series of Patients with Primary Immunodeficiencies. *Front. Immunol.* **2021**, *12*, 665322. [CrossRef]

80. Boyle, R.J.; Le, C.; Balloch, A.; Tang, M.L. The clinical syndrome of specific antibody deficiency in children. *Clin. Exp. Immunol.* **2006**, *146*, 486–492. [CrossRef]

81. Cheng, Y.K.; Decker, P.A.; O'Byrne, M.M.; Weiler, C.R. Clinical and laboratory characteristics of 75 patients with specific polysaccharide antibody deficiency syndrome. *Ann. Allergy Asthma Immunol.* **2006**, *97*, 306–311. [CrossRef]

82. Gennery, A.R.; Cant, A.J.; Jeggo, P.A. Immunodeficiency associated with DNA repair defects. *Clin. Exp. Immunol.* **2000**, *121*, 1–7. [CrossRef]

83. Vorechovsky, I.; Munzarova, M.; Lokaj, J. Increased bleomycin-induced chromosome damage in lymphocytes of patients with common variable immunodeficiency indicates an involvement of chromosomal instability in their cancer predisposition. *Cancer Immunol. Immunother.* **1989**, *29*, 303–306. [CrossRef]

84. Gantt, R.; Parshad, R.; Price, F.M.; Sanford, K.K. Biochemical evidence for deficient DNA repair leading to enhanced G2 chromatid radiosensitivity and susceptibility to cancer. *Radiat. Res.* **1986**, *108*, 117–126. [CrossRef]

85. Rothblum-Oviatt, C.; Wright, J.; Lefton-Greif, M.A.; McGrath-Morrow, S.A.; Crawford, T.O.; Lederman, H.M. Ataxia telangiectasia: A review. *Orphanet J. Rare Dis.* **2016**, *11*, 159. [CrossRef]

86. Luzi, G.; Zullo, A.; Iebba, F.; Rinaldi, V.; Sanchez Mete, L.; Muscaritoli, M.; Aiuti, F. Duodenal pathology and clinical-immunological implications in common variable immunodeficiency patients. *Am. J. Gastroenterol.* **2003**, *98*, 118–121. [CrossRef]

87. Malamut, G.; Verkarre, V.; Suarez, F.; Viallard, J.F.; Lascaux, A.S.; Cosnes, J.; Bouhnik, Y.; Lambotte, O.; Béchade, D.; Ziol, M. The enteropathy associated with common variable immunodeficiency: The delineated frontiers with celiac disease. *Am. J. Gastroenterol.* **2010**, *105*, 2262–2275. [CrossRef]

88. Daniels, J.A.; Lederman, H.M.; Maitra, A.; Montgomery, E.A. Gastrointestinal tract pathology in patients with common variable immunodeficiency (CVID): A clinicopathologic study and review. *Am. J. Surg. Pathol.* **2007**, *31*, 1800–1812. [CrossRef] [PubMed]

89. Westermann-Clark, E.; Meehan, C.A.; Meyer, A.K.; Dasso, J.F.; Amre, D.; Ellison, M.; Patel, B.; Betensky, M.; Hauk, C.I.; Mayer, J. Primary Immunodeficiency in Children with Autoimmune Cytopenias: Retrospective 154-Patient Cohort. *Front. Immunol.* **2021**, *12*, 649182. [CrossRef] [PubMed]

90. Mayor, P.C.; Eng, K.H.; Singel, K.L.; Abrams, S.I.; Odunsi, K.; Moysich, K.B.; Fuleihan, R.; Garabedia, E.; Lugar, P.; Ochs, H.D.; et al. Cancer in primary immunodeficiency diseases: Cancer incidence in the United States Immune Deficiency Network Registry. *J. Allergy Clin. Immunol.* **2018**, *141*, 1028–1035. [CrossRef] [PubMed]

91. Bonilla, F.A.; Barlan, I.; Chapel, H.; Costa-Carvalho, B.T.; Cunningham-Rundles, C.; de la Morena, M.T.; Espinosa-Rosales, F.J.; Hammarström, L.; Nonoyama, S.; Quinti, I. International Consensus Document (ICON): Common Variable Immunodeficiency Disorders. *J. Allergy Clin. Immunol. Pract.* **2016**, *4*, 38–59. [CrossRef] [PubMed]

92. Gangemi, S.; Allegra, A.; Musolino, C. Lymphoproliferative disease and cancer among patients with common variable immunodeficiency. *Leuk. Res.* **2015**, *39*, 389–396. [CrossRef]

93. Cohen, J.M.; Sebire, N.J.; Harvey, J.; Gaspar, H.B.; Cathy, C.; Jones, A.; Rao, K.; Cubitt, D.; Amrolia, P.J.; Davies, E.G.; et al. Successful treatment of lymphoproliferative disease complicating primary immunodeficiency/immunodysregulatory disorders with reduced-intensity allogeneic stem-cell transplantation. *Blood* **2007**, *110*, 2209–2214. [CrossRef]

94. Teachey, D.T.; Seif, A.E.; Grupp, S.A. Advances in the management and understanding of autoimmune lymphoproliferative syndrome (ALPS). *Br. J. Haematol.* **2010**, *148*, 205–216. [CrossRef]

95. Teachey, D.T.; Greiner, R.; Seif, A.; Attiyeh, E.; Bleesing, J.; Choi, J.; Manno, C.; Rappaport, E.; Schwabe, D.; Sheen, C.; et al. Treatment with sirolimus results in complete responses in patients with autoimmune lymphoproliferative syndrome. *Br. J. Haematol.* **2009**, *145*, 101–106. [CrossRef]

96. Neven, B.; Magerus-Chatinet, A.; Florkin, B.; Gobert, D.; Lambotte, O.; De Somer, L.; Lanzarotti, N.; Stolzenberg, M.C.; Bader-Meunier, B.; Aladjidi, N. A survey of 90 patients with autoimmune lymphoproliferative syndrome related to TNFRSF6 mutation. *Blood* **2011**, *118*, 4798–4807. [CrossRef]

97. Price, S.; Shaw, P.A.; Seitz, A.; Joshi, G.; Davis, J.; Niemela, J.E.; Perkins, K.; Hornung, R.L.; Folio, L.; Rosenberg, P.S. Natural history of autoimmune lymphoproliferative syndrome associated with *FAS* gene mutations. *Blood* **2014**, *123*, 1989–1999. [CrossRef]

98. Su, H.C.; Lenardo, M.J. Genetic defects of apoptosis and primary immunodeficiency. *Immunol. Allergy Clin. N. Am.* **2008**, *28*, 329–351. [CrossRef]

99. Oliveira, J.B. The expanding spectrum of the autoimmune lymphoproliferative syndromes. *Curr. Opin. Pediatr.* **2013**, *25*, 722–729. [CrossRef]

100. Pace, R.; Vinh, D.C. Autoimmune Lymphoproliferative Syndrome and Epstein-Barr Virus-Associated Lymphoma: An Adjunctive Diagnostic Role for Monitoring EBV Viremia? *Case Rep. Immunol.* **2013**, *2013*, 245893. [CrossRef]

101. Rao, V.K.; Carrasquillo, J.A.; Dale, J.K.; Bacharach, S.L.; Whatley, M.; Dugan, F.; Tretler, J.; Fleisher, T.; Puck, J.M.; Wilson, W. Fluorodeoxyglucose positron emission tomography (FDG-PET) for monitoring lymphadenopathy in the autoimmune lymphoproliferative syndrome (ALPS). *Am. J. Hematol.* **2006**, *81*, 81–85, Correction in *Am. J. Hematol.* **2006**, *81*, 389. [CrossRef]

102. Bride, K.; Teachey, D. Autoimmune lymphoproliferative syndrome: More than a FAScinating disease. *F1000Research* **2017**, *6*, 1928. [CrossRef]

103. Vajdic, C.M.; Mao, L.; van Leeuwen, M.T.; Kirkpatrick, P.; Grulich, A.E.; Riminton, S. Are antibody deficiency disorders associated with a narrower range of cancers than other forms of immunodeficiency? *Blood* **2010**, *116*, 1228–1234. [CrossRef]

104. Mellemkjaer, L.; Hammarstrom, L.; Andersen, V.; Yuen, J.; Heilmann, C.; Barington, T.; Bjorkander, J.; Olsen, J.H. Cancer risk among patients with IgA deficiency or common variable immunodeficiency and their relatives: A combined Danish and Swedish study. *Clin. Exp. Immunol.* **2002**, *130*, 495–500. [CrossRef]

105. Rezaei, N.; Hedayat, M.; Aghamohammadi, A.; Nichols, K.E. Primary immunodeficiency diseases associated with increased susceptibility to viral infections and malignancies. *J. Allergy Clin. Immunol.* **2011**, *127*, 1329–1343. [CrossRef]

106. Taskinen, M.; Ranki, A.; Pukkala, E.; Jeskanen, L.; Kaitila, I.; Mäkitie, O. Extended follow-up of the Finnish cartilage-hair hypoplasia cohort confirms high incidence of non-Hodgkin lymphoma and basal cell carcinoma. *Am. J. Med. Genet. A* **2008**, *146*, 2370–2375. [CrossRef]

107. Leechawengwongs, E.; Shearer, W.T. Lymphoma complicating primary immunodeficiency syndromes. *Curr. Opin. Hematol.* **2012**, *19*, 305–312. [CrossRef]

108. Ngalamika, O.; Zhang, Y.; Yin, H.; Zhao, M.; Gershwin, M.E.; Lu, Q. Epigenetics, autoimmunity and hematologic malignancies: A comprehensive review. *J. Autoimmun.* **2012**, *39*, 451–465. [CrossRef]

109. Dhalla, F.; da Silva, S.P.; Lucas, M.; Travis, S.; Chapel, H. Review of gastric cancer risk factors in patients with common variable immunodeficiency disorders, resulting in a proposal for a surveillance programme. *Clin. Exp. Immunol.* **2011**, *165*, 1–7. [CrossRef] [PubMed]

110. Quiding-Järbrink, M.; Sundström, P.; Lundgren, A.; Hansson, M.; Bäckström, M.; Johansson, C.; Enarsson, K.; Hermansson, M.; Johnsson, E.; Svennerholm, A.M. Decreased IgA antibody production in the stomach of gastric adenocarcinoma patients. *Clin. Immunol.* **2009**, *131*, 463–471. [CrossRef] [PubMed]

111. Cunningham-Rundles, C. The many faces of common variable immunodeficiency. *Hematol. Am. Soc. Hematol. Educ. Program* **2012**, *2012*, 301–305. [CrossRef]

112. Verma, N.; Thaventhiran, A.; Gathmann, B.; ESID Registry Working Party; Thaventhiran, J.; Grimbacher, B. Therapeutic management of primary immunodeficiency in older patients. *Drugs Aging* **2013**, *30*, 503–512. [CrossRef] [PubMed]

113. Ludvigsson, J.F.; Neovius, M.; Ye, W.; Hammarström, L. IgA deficiency and risk of cancer: A population-based matched cohort study. *J. Clin. Immunol.* **2015**, *35*, 182–188. [CrossRef]

114. De Miranda, N.F.; Björkman, A.; Pan-Hammarström, Q. DNA repair: The link between primary immunodeficiency and cancer. *Ann. N. Y. Acad. Sci.* **2011**, *1246*, 50–63. [CrossRef]

115. Gennery, A.R. Primary immunodeficiency syndromes associated with defective DNA double-strand break repair. *Br. Med. Bull.* **2006**, *77–78*, 71–85. [CrossRef]

116. Carroll, C.; Badu-Nkansah, A.; Hunley, T.; Baradaran-Heravi, A.; Cortez, D.; Frangoul, H. Schimke Immunoosseous Dysplasia associated with undifferentiated carcinoma and a novel SMARCAL1 mutation in a child. *Pediatr. Blood Cancer* **2013**, *60*, E88–E90. [CrossRef]

117. Baradaran-Heravi, A.; Raams, A.; Lubieniecka, J.; Cho, K.S.; DeHaai, K.A.; Basiratnia, M.; Mari, P.O.; Xue, Y.; Rauth, M.; Olney, A.H. SMARCAL1 deficiency predisposes to non-Hodgkin lymphoma and hypersensitivity to genotoxic agents in vivo. *Am. J. Med. Genet. A* **2012**, *158*, 2204–2213. [CrossRef]

118. Karalis, A.; Tischkowitz, M.; Millington, G.W. Dermatological manifestations of inherited cancer syndromes in children. *Br. J. Dermatol.* **2011**, *164*, 245–256. [CrossRef]

119. Spinner, M.A.; Sanchez, L.A.; Hsu, A.P.; Shaw, P.A.; Zerbe, C.S.; Calvo, K.R.; Arthur, D.C.; Gu, W.; Gould, C.M.; Brewer, C.C. GATA2 deficiency: A protean disorder of hematopoiesis, lymphatics, and immunity. *Blood* **2014**, *123*, 809–821. [CrossRef]

120. Sokolic, R. Neutropenia in primary immunodeficiency. *Curr. Opin. Hematol.* **2013**, *20*, 55–65. [CrossRef]

121. Meng, X.; Yang, B.; Suen, W.C. Prospects for modulating the CD40/CD40L pathway in the therapy of the hyper-IgM syndrome. *Innate Immun.* **2018**, *24*, 4–10. [CrossRef]

122. Jacobsohn, D.A.; Emerick, K.M.; Scholl, P.; Melin-Aldana, H.; O'Gorman, M.; Duerst, R.; Kletzel, M. Nonmyeloablative hematopoietic stem cell transplant for X-linked hyper-immunoglobulin m syndrome with cholangiopathy. *Pediatrics* **2004**, *113*, e122–e127. [CrossRef]

123. Hayward, A.R.; Levy, J.; Facchetti, F.; Notarangelo, L.; Ochs, H.D.; Etzioni, A.; Bonnefoy, J.Y.; Cosyns, M.; Weinberg, A. Cholangiopathy and tumors of the pancreas, liver, and biliary tree in boys with X-linked immunodeficiency with hyper-IgM. *J. Immunol.* **1997**, *158*, 977–983.

124. Hauck, F.; Voss, R.; Urban, C.; Seidel, M.G. Intrinsic and extrinsic causes of malignancies in patients with primary immunodeficiency disorders. *J. Allergy Clin. Immunol.* **2018**, *141*, 59–68.e4. [CrossRef]

125. De Flora, S.; Bonanni, P. The prevention of infection-associated cancers. *Carcinogenesis* **2011**, *32*, 787–795. [CrossRef]

126. Philip, M.; Rowley, D.A.; Schreiber, H. Inflammation as a tumor promoter in cancer induction. *Semin. Cancer Biol.* **2004**, *14*, 433–439. [CrossRef] [PubMed]

127. Leiding, J.W.; Holland, S.M. Warts and all: Human papillomavirus in primary immunodeficiencies. *J. Allergy Clin. Immunol.* **2012**, *130*, 1030–1048. [CrossRef] [PubMed]

128. Pablos, J.L.; Amara, A.; Bouloc, A.; Santiago, B.; Caruz, A.; Galindo, M.; Delaunay, T.; Virelizier, J.L.; Arenzana-Seisdedos, F. Stromal-cell derived factor is expressed by dendritic cells and endothelium in human skin. *Am. J. Pathol.* **1999**, *155*, 1577–1586. [CrossRef]

129. Grierson, H.L.; Skare, J.; Hawk, J.; Pauza, M.; Purtilo, D.T. Immunoglobulin class and subclass deficiencies prior to Epstein-Barr virus infection in males with X-linked lymphoproliferative disease. *Am. J. Med. Genet.* **1991**, *40*, 294–297. [CrossRef]

130. Rezaei, N.; Mahmoudi, E.; Aghamohammadi, A.; Das, R.; Nichols, K.E. X-linked lymphoproliferative syndrome: A genetic condition typified by the triad of infection, immunodeficiency and lymphoma. *Br. J. Haematol.* **2011**, *152*, 13–30. [CrossRef]

131. Conrado, L.A.; Marques, S.A.; Lastoria, J.C.; Cucé, L.C.; Marques, M.E.; Dillon, N.L. Keratitis-ichthyosis-deafness (KID) syndrome with squamous cell carcinoma. *Int. J. Dermatol.* **2007**, *46*, 403–406. [CrossRef]

132. Natsuga, K.; Akiyama, M.; Shimizu, H. Malignant skin tumours in patients with inherited icthyosis. *Br. J. Dermatol.* **2011**, *165*, 263–268. [CrossRef]

133. Filipovich, A.H.; Mathur, A.; Kamat, D.; Shapiro, R.S. Primary immunodeficiencies: Genetic risk factors for lymphoma. *Cancer Res.* **1992**, *52* (Suppl. S19), 5465s–5467s.

134. Broeks, A.; Urbanus, J.H.; Floore, A.N.; Dahler, E.C.; Klijn, J.G.; Rutgers, E.J.; Devilee, P.; Russell, N.S.; Van Leeuwen, F.E.; Van 't Veer, L.J. ATM-heterozygous germline mutations contribute to breast cancer-susceptibility. *Am. J. Hum. Genet.* **2000**, *66*, 494–500. [CrossRef]

135. Jerzak, K.J.; Mancuso, T.; Eisen, A. Ataxia-telangiectasia gene (*ATM*) mutation heterozygosity in breast cancer: A narrative review. *Curr. Oncol.* **2018**, *25*, e176–e180. [CrossRef]

136. Kiykim, A.; Eker, N.; Surekli, O.; Nain, E.; Kasap, N.; Aktürk, H.; Dogru, O.; Canbolat, A.; Somer, A.; Koc, A. Malignancy and lymphoid proliferation in primary immune deficiencies; hard to define, hard to treat. *Pediatr. Blood Cancer* **2020**, *67*, e28091. [CrossRef]

137. Seidemann, K.; Tiemann, M.; Henze, G.; Sauerbrey, A.; Müller, S.; Reiter, A. Therapy for non-Hodgkin lymphoma in children with primary immunodeficiency: Analysis of 19 patients from the BFM trials. *Med. Pediatr. Oncol.* **1999**, *33*, 536–544. [CrossRef]

138. Bomken, S.; Van der Werff Ten Bosch, J.; Attarbaschi, A.; Bacon, C.M.; Borkhardt, A.; Boztug, K.; Fischer, U.; Hauck, F.; Kuiper, R.P.; Lammens, T.; et al. Current Understanding and Future Research Priorities in Malignancy Associated with Inborn Errors of Immunity and DNA Repair Disorders: The Perspective of an Interdisciplinary Working Group. *Front. Immunol.* **2018**, *9*, 2912. [CrossRef]

139. Gompels, M.M.; Hodges, E.; Lock, R.J.; Angus, B.; White, H.; Larkin, A.; Chapel, H.M.; Spickett, G.P.; Misbah, S.A.; Smith, J.L.; et al. Lymphoproliferative disease in antibody deficiency: A multi-centre study. *Clin. Exp. Immunol.* **2003**, *134*, 314–320. [CrossRef]

140. Ren, A.; Yin, W.; Miller, H.; Westerberg, L.S.; Candotti, F.; Park, C.S.; Lee, P.; Gong, Q.; Chen, Y.; Liu, C. Novel Discoveries in Immune Dysregulation in Inborn Errors of Immunity. *Front. Immunol.* **2021**, *12*, 725587. [CrossRef]

141. Fournier, B.; Tusseau, M.; Villard, M.; Malcus, C.; Chopin, E.; Martin, E.; Jorge Cordeiro, D.; Fabien, N.; Fusaro, M.; Gauthier, A. DEF6 Deficiency, A Mendelian Susceptibility to EBV Infection, Lymphoma, and Autoimmunity. *J. Allergy Clin. Immunol.* **2021**, *147*, 740–743.e9. [CrossRef]

142. Somekh, I.; Thian, M.; Medgyesi, D.; Gülez, N.; Magg, T.; Gallón Duque, A.; Stauber, T.; Lev, A.; Genel, F.; Unal, E.; et al. CD137 Deficiency Causes Immune Dysregulation with Predisposition to Lymphomagenesis. *Blood* **2019**, *134*, 1510–1516. [CrossRef]

143. Salzer, E.; Zoghi, S.; Kiss, M.G.; Kage, F.; Rashkova, C.; Stahnke, S.; Haimel, M.; Platzer, R.; Caldera, M.; Ardy, R.C.; et al. The Cytoskeletal Regulator HEM1 Governs B Cell Development and Prevents Autoimmunity. *Sci. Immunol.* **2020**, *5*, eabc3979. [CrossRef]

144. Delmonte, O.M.; Castagnoli, R.; Calzoni, E.; Notarangelo, L.D. Inborn Errors of Immunity with Immune Dysregulation: From Bench to Bedside. *Front. Pediatr.* **2019**, *7*, 353. [CrossRef] [PubMed]

145. Perez, E. Future of Therapy for Inborn Errors of Immunity. *Clin. Rev. Allergy Immunol.* **2022**, 1–15. [CrossRef] [PubMed]

146. Hoyos-Bachiloglu, R.; Platt, C. Precision Medicine as Treatment for Primary Immunodeficiency and Immune Dysregulation. *Immunol. Genet. J.* **2019**, *2*, 153–172.

147. Rao, V.K.; Webster, S.; Dalm, V.; Šedivá, A.; Van Hagen, P.M.; Holland, S.; Rosenzweig, S.D.; Christ, A.D.; Sloth, B.; Cabanski, M. Effective "activated PI3Kδ syndrome"-targeted therapy with the PI3Kδ inhibitor leniolisib. *Blood* **2017**, *130*, 2307–2316. [CrossRef]

148. Riaz, I.B.; Faridi, W.; Patnaik, M.M.; Abraham, R.S. A Systematic Review on Predisposition to Lymphoid (B and T cell) Neoplasias in Patients with Primary Immunodeficiencies and Immune Dysregulatory Disorders (Inborn Errors of Immunity). *Front. Immunol.* **2019**, *10*, 777. [CrossRef]

some patients also have autoimmune diseases, alteration in lymphocyte populations, or cancer [3].

Currently, its diagnosis is complicated by the lack of standardized diagnostic criteria due to its great clinical and analytical heterogeneity, low awareness on the part of clinicians, and the similarity of the disease with other immune disorders. Even so, several guidelines have emerged that help in the diagnosis of CVID, including the International Consensus Document (ICON) guidelines from 2015 [4] and the European Society for Immunodeficiencies (ESID) Registry Working Definitions for the Clinical Diagnosis of Inborn Errors of Immunity from 2019, with several updates [5]. In the laboratory, determinations of serum levels of immunoglobulin (Ig)G, IgA, and IgM antibodies, the study of lymphocyte populations, and B-cell immunophenotyping are useful in evaluating the patient's condition [2].

It is well known that CVID patients present suboptimal responses to vaccines, some of them being potentially dangerous and even contraindicated (such as live vaccines). In fact, the current ESID guideline for CVID includes a poor antibody response to vaccines as a criterion [5], although several studies show that the response to vaccines is not uniform in patients with mild hypogammaglobulinemia [6]. Very few studies exist on humoral and cellular responses to COVID-19 vaccines in patients with CVID [7–9]. This is due, on the one hand, to the fact that SARS-CoV-2 infection is very recent and, on the other hand, because the response to immunization depends on the type of vaccine, the patient's immune defect, and the type of antigen studied [7]. Looking at the humoral and cellular response to other vaccines not related to SARS-CoV-2 in patients with CVID, we find studies that suggest that these patients have a lower humoral response to the vaccines compared to the control group while maintaining a good cellular response [10,11].

Regarding the etiology, most cases of CVID are idiopathic, but to date, between 10 and 35% of CVID patients present monogenic defects, predominantly autosomal dominant with incomplete penetrance or with late onset of symptoms. The development of high-throughput sequencing technologies has allowed the identification of several mutations in genes encoding proteins essential for immune function that could be involved in the development of CVID. Thus, panels of primary immunodeficiency-associated genes have been created, allowing rapid screening of identified mutations using genetic techniques. Some of the CVID-associated genes include: *ICOS* (OMIM: #604558), *CD19* (OMIM: #107265), *CD81* (OMIM: #186845), *MS4A1* (OMIM: #112210), *CR2* (OMIM: #120650), *TNFSF12* (OMIM: #602695), *CTLA4* (OMIM: #123890), *LRBA* (OMIM: #606453) *TNFRSF13B* (OMIM: #60907), *TNFRSF13C* (OMIM#606269), *NFKB1* (OMIM: #164011), *NFKB2* (OMIM: #164012), *IL21* (OMIM: #605384), *IRF2BP2* (OMIM: #615332), *PIK3CD* (OMIM: #602839), *STAT3* (OMIM: #102582), and others [3]. Alteration in any of these genes can cause perturbations of specific immune pathways, resulting in "unique phenotypes" that can aid in the diagnosis of CVID [2]. In fact, in the 2019 update of the classification of primary immunodeficiencies (PID) by the expert committee of the International Union of Immunological Societies (IUIS), CVID is included within the group of PID with a predominance of antibody deficiencies. Furthermore, information regarding clinical, analytical, and molecular tests is provided and that could help clinicians in the diagnosis and management of these patients and their relatives [12].

Here, we report a case of a 46-year-old patient diagnosed with CVID in current treatment with intravenous immunoglobulins and clinically stable. Because of his negative family history of primary immunodeficiencies and in the context of the SARS-CoV-2 pandemic, we reassessed the patient immunologically, employing an NGS analysis to try to clarify the genetic cause of his disease, and analyzed his serological and cellular response to the COVID-19 vaccine. We identified a heterozygous R162Q variant of the *IKZF1* gene in our patient that was not present in his parents. This variant was previously described as pathogenic, associated with CVID type 13 following an autosomal dominant inheritance and high penetrance. Therefore, we present for the first time a case of CVID associated with a de novo heterozygous R162Q variant in the *IKZF1* gene in a patient with a low antibody immune response and a positive cellular response to the complete COVID-19 vaccination program.

2. Materials and Methods

2.1. Subjects and Study Design

The family (2 parents and the proband) was recruited at the Clinical Immunological Department of the University Hospital Virgen de las Nieves, Granada, Spain, as part of a continuous and systematic program of phenotyping and genotyping focused on CVID. The patient was diagnosed with CVID at the age of 25 due to recurrent episodes of bronchiectasis and respiratory infections. Since then, he has been under treatment with intravenous immunoglobulins with a good response and is currently clinically stable. His phenotype was evaluated with blood tests such as cytometry, immunoturbidimetry, chemiluminescence microparticle immunoassay (CMIA), and enzyme-linked immunosorbent assay (ELISA), as well as genetic studies of the patient and, subsequently, his parents.

The timeline of the study design is represented in Figure 1. All family members provided their written informed consent. The study was approved by the ethics committee, via the Portal de Ética de la Investigación Biomédica de Andalucía, Junta de Andalucía (Code: 0297-N-21).

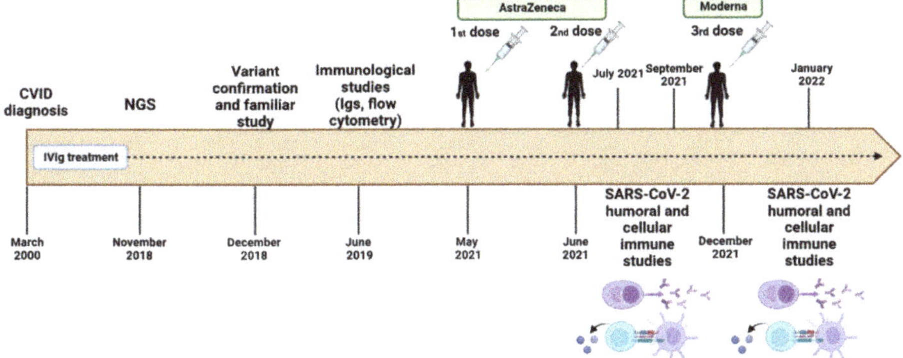

Figure 1. Study design. Timeline of the experimental plan and the schedule of the SARS-CoV-2 vaccine doses. CVID: common variable immunodeficiency; NSG: next-generation sequencing; IVIg: intravenous immunoglobulin treatment; Igs: immunoglobulins.

2.2. Blood Collection

Blood samples were obtained from the antecubital vein after an overnight fast of 12 h and under resting conditions (at least 10 min before) in the supine position and collected in Vacutainer SST 16 mm × 100 mm tubes for serum (Becton Dickinson, NJ, USA), Vacutainer ethylenediaminetetraacetic acid (EDTA K2) 13 mm × 75 mm tubes for anticoagulated whole blood, and Vacutainer lithium-heparin 16 mm × 100 mm tubes for plasma. Serum was obtained by centrifugation (4 min at $3000\times g$), aliquoted, and processed immediately. Plasma was obtained after pipetting the heparinized whole blood sample into the stimulation tubes, incubating them, and centrifuging for 10 min at $1300\times g$.

2.3. Serum Immunoturbidimetry

To assess the overall immune status of the patient, in addition to the study of lymphocyte subpopulations, we quantified the total IgG, IgA, and IgM antibody levels in the patient's serum by immunoturbidimetry using the automatic analyzer Alinity c system (Abbott Laboratories, Chicago, IL, USA). This procedure measures the increase in sample turbidity caused by the formation of insoluble immune complexes when the antibody is added to the sample. This antibody (reagent) consists of goat anti-human IgG, IgM, or IgA serum. Results are expressed in milligrams per deciliter (mg/dL).

2.4. Whole Blood Cytometry

To assess different lymphocyte subpopulations and perform the B-cell immunophenotyping, we used fresh whole blood. Both analyses were performed using the BD FACSCanto II Flow Cytometer (BD Biosciences, San Diego, CA, USA).

For the lymphocyte subpopulations, we used BD Trucount tubes and the BD Multitest 6 Color BTNK kit (BD Biosciences, San Diego, CA, USA), which included the following mixtures of fluorophore-conjugated monoclonal antibodies (mAb): anti-CD45-PerCP-Cy5.5, anti-CD3-FITC, anti-CD8-APC-Cy7, anti-CD4-PECy7, anti-CD19-APC, and anti-CD16+CD56-PE. Pre-sample preparation included the following steps: 20 µL of the antibody mix was added to a Trucount tube with 50 µL of patient blood. The tube was vortexed and incubated for 15 min at room temperature. Subsequently, 450 µL of FACS lysing solution (BD Biosciences) was added and left to act for 10 min. Finally, the cells were acquired on the flow cytometer and analyzed with the BD FACSCanto Clinical software (BD Biosciences, San Diego, CA, USA).

For B-cell immunophenotyping, we use an eight-color panel of the following mAb: anti-CD45-APH-7, anti-CD19-V500, anti-CD10-V450, anti-CD38-PECy7, anti-CD21-PE, anti-CD27-PerCP-Cy5, anti-IgD-FITC, and anti-IgM-APC (BD Biosciences, San Diego, CA, USA). The blood sample was stained with the mAb mixture in a test tube for 20 min at room temperature in the dark. After incubation, red cells were lysed with 1 mL of lysing solution (BD Pharm Lyse) for 10 min. Cells were then washed twice with PBA (1% bovine serum albumin in PBS) and fixed using 1% formalin in PBS. Finally, the cells were acquired on the flow cytometer, analyzed with the BD FACSDiva software (BD Biosciences, San Diego, CA, USA), and read with Infinicyt software (Cytognos, Salamanca, Spain).

2.5. Next-Generation Sequencing (NGS) and Sanger Validation

EDTA blood samples were subjected to an automated extraction and purification process to obtain genomic DNA using the QIAsymphony SP (Qiagen, Hilden, Germany). The proband sample was prepared for an NGS study. We performed targeted sequencing using a designed panel of 41 genes associated with primary antibody immunodeficiency (*AICDA, ATP6AP1, BLNK, BTK, CARD11, CD19, CD40, CD40LG, CD79A, CD79B, CD81, CR2, CTLA4, CXCR4, ICOS, IGLL1, IKZF1, IL21, INO80, IRF2BP2, LRBA, LRRC8A, MOGS, MS4A1, MSH6, NFKB1, NFKB2, PIK3CD, PIK3R1, PLCG2, PMS2, PTEN, SEC61A1, TCF3, TNFRSF13B, TNFRSF13C, TNFSF12, TNFSF13, TRNT1, TTC37, UNG*). The library was created using the SureSelect XT Reagent library preparation kit (Agilent Technologies, Inc., Santa Clara, CA, USA) for paired-end multiplexed sequencing of Illumina (Inc., San Diego, CA, USA). Target regions were enriched with the Custom SureSelect probe kit (Agilent). Cluster preparation was performed using the cBot device, and library sequencing was performed using the Illumina HiSeq1500 platform (Illumina, Inc., San Diego, CA, USA). Bioinformatic analysis was applied through an end-to-end in-house pipeline developed by Health in Code (A Coruña, Spain), in accordance with the best WES analysis practices. The identified pathogenic variant in the proband was confirmed by Sanger sequencing, by sequencing exon 5 of the *IKZF1* gene bidirectionally, with its intronic flanking regions. Sanger sequencing was also performed in the patient's parents.

2.6. SARS-CoV-2 Serological Study

In the context of the SARS-CoV-2 pandemic, the patient received the complete vaccination program against COVID-19: the first and second dose with the vector SARS-CoV-2 vaccine (Oxford-AstraZeneca AZD1222, ChAdOx1 nCoV-19, Vaxzevria) and the third dose (booster dose) with the messenger RNA SARS-CoV-2 vaccine (mRNA-1273, Moderna).

We analyzed the humoral immune response against the SARS-CoV-2 Spike (S) protein one and three months after the second dose with AstraZeneca and one month after the third dose with Moderna to ensure complete vaccination and subsequent booster. Quantitative determination of IgG against protein S was performed using the chemiluminescent microparticle immunoassay (CMIA) in the Alinity autoanalyzer (Abbott) following the manufacturer's instructions with the SARS-CoV-2 IgG II Quant Assay kit. Results were expressed in binding antibody units per milliliter (BAU/mL). A test was determined as positive if the signal was >7.5 BAU/mL.

2.7. SARS-CoV-2 Cellular Immunity Study

Cellular immunity was assessed by quantifying SARS-CoV-2-specific IFN-γ using the SARS-CoV-2 IGRA stimulation tube set (Euroimmun, Lüebeck, Germany), following the manufacturer's instructions. This stimulation is used for the treatment of whole blood to obtain plasma and contains: (1) CoV-2 IGRA BLANK (no activating component for T cells, used for the determination of the individual's INF-γ background); (2) CoV-2 IGRA TUBE (with components of the S1 domain of the SARS-CoV-2 Spike protein); and (3) CoV-2 IGRA STIM (coated with a mitogen to verify the sample quality). A 500 µL volume of heparinized whole blood was pipetted into each tube and incubated for 20–24 h at 37 °C. Subsequently, the concentration of the released interferon-gamma was measured in the plasma by enzyme-linked immunosorbent assay (ELISA) using the Interferon-Gamma ELISA kit (Euroimmun, Lüebeck, Germany, EQ 6841-9601).

Cellular response assays to the vaccine were performed at the same time as post-vaccination SARS-CoV-2 serology. The results were expressed in mUI/mL (milli-international unit per milliliter). The cutoff point for positivity was set at >200 mIU/mL for either of the two tubes (IGRA TUBE and IGRA STIM) of the technique. The mIU/mL value of each tube was calculated by subtracting the value obtained in IGRA TUBE and IGRA STIM minus the IGRA BLANK.

3. Results

3.1. Immunological Studies

The immunological features of the patient are shown in Table 1. The serum level of IgM was normal, whereas no IgA was detected. The IgG level was normal due to the immunoglobulin treatment. However, in the diagnostic sample, IgG was absent. The study of lymphocyte subpopulations showed a marked reduction in B cells, a reversed CD4/CD8 T-cell ratio, and normal NK cell levels. The B-cell immunophenotyping highlighted the absence of transitional B cells, plasmatocytes, and switched-memory B cells, whereas the percentages of unswitched-memory B cells, marginal-zone B cells, and CD21*low* B cells were elevated.

The study of the complete COVID-19 vaccination program showed a very low antibody response after the second dose of the AstraZeneca vaccine, with a decrease in antibody level of −23% at 3 months and a stronger response after the booster dose with the Moderna vaccine. The cellular immunity study by the quantification of the SARS-CoV-2-specific IFN-γ production showed a positive cellular response one month after the second dose, which, as in the case of antibodies, was lost at 3 months, with subsequent recovery of T-cell-mediated responses after the booster dose. All immunological values of the humoral and cellular responses are shown in Table 2. After the booster dose of the vaccine, the lymphocyte subpopulations remained stable, but we detected an increase in the percentage of switched-memory B cells (5.8%) and a decrease in unswitched-memory B cells (17%) in the immunophenotyping of B cells.

Table 1. Immunological features of the CVID patient in current treatment with intravenous immunoglobulins previous to the SARS-CoV-2 vaccination. Ig: immunoglobulin.

	Patient	Reference Values
Total serum immunoglobulins (mg/dL)		
IgM	79	22–240
IgA	<5	65–470
IgG	977	540–1822
Lymphocyte subsets (cells/μL—%)		
CD3+	1825 (82%)	960–2600 (61–84%)
CD3+CD4+	611 (27%)	540–1660 (32–60%)
CD3+CD8+	1164 (52%)	270–930 (13–40%)
CD4+/CD8+ ratio	0.53	0.9–4.5
CD19+	24 (1%)	122–632 (6–27%)
CD3−CD56+CD16+	358 (16%)	127–509 (10.1–20.9%)
B-cell immunophenotyping (%)		
Naive B cells (CD27−, IgD+, IgM+)	54.40%	53–86
Memory B cells (CD27+)	35.90%	9.1–33
Switched-memory B cells (CD21+, CD27+, IgM−, IgD−)	0%	4–22
Unswitched-memory B cells and marginal-zone B cells (CD21+, CD27+, IgM+, IgD+)	29.50%	3.3–12.8
Transitional B cells (CD38*high*, IgM*high*, CD21*low*)	0%	0.9–6.3
CD21*low* B cells (IgM+, CD38*low*, CD21*low*)	7.70%	0.4–4.5
Plasmatocytes (CD38+, CD138+)	0%	0.1–1.5

Table 2. Humoral and cellular responses to the SARS-CoV-2 vaccine in the CVID patient. We made the measurements one month and three months after the second dose of the vaccine and one month after the booster dose.

	Patient	Cut Off Values
Anti-S SARS-CoV-2 protein IgG (BAU/mL)		
After 2nd dose (1 month)	55.85	
After 2nd dose (3 month)	12.91	0–7.5
After 3rd dose (1 month)	180.03	
SARS-CoV-2-specific IFN-γ (mUI/mL)		
After 2nd dose (1 month)	420	Negative < 100
After 2nd dose (3 month)	14	Borderline 100–200
After 3rd dose (1 month)	638	Positive > 200

3.2. Genetic Study

The NGS test revealed a heterozygous substitution of guanine to adenine at position 485 of exon 5 of the *IKZF1* gene (chromosome 7p12.2) in the patient (c.485G>A) that causes a codon change of "CGG" to "CAG", which mean a non-synonymous switch from arginine to glutamine in the 162 protein residue (Arg162Gln). This variant was not identified in the DNA of the blood samples from the patient's parents by Sanger sequencing (Figure 2A–C).

3.3. Bioinformatic Studies

The heterozygous mutation (c.485G>A) causes a change in the physical–chemical properties of the protein because of the substitution of the amino acid arginine for glutamine (Arg162Gln). This variant affects the second C2H2 domain with the zinc-finger structure of IKAROS protein (ZF2, amino acids 145–167), a residue essential for DNA binding. All evaluated in silico predictors of protein damage determined a deleterious effect for the IKAROS protein (Table 3).

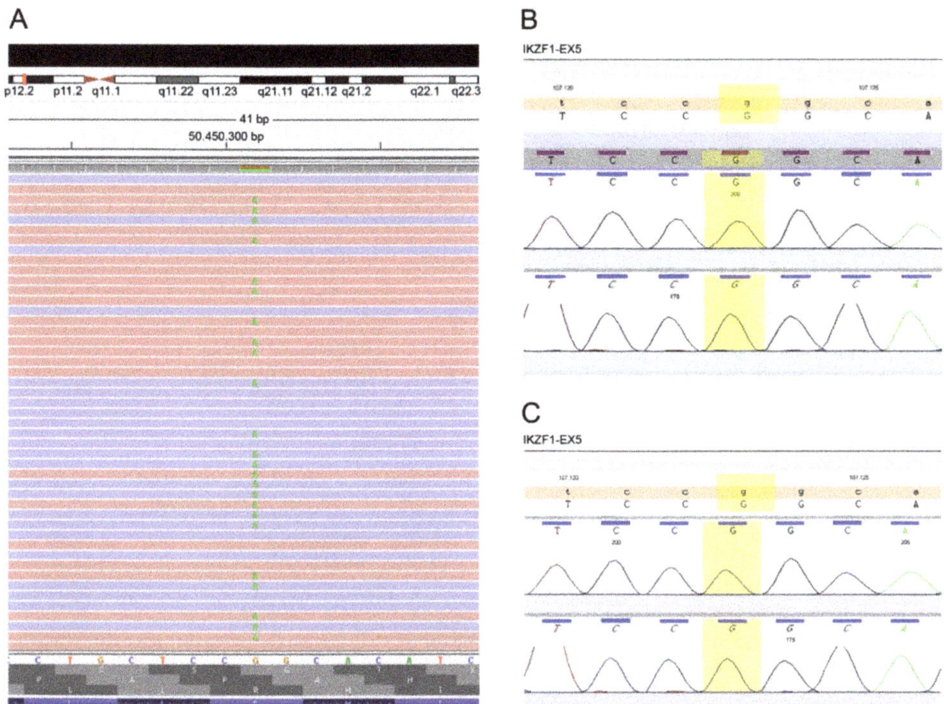

Figure 2. Genetic analysis of the *IKZF1* gene. (**A**) Next-generation sequencing of the patient revealed a heterozygous *missense* variant (c.485G>A) in exon 5 of the *IKZF1* gene, in chromosome 7p12.2, that results in a substitution of arginine to glutamine (p.Arg162Gln). Electropherograms of the Sanger sequencing performed in the patient's father (**B**) and mother (**C**) demonstrated homozygosity for G at position c.485 of exon 5 of *IKZF1*.

Table 3. Results of the in silico predictors of the effect of the variant. MutationTaster (values range from 0 to 1): probability close to 1 indicates greater confidence in the prediction. DANN (Deleterious Annotation of Genetic Variants using Neural Networks; values range from 0 to 1): the highest values are potentially the most pathogenic. FATHMM MKL (Functional Analysis through Hidden Markov Models; values range from 0 to 1): the highest values are potentially the most pathogenic. Coding and non-coding variants are scored independently.

	Prediction	Score
MutationTaster	Pathogenic	1
DANN		0.999552
FATHMM MKL Coding		0.98952
FATHMM MKL Non-Coding		0.99606

4. Discussion

The *IKZF1* gene (OMIM: #603023) encodes the IKAROS zinc-finger transcription factor. It is located on chromosome 7 (7q12.2) and has eight exons. The *IKZF1* gene can produce different isoforms due to alternative splicing, but the main "DNA-binding" form of IKAROS is isoform 1, whose structure is composed of an N-terminal DNA-binding domain, which is made up of four zinc-finger motifs (ZF1–ZF4), a central activation/repression domain, and a C-terminal dimerization domain (ZF5–ZF6) (Figure 3). IKAROS binds as homodimers or heterodimers to pericentric heterochromatin regions, promoting the expression of target genes that play important roles in lymphocyte development, differentiation and function, and myeloid cell development [13].

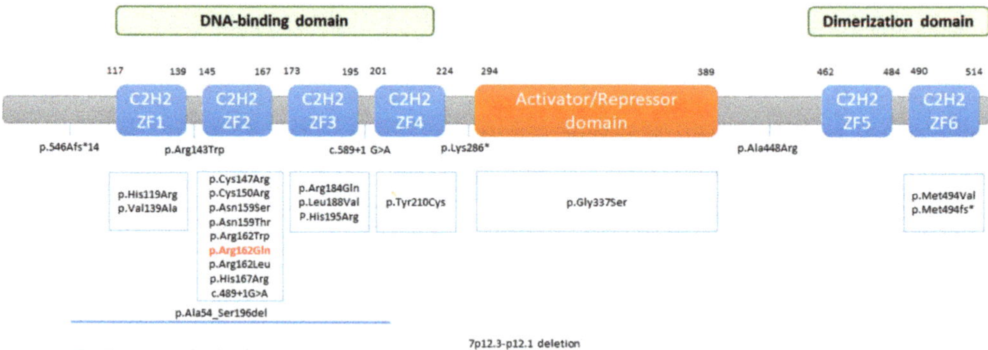

Figure 3. Schematic representation of the structure of human IKAROS protein encoded by the *IKZF1* gene. The DNA-binding domain consists of four zinc-finger motifs (ZF1–ZF4) and the dimerization domain of two (ZF5–ZF6). The p.Arg162Gln variant found in our patient is highlighted in red and affects the ZF2 domain. Other amino acid positions of IKAROS variants identified in CVID patients are indicated. ZF: zinc-finger.

The first *IKZF1* mutations that were linked to human pathologies were somatic and associated with a worse prognosis in B-cell acute lymphoblastic leukemia (B-ALL). However, more recently, heterozygous germline mutations have also been related to B-ALL development [14]. On the other hand, germline heterozygous mutations in the *IKZF1* gene have been detected in patients with primary immune deficiency or inborn errors of immunity. The first de novo heterozygous *missense* mutation in the *IKZF1* gene was described by Goldman et al. in 2012, in a 33-week preterm infant with pancytopenia and loss of B cells, which shared phenotypic similarities with the *IKZF1-null* mouse animal model and that affected the ZF4 domain of IKAROS [15].

Subsequently, various types of germline mutations in the *IKZF1* gene have been identified in patients with a CVID-like phenotype, including missense mutations (all of which have been classified as pathogenic or probably pathogenic), frame-shift mutations, nonsense mutations, and intragenic deletions of *IKZF1* [15–22] (Figure 3). These mutations can act in different ways, giving rise to haploinsufficiency (HI), dominant-negative (DN), or dimerization-defective (DD) effects. However, most patients carrying germline heterozygous missense mutations that act in a HI manner and affect the DNA binding domain manifest CVID when they have symptoms [23,24].

The heterozygous missense R162Q variant of the *IKZF1* gene present in our patient was discovered for the first time in 2016 by Kuehn et al., in a European family composed of three generations of affected individuals, where the variant cosegregated with the disease in seven of the twelve members of the family [17]. Later, in 2017, Hoshino et al. also reported the R162Q variant in three generations of a family with two affected individuals from three carriers [20]. Both studies revealed an autosomal dominant inheritance with incomplete but high penetrance. However, in our patient, it appears for the first time as a de novo mutation.

The pathogenicity of the variant was demonstrated by Kuehn et al. [17], through functional analysis and subsequently corroborated by Hoshino et al. [20]. Flow cytometry data of the Kuehn et al. study showed that the amount of IKAROS in the T and B cells of the patient with the R162Q mutation in *IKZF1* was equal to controls. To determine the effect of the R162Q mutation, they used transfected cells. They noticed that the R162Q variant produced stable proteins, which could dimerize with wild-type (WT) protein forms and migrate to the nucleus, but they were not capable of binding to the pericentromeric target DNA region. To mimic the heterozygous state, Kuehn et al. used transfected cells with vectors expressing 100% WT IKAROS and transfected cells with vectors expressing 50% WT and 50% mutant IKAROS. They found that DNA binding was reduced by 38–74% in cells transfected with 50% mutant vectors, compared to cells transfected with 100% WT vectors. The heterozygous state and the haploinsufficiency mechanism explain the high but incomplete penetrance of this mutation. Finally, based on our in silico analysis using MutationTaster and other prediction methods, the R162Q variant causes a deleterious effect on the mutated IKAROS protein, supporting the previous results.

In the Kuehn study, individuals with the heterozygous R162Q variant were characterized by recurrent infections, mainly respiratory tract infections or otitis, and hypogammaglobulinemia. The age onset was before the age of ten in most of them. One of the patients died of pneumonia at the age of 74 and another patient developed B-ALL and died from a relapse at the age of 5. All of the patients presented a reduction in B cells and most of them a reversed CD4/CD8 T-cell ratio [17]. In the Hoshino study, one of the symptomatic patients, at seven years of age, had bacterial pneumonia with agammaglobulinemia and IgA vasculitis and analytically low levels of B cells, normal NK cells, and low CD8+ T cells. The other symptomatic patient had a history of thrombocytopenic thrombotic purpura (ITP) without dysgammaglobulinemia, with normal levels of B and NK cells, an increase in T follicular helper cells, and a reduction in CD4/CD8 T-cell ratio [20]. Our patient was diagnosed at an older age compared to the patients previously described with this variant. Analytically, he is also characterized by recurrent infections, a marked reduction in B cells, a reversed CD4/CD8 T-cell ratio, and very low immunoglobulins levels at the diagnosis, except for IgM.

The humoral response to the complete vaccination schedule with AstraZeneca and the booster with Moderna in our patient is considered very low. The study carried out by Atmar et al. showed circulating antibody levels against the Spike protein of SARS-CoV-2 after 1, 15, and 29 days of the booster dose in healthy individuals. This study showed that there are no major differences between homologous booster vaccination and heterologous booster. The normal humoral response after the regimen of two doses of Janssen (viral vector) and the booster of Moderna is estimated at 4560 BAU/mL of circulating antibodies, which is much higher than our patient [25]. In the same way, Munro et al. published an article that compared the immunogenicity of seven booster vaccines after the two doses from AstraZeneca and Pfizer-BioNTech (COV-BOOST) by means of, among others, the levels of anti-spike IgG measured by ELISA [26]. In the 98 healthy cases studied that were vaccinated with two doses of AstraZeneca and subsequent booster with Moderna, a mean of 31,111 ELISA laboratory units per milliliter (ELU/mL) was found at day 29 post-boost. According to the WHO international standard for COVID-19 serological tests to harmonize anti-spike SARS-CoV-2 assays [27], taking as reference the ELISA technique, BAU/mL = 0.142 × AU/mL. Thus, the results obtained by Munro are similar to those obtained by Atmar (4417 BAU/mL) and are higher compared to our patient. On the other hand, Amodio et al. conducted the first study in a cohort of 21 European patients with CVID. They evaluated the humoral response 21 days after the first dose and 7 days after the second dose with the Pfizer-BioNTech vaccine and found that patients with CVID could generate an antibody response, but at a lower magnitude than healthy controls [8]. It should be noted that some of these patients had a genetic alteration, but none with a mutation in the *IKZF1* gene. In the same way, Hagin et al. observed a positive response, but at lower levels than healthy controls in a cohort of 13 CVID patients treated with immunoglobulins, measuring the SARS-CoV-2 anti-protein S antibody response 2 weeks after the second dose of the Pfizer-BioNTech vaccine [9]. In addition, in the group of patients with CVID, they found a greater response in patients under 50 years of age. Bergman et al. showed that CVID patients with a lower percentage of switched-memory B cells or an increased percentage of CD21*low* B cells have a poor response to the Pfizer-BioNTech mRNA vaccine [28]. These immunophenotypic characteristics and results were obtained in our patient, although after the booster dose of the SARS-CoV-2 vaccine, he generated a small increase in switched-memory B cells, coinciding with his greater production of antibodies after vaccination.

The T-cell response after the booster dose in our patient, despite being positive, is far from those found in the healthy population used as a control group in other studies and is similar to that of vulnerable and immunocompromised patients in whom the SARS-CoV-2-specific IFN-γ were quantified using the same commercial kits (Euroimmun) and techniques as in our case. Schwarz et al. quantified SARS-CoV-2-specific IFN-γ 7 days after the second dose of the Pfizer-BioNTech vaccine, finding that the mean concentration in people >70 years old was 707.3 mIU/mL, while in healthy young adults it was 2184 mIU/mL [29]. Ruether et al. compared the levels of SARS-CoV-2-specific IFN-γ 29 days after the second dose of the mRNA vaccine (Pfizer-BioNTech or Moderna) or vector-based vaccine (AstraZeneca) in 3 groups: 19 healthy controls (mean around 1000 mIU/mL), 26 patients with liver cirrhosis (mean between 100 and 200 mIU/mL or borderline), and 82 liver transplant patients (mean below 100 mIU/mL or negative response) [30]. Our patient presents a loss of cellular response 3 months after the second dose with the AstraZeneca vaccine, and although the response is good after the booster dose with Moderna, it is presumably lower than in the healthy population and will be lost over time. This hypothesis may be demonstrated in subsequent controls on the patient and corroborated by different studies that are still underway.

The molecular heterogeneity of patients with CVID, the use of different techniques to measure the immune response to SARS-CoV-2 vaccines, and the use of different vaccines and vaccine regimens make it difficult to compare the results between studies, but reinforces the importance of vaccinating this vulnerable population, emphasizing the need to measure not only the humoral response but also the cellular response in these patients [7–9].

5. Conclusions

We report a patient diagnosed with CVID due to his phenotypic and immunological features in which, among other findings, we found a positive cellular response to the complete COVID-19 vaccination schedule with a very low humoral response. In summary, vaccination against SARS-CoV-2 activates the adaptive immune response by inducing both the humoral response (specific antibodies against SARS-CoV-2) and the cellular response (specific T cells against SARS-CoV-2). However, in our patient, as well as in other immunosuppressed or vulnerable patients, we observed a low and less sustained immune response to the COVID-19 vaccines. Although there are not many studies on the effects of vaccines against SARS-CoV-2 in vulnerable patients (CVID, elderly, immunosuppression), it seems advisable to routinely include serological and/or cellular tests of response to vaccination, since this is usually suboptimal, and it would be necessary to implement additional booster doses (as many as necessary) in these patients with low or no response. The vaccine-induced T-cell response has a greater effect than the humoral response mediated by B cells, in addition to having a protective effect even in the absence of a humoral response by limiting the viral replication and supporting immunological memory in the timing of long-term vaccination. What is clear is that these patients require more research on the effects of vaccination in the short and medium term and closer monitoring to try to keep them adequately treated and protected.

Targeted sequencing using a panel of genes associated with primary antibody immunodeficiency was performed on our patient to elucidate the underlying cause of his disease. Our analysis revealed an R162Q variant, a heterozygous IKZF1 mutation, which was not present in his parents. Importantly, this is the first time this variant is described as a de novo mutation. The p.Arg162Gln variant has a deleterious effect on the IKAROS transcription factor, which in heterozygosity, generates a reduction of the binding affinity for DNA, altering the regulation of the targeted genes involved in lymphocyte differentiation, which supports the CVID phenotype of the patient. Despite all the studies carried out to date, it is still not fully understood how mutations in the *IKZF1* gene influence the etiopathogenesis of CVID. It is necessary to further investigate possible factors that could influence the clinical–immunological heterogeneity between patients, particularly in individuals with de novo mutations, exemplified by our patient, and asymptomatic patients. Genetic studies of these patients and their relatives must be included in the diagnostic algorithm for these pathologies in order to better understand the development of CVID, to find new pathological variants in *IKZF1* or other genes, or even in novel genes not yet described that could produce CVID and, perhaps in the future, to find a more effective and targeted treatment than the basic intravenous administration of immunoglobulins.

Author Contributions: Conceptualization, I.D.-A. and M.Á.L.-N.; Data curation, J.M.G.-A. and C.G.-G.; Investigation, I.D.-A. and A.E.-O.; Methodology, J.M.G.-A., C.G.-G., M.A.-F., A.S. and P.J.; Resources, M.Á.L.-N. and P.J.; Supervision, M.Á.L.-N. and P.J.; Visualization, I.D.-A.; Writing—original draft, I.D.-A. and A.E.-O.; Writing—review and editing, M.Á.L.-N. and P.J. All authors have read and agreed to the published version of the manuscript.

Funding: This study was partially financed by Palex Medical S.A.

Institutional Review Board Statement: The study was conducted in accordance with the Declaration of Helsinki, and approved by Portal de Ética de la Investigación Biomédica. Junta de Andalucía (Code: 0297-N-21).

Informed Consent Statement: Informed consent was obtained from all family members.

Acknowledgments: We thank Per Anderson for the English revision. This study is part of the doctoral thesis of Irene Díaz Alberola, within the program of Biomedicine, conducted at the University of Granada, Spain.

Conflicts of Interest: José María García-Aznar and Christian Ganoza-Gallardo are employees of Health in Code S. L. All authors declare no conflict of interest.

References

1. Ramirez, N.J.; Posadas-Cantera, S.; Caballero-Oteyza, A.; Camacho-Ordonez, N.; Grimbacher, B. There Is No Gene for CVID—Novel Monogenetic Causes for Primary Antibody Deficiency. *Curr. Opin. Immunol.* **2021**, *72*, 176–185. [CrossRef] [PubMed]
2. Aggarwal, V.; Banday, A.Z.; Jindal, A.K.; Das, J.; Rawat, A. Recent Advances in Elucidating the Genetics of Common Variable Immunodeficiency. *Genes Dis.* **2020**, *7*, 26–37. [CrossRef] [PubMed]
3. Yazdani, R.; Habibi, S.; Sharifi, L.; Azizi, G.; Abolhassani, H.; Olbrich, P.; Aghamohammadi, A. Common Variable Immunodeficiency: Epidemiology, Pathogenesis, Clinical Manifestations, Diagnosis, Classification, and Management. *J. Investig. Allergol. Clin. Immunol.* **2020**, *30*, 14–34. [CrossRef] [PubMed]
4. Bonilla, F.A.; Barlan, I.; Chapel, H.; Costa-Carvalho, B.T.; Cunningham-Rundles, C.; de la Morena, M.T.; Espinosa-Rosales, F.J.; Hammarström, L.; Nonoyama, S.; Quinti, I.; et al. International Consensus Document (ICON): Common Variable Immunodeficiency Disorders. *J. Allergy Clin. Immunol. Pract.* **2016**, *4*, 38–59. [CrossRef]
5. Seidel, M.G.; Kindle, G.; Gathmann, B.; Quinti, I.; Buckland, M.; van Montfrans, J.; Scheible, R.; Rusch, S.; Gasteiger, L.M.; Grimbacher, B.; et al. The European Society for Immunodeficiencies (ESID) Registry Working Definitions for the Clinical; Diagnosis of Inborn Errors of Immunity. *J. Allergy Clin. Immunol. Pract.* **2019**, *7*, 1763–1770. [CrossRef]
6. Ameratunga, R.; Longhurst, H.; Steele, R.; Lehnert, K.; Leung, E.; Brooks, A.E.S.; Woon, S.-T. Common Variable Immunodeficiency Disorders, T-Cell Responses to SARS-CoV-2 Vaccines, and the Risk of Chronic COVID-19. *J. Allergy Clin. Immunol. Pract.* **2021**, *9*, 3575–3583. [CrossRef]
7. Arroyo-Sánchez, D.; Cabrera-Marante, O.; Laguna-Goya, R.; Almendro-Vázquez, P.; Carretero, O.; Gil-Etayo, F.J.; Suàrez-Fernández, P.; Pérez-Romero, P.; Rodríguez de Frías, E.; Serrano, A.; et al. Immunogenicity of Anti-SARS-CoV-2 Vaccines in Common Variable Immunodeficiency. *J. Clin. Immunol.* **2022**, *42*, 240–252. [CrossRef]
8. Amodio, D.; Ruggiero, A.; Sgrulletti, M.; Pighi, C.; Cotugno, N.; Medri, C.; Morrocchi, E.; Colagrossi, L.; Russo, C.; Zaffina, S.; et al. Humoral and Cellular Response Following Vaccination with the BNT162b2 MRNA COVID-19 Vaccine in Patients Affected by Primary Immunodeficiencies. *Front. Immunol.* **2021**, *12*, 3947. [CrossRef]
9. Hagin, D.; Freund, T.; Navon, M.; Halperin, T.; Adir, D.; Marom, R.; Levi, I.; Benor, S.; Alcalay, Y.; Freund, N.T. Immunogenicity of Pfizer-BioNTech COVID-19 Vaccine in Patients with Inborn Errors of Immunity. *J. Allergy Clin. Immunol.* **2021**, *148*, 739–749. [CrossRef]
10. Hanitsch, L.G.; Löbel, M.; Mieves, J.F.; Bauer, S.; Babel, N.; Schweiger, B.; Wittke, K.; Grabowski, P.; Volk, H.-D.; Scheibenbogen, C. Cellular and Humoral Influenza-Specific Immune Response upon Vaccination in Patients with Common Variable Immunodeficiency and Unclassified Antibody Deficiency. *Vaccine* **2016**, *34*, 2417–2423. [CrossRef]
11. van Assen, S.; de Haan, A.; Holvast, A.; Horst, G.; Gorter, L.; Westra, J.; Kallenberg, C.G.M.; Telgt, D.S.C.; Palache, A.M.; Giezeman, K.M.; et al. Cell-Mediated Immune Responses to Inactivated Trivalent Influenza-Vaccination are Decreased in Patients with Common Variable Immunodeficiency. *Clin. Immunol.* **2011**, *141*, 161–168. [CrossRef] [PubMed]

12. Tangye, S.G.; Al-Herz, W.; Bousfiha, A.; Chatila, T.; Cunningham-Rundles, C.; Etzioni, A.; Franco, J.L.; Holland, S.M.; Klein, C.; Morio, T.; et al. Human Inborn Errors of Immunity: 2019 Update on the Classification from the International Union of Immunological Societies Expert Committee. *J. Clin. Immunol.* **2020**, *40*, 24–64. [CrossRef] [PubMed]

13. Kuehn, H.S.; Nunes-Santos, C.J.; Rosenzweig, S.D. Germline *IKZF1* Mutations and Their Impact on Immunity: IKAROS-Associated Diseases and Pathophysiology. *Expert Rev. Clin. Immunol.* **2021**, *17*, 407–416. [CrossRef] [PubMed]

14. Marke, R.; van Leeuwen, F.N.; Scheijen, B. The Many Faces of *IKZF1* in B-Cell Precursor Acute Lymphoblastic Leukemia. *Haematologica* **2018**, *103*, 565–574. [CrossRef]

15. Goldman, F.D.; Gurel, Z.; Al-Zubeidi, D.; Fried, A.J.; Icardi, M.; Song, C.; Dovat, S. Congenital Pancytopenia and Absence of B Lymphocytes in a Neonate with a Mutation in the Ikaros Gene. *Pediatr. Blood Cancer* **2012**, *58*, 591–597. [CrossRef]

16. Eskandarian, Z.; Fliegauf, M.; Bulashevska, A.; Proietti, M.; Hague, R.; Smulski, C.R.; Schubert, D.; Warnatz, K.; Grimbacher, B. Assessing the Functional Relevance of Variants in the IKAROS Family Zinc Finger Protein 1 (*IKZF1*) in a Cohort of Patients with Primary Immunodeficiency. *Front. Immunol.* **2019**, *10*, 568. [CrossRef]

17. Kuehn, H.S.; Boisson, B.; Cunningham-Rundles, C.; Reichenbach, J.; Stray-Pedersen, A.; Gelfand, E.W.; Maffucci, P.; Pierce, K.R.; Abbott, J.K.; Voelkerding, K.V.; et al. Loss of B Cells in Patients with Heterozygous Mutations in IKAROS. *N. Engl. J. Med.* **2016**, *374*, 1032–1043. [CrossRef] [PubMed]

18. Boutboul, D.; Kuehn, H.S.; Van de Wyngaert, Z.; Niemela, J.E.; Callebaut, I.; Stoddard, J.; Lenoir, C.; Barlogis, V.; Farnarier, C.; Vely, F.; et al. Dominant-Negative *IKZF1* Mutations Cause a T, B, and Myeloid Cell Combined Immunodeficiency. *J. Clin. Investig.* **2018**, *128*, 3071–3087. [CrossRef]

19. Van Nieuwenhove, E.; Garcia-Perez, J.E.; Helsen, C.; Rodriguez, P.D.; van Schouwenburg, P.A.; Dooley, J.; Schlenner, S.; van der Burg, M.; Verhoeyen, E.; Gijsbers, R.; et al. A Kindred with Mutant IKAROS and Autoimmunity. *J. Allergy Clin. Immunol.* **2018**, *142*, 699–702.e12. [CrossRef]

20. Hoshino, A.; Okada, S.; Yoshida, K.; Nishida, N.; Okuno, Y.; Ueno, H.; Yamashita, M.; Okano, T.; Tsumura, M.; Nishimura, S.; et al. Abnormal Hematopoiesis and Autoimmunity in Human Subjects with Germline *IKZF1* Mutations. *J. Allergy Clin. Immunol.* **2017**, *140*, 223–231. [CrossRef]

21. Chen, Q.Y.; Wang, X.C.; Wang, W.J.; Zhou, Q.H.; Liu, D.R.; Wang, Y. B-Cell Deficiency: A De Novo *IKZF1* Patient and Review of the Literature. *J. Investig. Allergol. Clin. Immunol.* **2018**, *28*, 53–56. [CrossRef] [PubMed]

22. Bogaert, D.J.; Kuehn, H.S.; Bonroy, C.; Calvo, K.R.; Dehoorne, J.; Vanlander, A.V.; De Bruyne, M.; Cytlak, U.; Bigley, V.; De Baets, F.; et al. A Novel IKAROS Haploinsufficiency Kindred with Unexpectedly Late and Variable B-Cell Maturation Defects. *J. Allergy Clin. Immunol.* **2018**, *141*, 432–435.e7. [CrossRef] [PubMed]

23. Kuehn, H.S.; Nunes-Santos, C.J.; Rosenzweig, S.D. IKAROS-Associated Diseases in 2020: Genotypes, Phenotypes, and Outcomes in Primary Immune Deficiency/Inborn Errors of Immunity. *J. Clin. Immunol.* **2021**, *41*, 1–10. [CrossRef] [PubMed]

24. Nunes-Santos, C.J.; Kuehn, H.S.; Rosenzweig, S.D. IKAROS Family Zinc Finger 1-Associated Diseases in Primary Immunodeficiency Patients. *Immunol. Allergy Clin. N. Am.* **2020**, *40*, 461–470. [CrossRef] [PubMed]

25. Atmar, R.L.; Lyke, K.E.; Deming, M.E.; Jackson, L.A.; Branche, A.R.; El Sahly, H.M.; Rostad, C.A.; Martin, J.M.; Johnston, C.; Rupp, R.E.; et al. Homologous and Heterologous COVID-19 Booster Vaccinations. *N. Engl. J. Med.* **2022**, *386*, 1046–1057. [CrossRef]

26. Munro, A.P.S.; Janani, L.; Cornelius, V.; Aley, P.K.; Babbage, G.; Baxter, D.; Bula, M.; Cathie, K.; Chatterjee, K.; Dodd, K.; et al. Safety and Immunogenicity of Seven COVID-19 Vaccines as a Third Dose (Booster) Following Two Doses of ChAdOx1 NCov-19 or BNT162b2 in the UK (COV-BOOST): A Blinded, Multicentre, Randomised, Controlled, Phase 2 Trial. *Lancet* **2021**, *398*, 2258–2276. [CrossRef]

27. Infantino, M.; Pieri, M.; Nuccetelli, M.; Grossi, V.; Lari, B.; Tomassetti, F.; Calugi, G.; Pancani, S.; Benucci, M.; Casprini, P.; et al. The WHO International Standard for COVID-19 Serological Tests: Towards Harmonization of Anti-Spike Assays. *Int. Immunopharmacol.* **2021**, *100*, 108095. [CrossRef]

28. Bergman, P.; Wullimann, D.; Gao, Y.; Wahren Borgström, E.; Norlin, A.-C.; Lind Enoksson, S.; Aleman, S.; Ljunggren, H.-G.; Buggert, M.; Smith, C.I.E. Elevated CD21(Low) B Cell Frequency Is a Marker of Poor Immunity to Pfizer-BioNTech BNT162b2 MRNA Vaccine Against SARS-CoV-2 in Patients with Common Variable Immunodeficiency. *J. Clin. Immunol.* **2022**, 1–12, online ahead of print. [CrossRef]

29. Schwarz, T.; Tober-Lau, P.; Hillus, D.; Helbig, E.; Lippert, L.; Thibeault, C.; Koch, W.; Landgraf, I.; Michel, J.; Bergfeld, L.; et al. Delayed Antibody and T-Cell Response to BNT162b2 Vaccination in the Elderly, Germany. *Emerg. Infect. Dis. J.* **2021**, *27*, 2174. [CrossRef]

30. Ruether, D.F.; Schaub, G.M.; Duengelhoef, P.M.; Haag, F.; Brehm, T.T.; Fathi, A.; Wehmeyer, M.; Jahnke-Triankowski, J.; Mayer, L.; Hoffmann, A.; et al. SARS-CoV2-Specific Humoral and T-Cell Immune Response After Second Vaccination in Liver Cirrhosis and Transplant Patients. *Clin. Gastroenterol. Hepatol. Off. Clin. Pract. J. Am. Gastroenterol. Assoc.* **2022**, *20*, 162–172.e9. [CrossRef]

Journal of
Clinical Medicine

Article

Clinical, Immunological, and Genetic Findings in a Cohort of Patients with the DiGeorge Phenotype without 22q11.2 Deletion

Antonino Maria Quintilio Alberio [1,†], Annalisa Legitimo [2,†], Veronica Bertini [3], Giampiero I. Baroncelli [1,‡], Giorgio Costagliola [1], Angelo Valetto [3] and Rita Consolini [2,*]

[1]   Pediatrics Unit, Department of Clinical and Experimental Medicine, University of Pisa, 56126 Pisa, Italy; antonino.alberio@gmail.com (A.M.Q.A.); g.baroncelli@med.unipi.it (G.I.B.); giorgio.costagliola@hotmail.com (G.C.)

[2]   Section of Clinical and Laboratory Immunology, Pediatric Unit, Department of Clinical and Experimental Medicine, University of Pisa, 56126 Pisa, Italy; a.legitimo@ao-pisa.toscana.it

[3]   Section of Cytogenetics, Department of Laboratory Medicine, Azienda Ospedaliero-Universitaria Pisana, 56126 Pisa, Italy; v.bertini@ao-pisa.toscana.it (V.B.); a.valetto@ao-pisa.toscana.it (A.V.)

*   Correspondence: rita.consolini@med.unipi.it

†   These authors contributed equally to this work.

‡   BOND-ERN Representative.

Abstract: Chromosome 22q11.2 deletion syndrome (22q11.2DS) is a primary immunodeficiency characterized by a broad and heterogeneous clinical presentation associated with various degrees of T-cell deficiency. We report the clinical, immunologic, and genetic findings of a cohort of eight patients presenting with a clinical phenotype that is highly suggestive of this syndrome but without the 22q11.2 deletion. The cardinal features of 22q11.2DS, such as congenital heart defects, hypoparathyroidism, and facial dysmorphisms, were observed in the majority of the patient cohort. The unusual features are described in detail. The immunologic assessment showed various degrees of immunodeficiency of the T-cell compartment, notably a reduction in the thymic output. Half of the patient cohort exhibited a reduction in total dendritic cells. Array comparative genomic hybridization (CGH) revealed six patients harboring copy number variations (CNVs) never reported in normal subjects. The gene content of these CNVs was carefully analyzed to understand the mechanisms leading to 22q11.2DS phenocopies. According to these results, we suggested that array-CGH should be used as a first-tier tool for patients resembling 22q11.2DS.

Keywords: DiGeorge syndrome; 22q11.2 deletion; thymic output; dendritic cells; immunodeficiency; autoimmunity; copy number variations; array-CGH

Citation: Alberio, A.M.Q.; Legitimo, A.; Bertini, V.; Baroncelli, G.I.; Costagliola, G.; Valetto, A.; Consolini, R. Clinical, Immunological, and Genetic Findings in a Cohort of Patients with the DiGeorge Phenotype without 22q11.2 Deletion. *J. Clin. Med.* **2022**, *11*, 2025. https://doi.org/10.3390/jcm11072025

Academic Editor: Takao Fujisawa

Received: 26 February 2022
Accepted: 31 March 2022
Published: 5 April 2022

Publisher's Note: MDPI stays neutral with regard to jurisdictional claims in published maps and institutional affiliations.

1. Introduction

The 22q11.2 genomic region is prone to meiotic errors due to the presence of several large blocks of low-copy repeats (LCRs) [1,2]. 22q11.2 deletion is causative of DiGeorge syndrome (DGS) and other clinical conditions, previously described separately, such as velocardiofacial syndrome (MIM #192430), conotruncal anomaly face syndrome (MIM #217095) (or Takao syndrome), (CTAFS), Opitz G/BBB syndrome (MIM #145410), and Cayler cardiofacial syndromes. All these conditions are now collected under the definition of "22q11.2 deletion syndrome" (22q11.2DS) according to the common genetic etiology [3]. 22q11.2DS has an estimated incidence of 1:4000 live births, with approximately 80–90% of cases presenting with de novo inheritance [2]. The 22q11.2 typical deleted region is approximately 3 Mb in size and harbors more than 40 protein-coding genes, seven microRNAs (miRNAs), and ten non-coding RNAs (according to build GRCh37). Different sets of genes are involved, such as *TBX1*, *HIRA*, and *COMT* [3], showing the great phenotypic variability that makes this pathology a classic example of a syndrome with variable expressivity and incomplete penetrance. The clinical phenotype is mainly characterized by congenital heart disease (CHD), palatal and craniofacial abnormalities, hypoparathyroidism, immune deficiencies or autoimmune diseases (related to thymic a/hypoplasia), and neurocognitive impairment [4,5]. The severity of symptoms is also variable, ranging from quite severe to near-normal life conditions [6]. After the introduction of array comparative genomic hybridization (CGH) technology, further copy number variations (CNVs) have been identified, which are associated with clinical pictures resembling 22q11.2DS [7–10]. However, in 6% to 17% of patients, the identification of a genetic cause remains unknown, with serious consequences for their therapeutic management.

In this paper, we describe the clinical picture, the immunological abnormalities, and the genomic alterations of a cohort of patients with highly evocative DGS phenotype without 22q11.2 deletion. This may contribute to the diagnosis of patients presenting with primary immunodeficiency and developmental defects of unknown etiology.

2. Materials and Methods

2.1. Study Design and Data Collection

We enrolled eight patients (four females and four males), who were followed at a single pediatric center for primary immunodeficiency (at the University of Pisa), presenting with a highly evocative clinical phenotype for 22q11.2DS. The Tobias criteria were used to consider these patients as susceptible to genetic analysis for 22q11.2DS [11]. The study was conducted according to the Declaration of Helsinki II. Informed consent was signed prior to performing the genetic analyses. Written and informed consent to report the clinical data and the publication of the genetic analysis was obtained from all patients' parents or legal guardians. Patient data were retrospectively retrieved from the clinical records and anonymously entered into a database. The cohort of patients was composed of six children and two adults currently followed in our center. Physical phenotypes, including auxologic features, behavioral or psychiatric disorders, immunological profile, and genomic analysis were evaluated. Frequent morbidity was reported using recurrent respiratory infections (RRIs), according to the previously described RRI criteria [12]. The analysis of 25-hydroxyvitamin D (25OHD) levels was limited to patients who did not initially receive vitamin D supplementation; 25OHD levels were considered deficient for values < 20 ng/mL, according to the Institute of Medicine (IOM), the American Academy of Pediatrics (AAP), and the European Society for Paediatric Gastroenterology, Hepatology, and Nutrition (ESPGHAN) recommendations [13–15]. Auxological parameters of weight and height were expressed in standard deviation (SD) scores, using growth charts as previously described [16]. The measurements of height were performed at time points T0, T1, and T2; T1 and T2 were related to the measurement of the height after 2 and 4 years of follow-up, respectively.

2.2. Flow Cytometry and Immunological Assessment

None of the patients had acute infections at the time of sample collection for the immunological evaluation. Lymphocyte counts, serum immunoglobulin concentration, and serum immunoglobulin subclasses were evaluated through standard methods and compared with age-related normal values. An extended immunological phenotype was performed in all patients and the data were compared with age-matched normal values [17–20]. Eight-color flow cytometric analysis was performed on fresh peripheral whole blood anticoagulated with ethylenediaminetetracetic acid (EDTA), according to standard protocols, to determine the following cell subpopulations: T lymphocytes ($CD3^+$), helper T lymphocytes ($CD3^+CD4^+$), cytotoxic T lymphocytes ($CD3^+CD8^+$), B lymphocytes ($CD19^+$), and natural killer (NK) cells ($CD16^+/56^+$). Helper and cytotoxic T lymphocytes were also analyzed for the expressions of CD45RA, CD62L, and CD31 to identify naïve ($CD45RA^+CD62L^+$), central memory (CM: $CD45RA^-CD62L^+$), effector memory (EM: $CD45RA^-CD62L^-$), terminal effector memory re-expressing CD45RA (TEMRA: $CD45RA^+CD62L^-$), and recent thymic emigrants (RTEs: $CD45RA^+CD62L^+CD31^+$) [21]. Circulating Treg cells were identified as a $CD4^+CD25^+CD127^-$ cell population, as previously described [22]. The expression of CD45RA was evaluated to estimate the amount of naïve Treg cells. The expression of CD185 (CXCR5) was analyzed on memory T helper cells (CD3$^+$CD4$^+$CD45RO$^+$) to identify follicular T cells. We defined naïve B cells as CD19$^+$CD27$^-$IgD$^+$ and switched memory B cells as CD19$^+$CD27$^+$IgD$^-$IgM$^-$. Circulating dendritic cells (DCs) were enumerated and phenotypically characterized directly into the two major subsets, namely, myeloid (mDCs) and plasmacytoid (pDCs), as previously described [22]. Due to the lack of a specific marker to detect DCs, we used a mixture of monoclonal antibodies specifically established to identify DCs, purchased from Immunotech (Beckman Coulter Inc., Brea, CA, USA). Cells were stained with the following antibodies: CD14, CD16, CD85k, CD33, or CD123 for the mDC and pDC subsets, respectively. Dendritic cells were identified as CD14$^{low/-}$CD16$^{low/-}$CD85k$^+$ and CD33$^+$ or CD123$^+$. The absolute numbers of DCs were estimated by multiplying the percentage of DCs in the mononuclear cell (MNC) gate by the absolute peripheral blood MNC count determined using a standard hemocytometer (Abbott Laboratories, Abbott Park, IL, USA). DC data were compared with our laboratory age-related normal values [22]. Data acquisition and analysis were performed on a dual laser BD FACSCanto (Becton Dickinson Immunocytometry Systems, San Jose, CA, USA) using the FACSDiva software (San Jose, CA, USA).

2.3. Genomic Analysis

Karyotyping was performed according to standard methods. The commercially available D22S75/N25 probe (Cytocell, Cambridge, UK) was used, according to the manufacturer's instructions, to perform fluorescent in situ hybridization (FISH) analysis.

Genomic DNA of the patients (tests) was isolated from peripheral blood using standard methods; DNA from healthy subjects (reference) was used as controls (Agilent Technologies, Santa Clara, CA, USA); tests and reference DNA were differentially labeled with Cy5-dCTP or with Cy3-dCTP using random primer labeling and applied to 60K arrays, according to the manufacturer's protocol (Agilent, Santa Clara, CA, USA). Quality slide evaluation was performed using the Agilent dedicated software (Feature Extraction, Agilent). We elaborated only on those experiments that met the "excellent" criteria, as determined by the QC report (Cytogenomic software, Agilent). In particular, the derivative log ratio spread (DLRS) was the main value considered for further analysis of the data: when >0.16, the experiment was discarded and repeated. CNVs were identified with Cytogenomics 4.0.3.12 (Agilent) using the ADM-2 (aberration detection method 2) algorithm. The threshold was set to a minimum of 6 with the minimum number of 3 probes required in a region and a minimum absolute log ratio of 0.25. We analyzed all the CNVs with 3 or more contiguous probes for deletions and duplications. The CNVs reported in the Database of Genomic Variants (http://projects.tcag.ca/variation/ (accessed on 25 February 2022)) were classified as benign and not further analyzed. All the other genomic imbalances were compared with those collected in DECIPHER (https://decipher.sanger.ac.uk/ (accessed on 25 February 2022)) and ClinVar (https://www.ncbi.nlm.nih.gov/clinvar/ (accessed on 25 February 2022)). Data about the biological function of genes and their interactions were retrieved from UCSC Genome Browser (http://genome.ucsc.edu/ (accessed on 25 February 2022)), PubMed (https://www.ncbi.nlm.nih.gov/pubmed/ (accessed on 25 February 2022)), and OMIM (https://www.omim.org/ (accessed on 25 February 2022)).

3. Results

3.1. Patients Characteristics

Two patients (25%) met one of the A Tobias criteria, while six patients (75%) met at least two of the B Tobias criteria, along with a combination of C criteria. The mean age of patients was 6.7 years (8 months–15.7 years) at diagnosis and 13.7 years (2.2–23.7 years) at the time of the study. The mean follow-up time of the cohort was 55.2 months (SD $\pm$ 31.6). Demographic and clinical features of the cohort are described in Table 1. No exposure to tobacco, alcohol, or teratogenic drugs during pregnancy was reported; no perinatal information was available for the two sisters P4 and P5. Patient P2 was born from an emergency cesarean section at 32 weeks of gestational age for maternal HELLP (hemolysis, elevated liver enzymes, low platelets) syndrome. Subject P6 was born to a human immunodeficiency virus (HIV)-positive mother, small for gestational age, and successfully received the prevention protocol of vertical transmission. The overall cases were sporadic, without familial history of severe chronic diseases, immunodeficiencies, or inherited pathologic conditions. Subject P2 presented with syndactyly like her maternal grandmother. The neonatal period of subject P1 was complicated by seizures, hypotonia, and sucking difficulty. Congenital heart defects were detected in 87.5% of the cohort ($n = 7$); two conotruncal anomalies and four non-conotruncal defects (patent oval foramen, patent ductus arteriosus, atrial and/or ventricular septum) were observed. A total of 71% of them underwent corrective or palliative cardiac surgery in the first year of life, with excellent outcomes. Subject P3 presented with persistent left superior vena cava and percutaneous cardiac catheterization was performed at 5 months for aortic re-coarctation, with an absence of residual obstruction. Otolaryngologic malformations were detected and successfully corrected in two patients (25%). The overall cohort exhibited mostly mild facial dysmorphisms. Only toddler P8 presented with a gastrointestinal malformation (esophageal atresia). Concerning congenital renal disorders, medullary sponge kidney was found in subject P4 and hypospadias in P8. Noteworthy, language disorder with speech delay was observed in six subjects (75%) and psychomotor delay in 25%. One patient (P7) suffered from attention-deficit hyperactivity disorder, mixed anxiety disorder with an obsessive-compulsive component, and sleep disturbance; her nuclear magnetic resonance (NMR) showed widening of the fourth ventricle associated with hypoplasia of the lower cerebellar vermis. Interestingly, psychiatric involvement in patient P3 developed after many years from neurological manifestations. A total of three patients (37.5%) had hypoacusia that was conductive in nature, with a hearing aid being necessary only in one patient (P7). The two sisters P4 and P5 suffered from monthly headache episodes. Other anomalies, such as myopia (P4 and P6) and hypermetropia (P5, P6, and P7), were found in 25% and 37.5% of patients, respectively. Moreover, subject P6 showed left cryptorchidism and developed a limitation in pronation movement of the right upper limb in the last year of follow-up, which is still under diagnostic investigation.

3.2. Infections and Autoimmunity

Recurrent upper respiratory infections were detected in the majority of the cohort, including otitis and sinusitis (75%); lower respiratory tract infections occurred in two patients (25%), requiring hospitalization. No patient developed bronchiectasis. A total of 37.5% (three out of eight subjects) were affected by recurrent periodic fevers, without genetic features of monogenic autoinflammatory syndromes, and subject P2 was affected by recurrent urinary infections. Subjects P2 and P3 presented with recurrent herpes simplex infections and aphthous stomatitis. No fungal or opportunistic infections were identified during the follow-up, except for one episode of oral candidiasis in patient P2, requiring fluconazole prophylaxis. Regarding severe infections, an episode of mediastinitis in the context of the sternal surgical wound was described in child P8, with a good outcome.

Concerning autoimmune manifestations, juvenile idiopathic arthritis with oligoarticular phenotype was diagnosed in patient P3. Joint involvement (the right knee, right temporomandibular, and proximal interphalangeal of the fifth finger of the right hand) required a step-up therapy with frequent evacuative and infiltrative (triamcinolone acetonamide) arthrocentesis, use of modifiers of the biologic response (methotrexate), and biologic drugs (anti-tumor necrosis factors). Moreover, the child was unresponsive and developed a chronic inflammatory process that resulted in deformity of the hand joint and residual synovial inflammation of the right temporomandibular joint, as detected using nuclear magnetic resonance, with functional limitation.

Table 1. Clinical features of the patients.

P	Sex	Age at Diagnosis (Years)	Frequent Morbidity	Autoimmune Disorders	Cardiac Malformations	Otolaryngologic Involvement	Neuro-Behavioural and Psychiatric Involvement	Endocrine Involvement	Dysmorphic Features and Dental Issues	Skeletal Abnormalities
1	F	15.5	Ear infections and sinusitis	nr	VSD, ASD		Psychomotor and language delay, epilepsy, moderate cognitive impairment		Right eye exophoria	Left flat foot, cleft posterior arch in cervical vertebrae C1
2	M	1.7	Upper and lower respiratory tract infections, urinary infections	nr	VSD, ASD, CoA	Adenoid hypertrophy	Language delay, moderate cognitive impairment	25OHD deficiency	Long face, hypertelorism, low and flat nasal bridge, low and retracted ears	Syndactyly IV and V finger hands
3	M	7.7	nr	JIA	HLHS, CoA, BAV	Short lingual frenulum	Motor and language delay, anxiety disorder with an obsessive-compulsive component, vocal tics		Wide ear pad, supernumerary and ectopic teeth in the hard palate	Bilateral clinodactyly V finger, lumbar scoliosis with right dorsal hump, mild leg dysmetria, right leg hypotrophy, valgus right foot, mild retro-tibial torsion
4	F	10.5	Ear infections and sinusitis	Chronic autoimmune thyroiditis				Hypo-parathyroidism	Low set ears, preauricular appendix	
5	F	2.5	Ear infections and sinusitis	nr	PFO	Conductive hearing loss		Hypo-parathyroidism	Low set ears	
6	M	0.8	Upper respiratory tract infections	nr	TOF	Conductive hearing loss, adenoid hypertrophy	Language delay	25OHD deficiency	Anteroverse ears, bilateral epicanthal folds	
7	F	13.4	Upper and lower respiratory tract infections, urinary infections	nr	PDA	Labiopalatoschisis, mild and predominantly conductive mixed hearing loss	Psychomotor, cognitive and language delay, attention-deficit hyperactivity disorder, mixed anxiety disorder with an obsessive-compulsive component, sleep disturbance	25OHD deficiency	Eyes with elongated and upward rhymes, pyriform aspect of the nose with prominent tip and widened nostrils, lower lip with a thickened edge, wide ear pad with an antiverse and low implantation	Hip dysplasia, hindfoot pronation, scoliosis
8	M	1.4	nr	nr	TA type 2, RAA		Motor and language delay and mild axial hypotonia		Micrognathia, hypertelorism, long palpebral fissures, and low and depressed nasal bridge	

nr: not reported; VSD: ventricular septal defect; ASD: atrial septal defect; CoA: aortic coarctation; HLHS: hypoplastic left heart syndrome; BAV: bicuspid aortic valve; JIA: juvenile idiopathic arthritis; PFO: patent foramen ovale; TOF: tetralogy of Fallot; PDA: patent ductus arteriosus; TA: truncus arteriosus; RAA: right-sided aortic arch.

3.3. Auxological and Endocrine Features

Weight and head circumference were within normal percentiles for the overall cohort, while a delay in height growth was remarked. No measurements were reported for subject P8 and, consequently, he was excluded. A reduction in the height (between -1 and -2 SD) was reported in almost all of the cohort, prevalently in childhood. No case of short stature (height < -2 SD) was registered. Of note, only the two sisters P4 and P5 presented within a height normal distribution (Supplementary Figure S1).

We focused on the endocrine features of the cohort by investigating the 25OHD level and its association with immunological parameters, as we previously published for a cohort of DGS patients [23]. Therefore, we identified three subgroups according to 25OHD levels: group A (normal values), group B (25OHD deficiency), and group C (patients with hypoparathyroidism). Toddler P8 was excluded due to the absence of 25OHD determination. Group B was composed of patients P2, P6, and P7. The vitamin D status was performed at diagnosis and showed the following 25OHD mean values: 30.23 ng/mL, 17.43 ng/mL, and 14.0 ng/mL in the A, B, and C groups, respectively (normal range < 20 ng/mL). Both the A and B groups presented with parameters of phospho-calcium metabolism (parathyroid hormone, PTH, calcium, phosphate, alkaline phosphatase) in the normal range for the patients' ages. Patients belonging to group B received oral supplementation with cholecalciferol at the dosage of 25,000 IU/month. An increase in 25OHD values was observed in P2 (46.9 ng/mL vs. basal value 16.8 ng/mL) and P6 (35.5 ng/mL vs. basal value 17.63 ng/mL).

Patients P4 and P5 (25%) received a diagnosis of congenital hypoparathyroidism. Both presented with neonatal hypocalcemia, elevated serum phosphorus concentration, PTH values constantly suppressed or undetectable (<3 ng/mL), and low vitamin D values (13 ng/mL and 15 ng/mL in P4 and P5 patients, respectively). They received treatment with calcium carbonate (50–100 mg/kg/day) and calcitriol (20–40 ng/kg/day) for at least 6 years, with good outcomes. Moreover, the vitamin D values returned to within the normal range only for a short time and subject P4 suffered from two hypercalcemia episodes during the follow-up.

However, we did not observe any difference between vitamin D levels and the immunological parameters examined in the various groups.

3.4. Immunological Features

The lymphocyte and lymphocyte subpopulation values of patients, compared with age-matched normal controls, are shown in Tables 2 and 3. Lymphocytopenia was reported in 62.5% of the cohort (five out of eight patients). A low absolute count of CD3$^+$ lymphocytes and CD4$^+$ and CD8$^+$ cells were observed in 75% ($n = 6$), 87.5% ($n = 7$), and 75% ($n = 6$) of subjects, respectively. The B cell absolute number was low in two lymphopenic patients, with them being normal if expressed as a percentage. Similarly, the NK cell count exhibited the same behavior in patient P2; conversely, high relative numbers of NK cells were observed in P3, P6, P7, and P8. Furthermore, extensive phenotyping was performed due to the suspicion of a putative thymic a/hypoplasia. RTEs were lower than normal age-matched values in six out of eight patients, in particular, P1, P2, and P3 < 15%. Naïve helper T cells were reduced in P2 and severely low in both patients P1 and P3 (8.8% and 12.7%, respectively). Naïve cytotoxic T cells were reduced in two subjects (P3, P6). Central memory cytotoxic T cells were low only in P8, whereas they were higher than the normal range in P1, P6, and P7. Both effector memory CD8$^+$ T cells (EM and TEMRA) were reduced in patients P5 and P7, while EM was only low in P4 and TEMRA only in P1. The overall cohort showed normal values of follicular T cells. Regulatory T cells were increased in patient P2 and reduced in patient P6. The analysis of the B cell compartment (naïve and switched memory cells) revealed mostly normal values, with the exception of two patients who showed a reduction of naïve B cells (P6 and P7) and an increase of switched memory cells (P6).

To explore the interface between innate and adaptative immunity, we investigated the dendritic cells, as shown in Table 4. A total of 50% of patients (P2, P3, P5, P7) exhibited a reduction in total DCs in comparison to healthy controls, whereas an increase was observed in patient P8. The observed reduction affected mainly the myeloid subset (37.5%), except for patient P2, who showed a reduction in both subsets.

Serum immunoglobulin levels are shown in Table 5. Two children (P2 and P6) had partial selective IgM deficiency (between −1 and −2 SD). Patients P3 and P5 (25% of the cohort) had a reduction in both IgG and IgA, with the IgA values being severely low (>−2 SD) in P5, whereas in P8, a reduction in both IgG and IgM (between −1 and −2 SD) was observed. P4 showed hypogammaglobulinemia (both IgG and IgA, between −2 and −3 SD), along with a compensatory increase (>+3 SD) in IgM levels, requiring immunoglobulin replacement therapy. Conversely, P1 had an increase in IgM (>+2 SD). IgE levels were normal in the overall cohort. The immunoglobulin subclass levels were in the normal range in the overall population, except for the IgG3 subtype, which was reduced in patient P8.

None of the patients had neutropenia. Mild eosinophilia (760/μL) was observed in subject P3.

Table 2. Lymphocyte subsets of the patients.

	P1	P2	P3	P4	P5	P6	P7	P8
Age (years) *	21.5	5.8	13.6	22.8	14.8	7.7	16	1.6
Lymphocyte ($\times 10^3$/μL)	1.27 *1.8 (0.9–4.5)*	0.92 *3.8 (2.3–6.1)*	1.72 *2.3 (1.3–3.2)*	1.45 *1.8 (0.9–4.5)*	1.20 *2.3 (1.3–3.2)*	1.44 *2.5 (1.7–3.4)*	1.1 *2.3 (1.3–3.2)*	2.89 *4.7 (3.9–6.1)*
T cells ($\times 10^3$/μL)	0.89 *1.5 (0.78–3.0)*	0.46 *2.6 (1.6–3.7)*	0.94 *1.6 (0.95–2.3)*	0.91 *1.5 (0.78–3.0)*	0.77 *1.6 (0.95–2.3)*	0.78 *1.8 (1.2–2.6)*	0.69 *1.6 (0.95–2.3)*	1.27 *3.1 (2.5–4.9)*
Helper T cells ($\times 10^3$/μL)	0.46 *1.0 (0.5–2.0)*	0.31 *1.4 (0.8–2.1)*	0.47 *0.9 (0.6–1.4)*	0.60 *1.0 (0.5–2.0)*	0.46 *0.9 (0.6–1.4)*	0.40 *1.0 (0.6–1.5)*	0.39 *0.9 (0.6–1.4)*	0.95 *1.8 (1.6–2.9)*
Cytotoxic T cells ($\times 10^3$/μL)	0.36 *0.5 (0.2–1.2)*	0.14 *0.8 (0.4–1.1)*	0.41 *0.5 (0.3–0.7)*	0.17 *0.5 (0.2–1.2)*	0.21 *0.5 (0.3–0.7)*	0.22 *0.6 (0.3–0.9)*	0.28 *0.5 (0.3–0.7)*	0.17 *0.9 (0.6–1.4)*
B cells ($\times 10^3$/μL)	0.18 *0.23 (0.06–0.8)*	0.27 *0.73 (0.4–1.2)*	0.37 *0.32 (0.2–0.7)*	0.21 *0.23 (0.06–0.8)*	0.20 *0.32 (0.2–0.7)*	0.21 *0.40 (0.3–0.6)*	0.94 *0.32 (0.2–0.7)*	1.03 *0.29 (0.19–0.7)*
NK cells ($\times 10^3$/μL)	0.20 *0.34 (0.10–1.2)*	0.13 *0.29 (0.16–0.6)*	0.40 *0.23 (0.09–0.5)*	0.27 *0.34 (0.1–1.2)*	0.19 *0.23 (0.09–0.5)*	0.44 *0.26 (0.12–0.5)*	0.30 *0.23 (0.09–0.5)*	0.58 *0.29 (0.19–0.7)*

The absolute numbers of cell subsets are indicated for each patient (upper line). Lower lines indicate normal values for age (median (10–90th percentile)). * Age at immunological evaluation; NK: natural killer.

Table 3. Advanced phenotypes of the patients.

	P1	P2	P3	P4	P5	P6	P7	P8
Age (years) *	21.5	5.8	13.6	22.8	14.8	7.7	16	1.6
T cells (%) [a]	69.1 *67 (50–91)*	50.2 *69 (60–77.6)*	54.6 *73 (62.6–80.4)*	62.5 *67 (50–91)*	64.3 *73 (62.6–80.4)*	54.0 *72 (63.2–77.8)*	63.0 *73 (62.6–80.4)*	44.0 *68 (60.7–75.8)*
Helper T cells (%) [a]	36.5 *42 (28–64)*	33.2 *38 (31.1–47.4)*	27.2 *44 (32.6–51.5)*	41.7 *42 (28–64)*	38.4 *44 (32.6–51.5)*	27.5 *40 (31.7–47)*	28.1 *44(32.6–51.5)*	33.0 *41 (35–52)*
Cytotoxic T cells (%) [a]	28.2 *22 (12–40)*	14.8 *21 (16–27)*	24.0 *23 (19–29)*	12.0 *22 (12–40)*	17.2 *23 (19–29)*	15.0 *24 (17.1–30)*	25.3 *23 (19–29)*	6.0 *19.3 (16.1–29.4)*
B cells (%) [a]	13.8 *10 (4–28)*	29.0 *22 (13–29.2)*	21.5 *14 (12–21)*	14.3 *10 (4–28)*	16.3 *14 (12–21)*	14.5 *15.6 (12–34)*	8.5 *14 (12–21)*	35.5 *24 (14.3–28.2)*
NK cells (%) [a]	15.5 *15 (4–24.6)*	14.0 *8 (4.7–16.2)*	23.5 *11.7 (4.3–16.2)*	18.3 *15 (4–24.6))*	15.5 *11.7 (4.3–16.2)*	30.7 *9.8 (5.4–18.6)*	27.6 *11.7 (4.3–16.2)*	20.0 *6.8 (4–13.8)*
Naïve helper T cells (%) [b]	8.8 *46 (16–100)*	32.6 *70 (50–85)*	12.7 *51 (31–65)*	32.1 *46 (16–100)*	42.4 *51 (31–65)*	51.2 *58 (42–74)*	46.5 *51 (31–65)*	85.2 *79 (62–90)*
RTE (%) [b]	2.7 *33 (7–100)*	13.0 *58 (41–81)*	2.9 *50 (31–81)*	20.6 *33 (7–100)*	27.0 *50 (31–81)*	26.6 *58 (41–81)*	41.0 *50 (31–81)*	39.9 *66 (40–100)*
CM helper T cells(%) [b]	58.5 *42 (18–95)*	45.7 *18 (0.35–100)*	44.2 *32 (13–76)*	45.4 *42 (18–95)*	45.0 *32 (13–76)*	35.0 *18 (0.35–100)*	36.4 *32 (13–76)*	1.5 *10 (0.09–40)*
EM helper T cells (%) [b]	32.5 *5 (1–23)*	20.0 *2 (0.27–18)*	39.0 *3 (0.49–25)*	22.0 *5 (1–23)*	12.2 *3 (0.49–25)*	12.7 *2 (0.27–18)*	16.8 *3 (0.49–25)*	1.4 *0.67 (0.024–4.7)*
TEMRA helper cells (%) [b]	0.1 *0.35 (0.008–6.8)*	1.6 *0.1 (0.003–1.8)*	4.1 *0.17 (0.004–5.8)*	0.6 *0.35 (0.008–6.8)*	0.3 *0.17 (0.004–5.8)*	1.1 *0.1 (0.003–1.8)*	0.3 *0.17 (0.004–5.8)*	11.9 *0.1 (0.0–4.1)*
Naïve cytotoxic T cells (%) [c]	10.5 *29 (6–100)*	53.9 *64 (42–81)*	9.0 *56 (42–73)*	17.8 *29 (6–100)*	78.8 *56 (42–73)*	30.6 *58 (39–73)*	75.2 *56 (42–73)*	73.1 *71(46–85)*
CM cytotoxic T cells (%) [c]	35.2 *5 (1–20)*	4.0 *3 (1–6)*	10.4 *3 (0.4–18)*	17.7 *5 (1–20)*	10.3 *3 (0.4–18)*	10.7 *3 (1–6)*	24.8 *3 (0.4–18)*	0.5 *3 (1–8)*
EM cytotoxic T cells (%) [c]	47.4 *36 (14–98)*	6.1 *24 (5–100)*	33.2 *22 (4–100)*	13.3 *36 (14–98)*	3.3 *22 (4–100)*	35.3 *24 (5–100)*	0.1 *22 (4–100)*	2.1 *15 (2–100)*
TEMRA cytotoxic T cells (%) [c]	6.8 *19 (7–53)*	22.8 *25 (15–41)*	47.3 *24 (9–65)*	17.8 *19 (7–53)*	7.6 *24 (9–65)*	23.3 *25 (15–41)*	0.10 *24 (9–65)*	24.3 *24 (8–71)*
Treg (%) [b]	10 *8 (4–17)*	15.8 *8 (4–14)*	4.4 *9 (4–20)*	8.6 *8 (4–17)*	9.8 *9 (4–20)*	3.4 *8 (4–14)*	13.2 *9 (4–20)*	6.2 *9 (6–13)*
Follicular T helper cells (%) [d]	27.4 *17 (5–56)*	36.5 *24 (7–85)*	22.7 *18 (7–47)*	25.9 *17 (5–56)*	26.2 *18 (7–47)*	44.8 *24 (7–85)*	28.6 *18 (7–47)*	27.1 *20 (8–51)*
Naïve B cells (%) [e]	53.8 *63 (33–100)*	82.0 *76 (62–94)*	91.4 *74 (49–100)*	78.1 *63 (33–100)*	89.1 *74 (49–100)*	61.9 *76 (62–94)*	47.2 *74 (49–100)*	97.4 *88 (78–99)*
Switched memory B cells (%) [e]	14.6 *12 (3–46)*	4.0 *7 (3–18)*	2.0 *8 (1–43)*	7.2 *12 (3–46)*	2.5 *8 (1–43)*	24.0 *7 (3–18)*	30.4 *8 (1–43)*	1.04 *3 (0.3–20)*

The frequency of cell subsets is indicated for each patient (upper line). Lower lines indicate normal values for age (median (10–90th percentile)). * Age at immunological evaluation. [a] % of total peripheral lymphocyte population; [b] % of helper T lymphocyte population; [c] % of cytotoxic T lymphocyte population; [d] % of CD4$^+$CD45RO$^+$ T lymphocytes; [e] % of B lymphocyte population; NK: natural killer; TEMRA: effector memory T cells re-expressing CD45RA; CM: central memory; EM effector memory; RTE: recent thymic emigrants; Treg: regulatory T cells.

Table 4. Absolute and relative numbers of the dendritic cells in the cohort.

	P1	P2	P3	P4	P5	P6	P7	P8
Age (years) *	21.5	5.8	13.6	22.8	14.8	7.7	16	1.6
DCtot/µL	31.71 24 (10.7–35.6)	10.69 41.4 (28.6–69.5)	26.41 36 (27.1–43.7)	18.15 24 (10.7–35.6)	20.74 36 (27.1–43.7)	32.61 41.4 (28.6–69.5)	14.3 36 (27.1–43.7)	59.07 53.6 (44.7–58.9)
DCtot (%) [f]	0.48 0.43 (0.22–0.69)	0.25 0.56 (0.39–0.68)	0.56 0.61 (0.5–0.72)	0.44 0.43 (0.22–0.69)	0.46 0.61 (0.5–0.72)	0.72 0.56 (0.39–0.68)	0.23 0.61 (0.5–0.72)	1.12 0.63 (0.48–0.89)
mDC/µL	18.29 14.7 (7.6–21.1)	7.33 25.5 (12.4–48.0)	17.98 23.5 (18.4–30.9)	11.37 14.7 (7.6–21.1)	11.36 23.5 (18.4–30.9)	14.6 25.5 (12.4–48.0)	5.69 23.5 (18.4–30.9)	27.62 32.8 (26.9–39.5)
mDC (%) [f]	0.28 0.26 (0.1–0.4)	0.17 0.34 (0.2–0.5)	0.38 0.40 (0.3–0.5)	0.28 0.26 (0.1–0.4)	0.25 0.40 (0.3–0.5)	0.32 0.34 (0.2–0.5)	0.09 0.40 (0.3–0.5)	0.52 0.38 (0.3–0.5)
pDC/µL	13.42 9.4 (3.2–17.0)	3.36 15.9 (8.6–23.6)	8.43 12.5 (5.4–18.8)	6.78 9.4 (3.25–17.0)	9.38 12.5 (5.4–18.8)	18.01 15.9 (8.6–23.6)	8.61 12.5 (5.4–18.8)	31.45 20.8 (12.6–30.8)
pDC (%) [f]	0.20 0.17 (0.07–0.3)	0.08 0.22 (0.1–0.4)	0.18 0.21 (0.1–0.3)	0.16 0.17 (0.07–0.3)	0.21 0.21 (0.1–0.3)	0.4 0.22 (0.1–0.4)	0.14 0.21 (0.1–0.3)	0.6 0.25 (0.1–0.4)

The frequency and absolute numbers of cell subsets are indicated for each patient (upper line). Lower lines indicate normal values for age (mean (10–90th percentile)). * Age at immunological evaluation; [f] % of WBC; WBC: white blood cells; DC: dendritic cells; mDC: myeloid dendritic cells; pDC: plasmacytoid dendritic cells.

Table 5. Immunoglobulins and their subclasses in the cohort.

Immunoglobulins	P1	P2	P3	P4	P5	P6	P7	P8
Age *	21.5	5.8	13.6	22.8	14.8	7.7	16	1.6
IgG (mg/dL)	1440 (1116 ± 208)	979 (1007 ± 236)	841 (1116 ± 208)	557 (1116 ± 208)	697 (1116 ± 208)	1000 (1040 ± 223)	962 (1116 ± 208)	361 (655 ± 176)
IgM (mg/dL)	256 (92 ± 34)	57 (87 ± 27)	96 (92 ± 34)	304 (92 ± 34)	91 (92 ± 34)	49 (90 ± 27)	127 (92 ± 34)	21 (67 ± 29)
IgA (mg/dL)	138 (189 ± 67)	84 (123 ± 41)	113 (189 ± 67)	59 (189 ± 67)	48 (189 ± 67)	163 (136 ± 48)	155 (189 ± 67)	35 (42 ± 23)
IgG1 (mg/dL)	844 (490–1140)	664 (370–1000)	NA	523 (490–1140)	535 (490–1140)	730 (370–1000)	588 (370–1280)	310 (200–770)
IgG2 (mg/dL)	511 (150–640)	181 (72–340)	NA	196 (150–640)	152 (150–640)	213 (72–340)	266 (106–610)	83 (34–230)
IgG3 (mg/dL)	64 (20–110)	77 (13–133)	NA	23 (20–110)	30 (20–110)	59 (13–133)	57 (18–263)	13 (15–97)
IgG4(mg/dL)	93 (8–140)	2 (0.01–158)	NA	0.0 (8–140)	9 (8–140)	24 (0.01–158)	25 (4–230)	2 (0.01–43)

The frequency of immunoglobulins and IgG subclasses is indicated for each patient (upper line). Lower lines indicate normal values for age expressed as mean ± SD or 10–90th percentile; NA, not available. * Age at immunological evaluation.

3.5. Genomic Features

The genetic assessment of all patients did not show any deletion in the 22q11.2 and 10p13-14 regions. Among the eight patients tested using array-CGH, two subjects did not show any pathological CNVs, according to the DGV database (http://dgv.tcag.ca/variation (accessed on 25 February 2022)), which collects variations reported in normal subjects. Table 6 shows the CNVs detected in the remaining six patients, along with the positions of the first and last abnormal probes and the extent and gene content of each CNV. The pattern of inheritance was assessed in four patients in which CNVs were inherited from the mother.

Table 6. Genetic abnormalities on array-CGH of the cohort.

Subjects	Position (GRCh37/hg19)	Extent (kb)	NCBI RefSeq Genes (UCSC)	Inheritance
P1	2q24.1 (156,761,199_157,075,778)x3	314	LINC01876	Maternal
P2	arr(X,Y)x1,(1-22)x2 *			
P3	arr(X,Y)x1,(1-22)x2 *			
P4	20p11.22 (21,419,411_21,784,484)x3	365	**PAX1**, NKX2-2, LINC01727,LINC01726	Maternal
P5	20p11.22 (21,419,411_21,784,484)x3	365	**PAX1**, NKX2-2, LINC01727,LINC01726	Maternal
P6	11p15.5 (723,382_917,649)x3	194	**EPS8L2**, **TALD01**, GATD1, LOC171391, CEND1, **SLC25A22**, PIDD1, RPLP2, SNORA52, **PNPLA2**, CRACR2B, **CD151**, POLR2L, TSPAN4, CHID1	NA
	Xp22.33 or Yp11.32 (61,091_658,258 or 11,091_608,258)x2	597	PLCXD1, GTPBP6, LINC00685, PPP2R3B, **SHOX**	
P7	17q21.31 (43,717,703_44,210,822)x1	493	LINC02210, LINC02210-CRHR1, CRHR1, MAPT-AS1, SPPL2C, **MAPT**, MAPT-TT1, STH, **KANSL1**	NA
P8	17p13.2(5882589_6140992)x1	258	WSCD1	Maternal
	Xq24(118647205_118715504)x0	68	**CXorf56, UBE2A**	

OMIM genes are in bold; NA: not available; * negative array-CGH.

Patient P1 showed a duplication of 314 kb in 2q24.1, inherited from her mother, that harbored the long intergenic non-protein coding RNA 1876. No overlapping duplications were reported in the Decipher database (https://decipher.sanger.ac.uk/ (accessed on 25 February 2022)), with this CNV classified as a variant of unknown significance (VOUS).

P4 and P5 were two sisters who showed a duplication in 20p11.22 inherited from their mother. This CNV has never been reported in individuals with a pathological phenotype (Decipher). This region harbors two non-coding protein LINC01727, LINC01726 and two coding protein PAX1 and NKX2 genes. PAX1 (*167411) encodes a transcription factor implicated in embryogenesis in vertebrates and plays an important role in segmental spine formation and thymus organogenesis. In humans, the phenotypic effects of the PAX1 duplication have not been yet described, whereas homozygous mutations of this gene have been associated with otofaciocervical syndrome 2 with T-cell deficiency (OTFCS2) (#615560) [24–26]. NKX2 contains a homeobox domain. It is highly expressed in the central nervous system and encodes for a protein that is likely a nuclear transcription factor involved in the morphogenesis of this system. According to these data, this CNV is likely pathogenetic.

In patient P6, two duplicated regions were detected in 11p15.5 and the pseudoautosomal region Xp22.33/Yp11.32. The duplication in 11p15.5 encompassed a total of 15 genes (Supplementary Table S1), among which, EPS8L2, TALD01, SLC25A22, PNPLA2, and CD151 are reported in the OMIM database (https://www.omim.org (accessed on 25 February 2022)). Apparently, none of these five genes appeared causative of the clinical features of this patient (Supplementary Table S1). Moreover, we highlight that the pathological phenotype described in OMIM is related to homozygous loss-of-function mutations. No overlapping duplications are reported neither in the literature nor in Decipher; thus, according to the few data available, this variant could be classified as VOUS. The duplication in Xp22.33/Yp11.32 harbors one non-coding gene (LINC00685) and four coding genes, including SHOX (*312865/*400020) (Supplementary Table S1). SHOX/SHOX enhancer deletions cause short stature and skeletal abnormalities (#249700, #127300, #300582); microduplications in the pseudoautosomal region including SHOX appear to be rare and have been related to autism spectrum disorders and neurodevelopmental pathologic conditions [27].

Patient P7 showed a deletion in 17q21.31 of about 493 kb that was causative of Koolen–De Vries syndrome (KDVS) (#610443). Moderate-to-severe intellectual disability, hypotonia, and a characteristic face represent the core phenotype of this syndrome. More variable features include cardiac, genitourinary anomalies, seizures, nasal speech, and a friendly demeanor [28]. Haploinsufficiency of KANSL1 (*612452) appears causative of the syndrome, as the clinical phenotype does not substantially differ between patients with 17q21.21 microdeletion encompassing KANSL1 and patients with a de novo heterozygous mutation in this gene. KANSL1 encodes a nuclear protein that plays a role in chromatin modification. It is a member of a histone acetyltransferase (HAT) complex [29].

Patient P8 showed two deleted regions in 17q13.2 and Xq24, with sizes of 258 kb and 68 kb, respectively, both inherited from his mother. The deletion in 17p13.2 harbors WSCD1, which encodes for a membrane protein with a sulfotransferase activity, whose role is not yet known. This CNV can be classified as a VOUS. The deletion in Xq24 encompasses two genes, namely, CXorf56 and UBE2A, and the phenotypic effects of their loss-of-function in males are reported in OMIM. CXorf56 is related to "Intellectual development disorder, X-linked 107" (#301013). UBE2A encodes a member of the E2 ubiquitin-conjugating enzyme family that is required for post-replicative DNA damage repair and may play a role in transcriptional regulation; its loss-of-function is causative of "UBE2A deficiency syndrome" or "X-linked Nascimento-type intellectual disability syndrome" (#300860). According to these data, this CNV is pathogenetic.

4. Discussion

We described the clinical phenotype of a cohort of eight patients that were highly suggestive of 22q11.2DS without harboring genomic aberrations of chromosome 22. The cardinal features of 22q11.2DS, such as congenital heart defects, hypoparathyroidism, facial dysmorphisms, and immunological abnormalities, were observed in this cohort of patients.

The typical cardiac defects, such as tetralogy of Fallot, right-sided aortic arch, and truncus arteriosus, were identified in a few patients of our cohort, whereas minor cardiac anomalies were observed with higher frequency and uncommon defects, such as aortic coarctation and hypoplastic left heart syndrome, were also described. Peculiar facial dysmorphic features, such as hypertelorism, narrow palpebral fissures, epicanthal folds, and micrognathia, were observed in all the patients of the study; furthermore, some uncommon traits, such as wide ear pad, low set ears, flat nasal bridge, and enlarged nasal, were also found. Hypoparathyroidism was not associated with cardiac defects, contrary to what is reported in 22q11.2DS. Concerning the neurodevelopmental disorders, they did not differ from 22q11.2DS, with speech and psychomotor delays being the most frequent manifestations identified during the follow-up. The frequency of the other phenotypic features was similar to that of typical 22q11.2DS [2,4,10,30].

The immunological profile did not substantially differ from typical 22q11.2DS, showing various degrees of mild or moderate immunodeficiency, mainly related to the cell-mediated compartment. We reported a reduced thymic output entailing low RTEs and reduced numbers of both $CD4^+$ and $CD8^+$ naïve T lymphocytes. The decline in naïve and the increase in memory T-cell populations observed in our cohort could have been due not only to the impaired thymic output but also to the accelerated conversion of naïve to memory phenotype, secondary to multiple mechanisms, such as infectious exposures or homeostatic expansion, as we previously reported [31]. No severe phenotype resembling a leaky SCID was observed in our cohort. Thymic hypoplasia and, more recently, the immature status of the thymus (mainly referred to as an impairment of the epithelium function) described in the majority of cases of 22q11.2DS [5,32] may have underlain the immunological abnormalities observed in our population. We previously described a defective Tregs number in 22q11.2DS [23] that, when associated with an impaired expression of AIRE-dependent tissue-restricted antigens, leads to an impaired peripheral tolerance and consequent escape of autoreactive T cells [32]. Conversely, the Tregs number was normal in our patients, as well as the number of switched memory B cells, which were reported (together with a low level of naïve T helper cells) as strong predictors for the development of autoimmune disorders in DGs patients, particularly in their adult life [33]. Interestingly, a low level of DCs was observed in half of our patients, as we recently reported in 22q11.2DS [23], most prevalently in the mDC compartment. We argue that this finding may contribute to the observed high susceptibility toward developing infectious diseases and autoimmune manifestations of these patients. Concerning vitamin D status, the limited population analyzed did not allow for investigating its relationship with the immunological parameters, as we previously reported in 22q11.2DS. Furthermore, due to the recognized immunomodulatory role of vitamin D, it appears reasonable to suggest its supplementation also in these patients, as infections and susceptibility to autoimmune diseases represent their major concern, similarly to 22q11.2DS.

Genetic assessment in our cohort revealed six patients harboring CNVs that are never reported in normal subjects (Table 6). The genes content of these CNVs was carefully analyzed for their correlation with the phenotypic features reminiscent of 22q11.2DS.

In subject P1, a 314 kb duplication in the 2p24.1 region, where the long intergenic non-protein coding RNA 1876 (LINC1876) is harbored, was identified. According to Gene-Hancer (https://genome.ucsc.edu (accessed on 25 February 2022)), LINC1876 regulates the expression of NR4A2, a gene located distally to the duplicated region. NR4A2 encodes a steroid–thyroid hormone-retinoid receptor, acting as a nuclear receptor (NR) transcription factor, and is mainly expressed in neurons of several areas of the CNS where it is specifically required for the development and function of the neurons. The dysregulation of this gene has been associated with neurodevelopmental delay and intellectual disability with or without epilepsy [34]. Altered NR4A2 expression is thought to have caused the neurological phenotype of P1, characterized by epilepsy with psychomotor delay and cognitive impairment. It was demonstrated that the NR4A family has a role in T cell development from thymic differentiation to peripheral response against infections and cancer; the overexpression of NR4A1 and NR4A3, but not NR4A2, induces thymocyte apoptosis in vivo [35]. Although NR4A2 appears not to be involved in this mechanism, its altered expression in our patient might contribute to determining his severe deficiency of thymic output.

P4 and P5 patients were two sisters who exhibited a 365 kb duplication in the 20p11.22 region, including the PAX1 gene. PAX1 is a member of the paired box (PAX) family of transcription factors that plays a critical role in human embryogenesis at the level of pharyngeal pouches, involving the development of the thymus, tonsils, parathyroid glands, thyroid, and middle ear [26,36,37]. PAX1 homozygous loss-of-function variants are causative of otofaciocervical syndrome 2 with T-cell deficiency (OTFCS2) (#615560 OMIM), which may include a severe combined immunodeficiency (SCID) due to abnormal thymic epithelium development [26]. The main aspects of this syndrome are facial anomalies, cup-shaped low-set ears, preauricular fistulas, hearing loss, branchial defects, skeletal anomalies, and mild intellectual disability [24,25]. Although the effects of PAX1 duplication are still unknown, it might alter the embryonic development of the pharyngeal region. It is possible to hypothesize that PAX1 dosage alteration can contribute specifically to the otolaryngological manifestations, such as the hearing loss observed in patient P5 and to the dysmorphic auricular appendix in patient P4, as well to the hypoparathyroidism and immunological alterations found in both sisters, which is usually observed in 22.11.2DS.

In patient P6, the coexistence of two CNVs, located on 11p15.5 and the pseudoautosomal regions Xp22.33/Yp11.32 was detected. The 11p15.5 duplication was rich in genes, five of which are related to known syndromes with an autosomal recessive inheritance (#617637, #606003, #609304, #610717, #609057 OMIM). The effects of their duplication are unknown, but, according to their expression and function, none of them seem to influence the patient's phenotype.

The duplication in the pseudoautosomal region Xp22.33/Yp11.32 included SHOX, encoding a homeodomain transcription factor involved in cell cycle and growth regulation. SHOX deletions cause well-defined pathologic phenotypes, mainly including short stature and skeletal abnormalities (#249700, #127300, #300582 OMIM). Recently, it was highlighted that microdeletions encompassing this gene are a risk factor for autism spectrum disorders and neurodevelopmental defects [27]. We outline the possible role of this variant in determining the speech delay observed in the patient. In the human embryo, SHOX is expressed both in the limbs and in the first and second pharyngeal arches, from which originate the maxilla, mandible, and several bony elements of the external and middle ear [38]. It could be speculated that conductive hearing loss may be attributed to the dysregulation of this gene.

In P7, a 17q21.13 deletion was identified, which is causative of the Koolen–De Vries syndrome (KDVS #610443). KDVS has variable expressivity and a wide clinical spectrum that can overlap with DGS. The pathologic features of the patient include neurodevelopmental delay, anxiety disorder with psychotic signs, as well as facial dysmorphisms (long face, malar flatness, hooded eyelids resulting in the appearance of narrow palpebral fissures, and bulbous nasal tip), which overlap between both syndromes [39]. The other clinical features, such as skeletal anomalies and otolaryngological manifestations, are less common in KDVS. Interestingly, a patient with KDVS was also identified in a previously reported cohort of DGS without a 22q11.2 deletion [8]. The overlapping phenotypes of these two syndromes could be attributed to an underlying common genetic pathway. The master gene of KDVS is KANSL1, a protein-coding gene that belongs to a histone acetyltransferase (HAT) complex. Even if direct interactions between KANSL1 and the protein-coding genes located in the 22q11.2-deleted region have not been demonstrated, a common miRNA regulatory network has been identified. miRNAs play a role in 22q11.2DS [40]; interestingly, it was highlighted that specific miRNAs (such as miR-106b-5p, miR-148a-3p, miR-23b-3p, miR-17-5p, miR-149-5p, and miR-130b-3p) involved in the KANSL1 regulation also regulate DGCR14, DGCR2, TXNRD2, MRPL40, and CRKL genes, which are included in the 22q11.2-deleted region [41].

Patient P8 showed two deletions: the first in 17p13.2 harboring the WSCD1 gene and the second one in Xq24 encompassing the CXorf56 and UBE2A genes. The clinical effects of WSCD1 haploinsufficiency are unknown, whereas the deletion of CXorf56 and UBE2A was related to "Intellectual developmental disorder, X-linked 107" (#301013) and to "UBE2A deficiency syndrome" or "X-linked Nascimento-type intellectual disability syndrome" (#300860), respectively. Developmental delay, motor delay, impaired speech, and mild axial hypotonia, all present in P8, could be related to this Xq24 deletion. The phenotypic spectrum of UBE2A deficiency syndrome was recently expanded, and cardiac defects, craniofacial dysmorphisms, urogenital malformations, and hypogammaglobulinemia emerged as frequent features [42,43]. Therefore, the clinical spectrum of our patient, including cardiologic and urogenital defects and facial dysmorphic features, together with hypogammaglobulinemia, could be attributed to this deletion.

5. Conclusions

We outline that array-CGH should be used as a first-tier tool in the diagnostic work-up of patients presenting with a phenotype resembling the 22q11.2DS. This technique allows for identifying CNVs, whose altered gene content should be carefully examined to understand the mechanisms leading to 22q11.2DS phenocopies. Further analysis, such as whole exome sequencing and methylome analyses, could be considered in DGS patients in the case of normal array-CGH.

Supplementary Materials: The following supporting information can be downloaded at https://www.mdpi.com/article/10.3390/jcm11072025/s1, Table S1: Gene content in the CNVs; Figure S1: Auxological findings of height expressed in standard deviations.

Author Contributions: Conceptualization, R.C.; Methodology, A.L., V.B. and A.V.; Software, V.B. and A.V.; Validation, A.L., V.B. and A.V.; Formal Analysis, A.M.Q.A., A.L., V.B. and A.V.; Investigation, A.M.Q.A., A.L., V.B., G.C., G.I.B., A.V. and R.C.; Data Curation, A.M.Q.A., A.L., V.B., G.C., G.I.B., A.V. and R.C.; Writing—Original Draft Preparation, A.M.Q.A., A.L. and V.B.; Writing—Review and Editing, A.L., V.B., A.V. and R.C.; Visualization, R.C. and A.V.; Supervision, R.C. and A.V. All authors have read and agreed to the published version of the manuscript.

Funding: This research received no external funding.

Institutional Review Board Statement: Ethical review and approval were waived by our institutional review board because of the retrospective chart review study design.

Informed Consent Statement: Informed consent was obtained from all subjects involved in the study. The study was conducted in accordance with the Declaration of Helsinki.

Data Availability Statement: The data presented in this study are available on reasonable request from the corresponding author.

Conflicts of Interest: The authors declare no conflict of interest.

References

1. Morrow, B.E.; McDonald-McGinn, D.M.; Emanuel, B.S.; Vermeesch, J.R.; Scambler, P.J. Molecular genetics of 22q11.2 deletion syndrome. *Am. J. Med. Genet. A* **2018**, *176*, 2070–2081. [CrossRef] [PubMed]
2. McDonald-McGinn, D.M.; Sullivan, K.E.; Marino, B.; Philip, N.; Swillen, A.; Vorstman, J.A.; Zackai, E.H.; Emanuel, B.S.; Vermeesch, J.R.; Morrow, B.E.; et al. 22q11.2 deletion syndrome. *Nat. Rev. Dis. Primers* **2015**, *1*, 15071. [CrossRef] [PubMed]
3. Burnside, R.D. 22q11.21 Deletion Syndromes: A Review of Proximal, Central, and Distal Deletions and Their Associated Features. *Cytogenet. Genome Res.* **2015**, *146*, 89–99. [CrossRef] [PubMed]
4. McDonald-McGinn, D.M.; Sullivan, K.E. Chromosome 22q11.2 deletion syndrome (DiGeorge syndrome/velocardiofacial syndrome). *Medicine (Baltimore)* **2011**, *90*, 1–18. [CrossRef] [PubMed]
5. Cancrini, C.; Puliafito, P.; Digilio, M.C.; Soresina, A.; Martino, S.; Rondelli, R.; Consolini, R.; Ruga, E.M.; Cardinale, F.; Finocchi, A.; et al. Clinical features and follow-up in patients with 22q11.2 deletion syndrome. *J. Pediatr.* **2014**, *164*, 1475–1480.e2. [CrossRef]
6. Bassett, A.S.; McDonald-McGinn, D.M.; Devriendt, K.; Digilio, M.C.; Goldenberg, P.; Habel, A.; Marino, B.; Oskarsdottir, S.; Philip, N.; Sullivan, K.; et al. Practical guidelines for managing patients with 22q11.2 deletion syndrome. *J. Pediatr.* **2011**, *159*, 332–339.e1. [CrossRef]
7. Bernstock, J.D.; Totten, A.H.; Elkahloun, A.G.; Johnson, K.R.; Hurst, A.C.; Goldman, F.; Groves, A.K.; Mikhail, F.M.; Atkinson, T.P. Recurrent microdeletions at chromosome 2p11.2 are associated with thymic hypoplasia and features resembling DiGeorge syndrome. *J. Allergy Clin. Immunol.* **2020**, *145*, 358–367.e2. [CrossRef]
8. Koczkowska, M.; Wierzba, J.; Śmigiel, R.; Sąsiadek, M.; Cabała, M.; Ślężak, R.; Iliszko, M.; Kardaś, I.; Limon, J.; Lipska-Ziętkiewicz, B.S. Genomic findings in patients with clinical suspicion of 22q11.2 deletion syndrome. *J. Appl. Genet.* **2017**, *58*, 93–98. [CrossRef]
9. Daw, S.C.; Taylor, C.; Kraman, M.; Call, K.; Mao, J.; Schuffenhauer, S.; Meitinger, T.; Lipson, T.; Goodship, J.; Scambler, P. A common region of 10p deleted in DiGeorge and velocardiofacial syndromes. *Nat. Genet.* **1996**, *13*, 458–460. [CrossRef]
10. Cirillo, E.; Prencipe, M.R.; Giardino, G.; Romano, R.; Scalia, G.; Genesio, R.; Nitsch, L.; Pignata, C. Clinical Phenotype, Immunological Abnormalities, and Genomic Findings in Patients with DiGeorge Spectrum Phenotype without 22q11.2 Deletion. *J. Allergy Clin. Immunol. Pract.* **2020**, *8*, 3112–3120. [CrossRef]
11. Tobias, E.S.; Morrison, N.; Whiteford, M.L.; Tolmie, J.L. Towards earlier diagnosis of 22q11 deletions. *Arch. Dis. Child.* **1999**, *81*, 513–514. [CrossRef] [PubMed]
12. Chiappini, E.; Santamaria, F.; Marseglia, G.L.; Marchisio, P.; Galli, L.; Cutrera, R.; de Martino, M.; Antonini, S.; Becherucci, P.; Biasci, P.; et al. Prevention of recurrent respiratory infections: Inter-society Consensus. *Ital. J. Pediatrics* **2021**, *47*, 211. [CrossRef] [PubMed]
13. Ross, A.C.; Manson, J.E.; Abrams, S.A.; Aloia, J.F.; Brannon, P.M.; Clinton, S.K.; Durazo-Arvizu, R.A.; Gallagher, J.C.; Gallo, R.L.; Jones, G.; et al. The 2011 report on dietary reference intakes for calcium and vitamin D from the Institute of Medicine: What clinicians need to know. *J. Clin. Endocrinol. Metab.* **2011**, *96*, 53–58. [CrossRef] [PubMed]
14. Wagner, C.L.; Greer, F.R. Prevention of rickets and vitamin D deficiency in infants, children, and adolescents. *Pediatrics* **2008**, *122*, 1142–1152. [CrossRef]
15. Braegger, C.; Campoy, C.; Colomb, V.; Decsi, T.; Domellof, M.; Fewtrell, M.; Hojsak, I.; Mihatsch, W.; Molgaard, C.; Shamir, R.; et al. Vitamin D in the healthy European paediatric population. *J. Pediatr. Gastroenterol. Nutr.* **2013**, *56*, 692–701. [CrossRef]
16. Cacciari, E.; Milani, S.; Balsamo, A.; Spada, E.; Bona, G.; Cavallo, L.; Cerutti, F.; Gargantini, L.; Greggio, N.; Tonini, G.; et al. Italian cross-sectional growth charts for height, weight and BMI (2 to 20 yr). *J. Endocrinol. Invest.* **2006**, *29*, 581–593. [CrossRef]
17. Schatorjé, E.J.; Gemen, E.F.; Driessen, G.J.; Leuvenink, J.; van Hout, R.W.; de Vries, E. Paediatric reference values for the peripheral T cell compartment. *Scand. J. Immunol.* **2012**, *75*, 436–444. [CrossRef]
18. Schatorjé, E.J.; Gemen, E.F.; Driessen, G.J.; Leuvenink, J.; van Hout, R.W.; van der Burg, M.; de Vries, E. Age-matched reference values for B-lymphocyte subpopulations and CVID classifications in children. *Scand. J. Immunol.* **2011**, *74*, 502–510. [CrossRef]
19. Shearer, W.T.; Rosenblatt, H.M.; Gelman, R.S.; Oyomopito, R.; Plaeger, S.; Stiehm, E.R.; Wara, D.W.; Douglas, S.D.; Luzuriaga, K.; McFarland, E.J.; et al. Lymphocyte subsets in healthy children from birth through 18 years of age: The Pediatric AIDS Clinical Trials Group P1009 study. *J. Allergy Clin. Immunol.* **2003**, *112*, 973–980. [CrossRef]
20. Tosato, F.; Bucciol, G.; Pantano, G.; Putti, M.C.; Sanzari, M.C.; Basso, G.; Plebani, M. Lymphocytes subsets reference values in childhood. *Cytom. A* **2015**, *87*, 81–85. [CrossRef]
21. Kohler, S.; Thiel, A. Life after the thymus: CD31+ and CD31− human naive CD4+ T-cell subsets. *Blood* **2009**, *113*, 769–774. [CrossRef] [PubMed]

22. Orsini, G.; Legitimo, A.; Failli, A.; Massei, F.; Biver, P.; Consolini, R. Enumeration of human peripheral blood dendritic cells throughout the life. *Int. Immunol.* **2012**, *24*, 347–356. [CrossRef] [PubMed]
23. Legitimo, A.; Bertini, V.; Costagliola, G.; Baroncelli, G.I.; Morganti, R.; Valetto, A.; Consolini, R. Vitamin D status and the immune assessment in 22q11.2 deletion syndrome. *Clin. Exp. Immunol.* **2020**, *200*, 272–286. [CrossRef] [PubMed]
24. Pohl, E.; Aykut, A.; Beleggia, F.; Karaca, E.; Durmaz, B.; Keupp, K.; Arslan, E.; Palamar, M.; Yigit, G.; Ozkinay, F.; et al. A hypofunctional PAX1 mutation causes autosomal recessively inherited otofaciocervical syndrome. *Hum. Genet.* **2013**, *132*, 1311–1320. [CrossRef]
25. Paganini, I.; Sestini, R.; Capone, G.L.; Putignano, A.L.; Contini, E.; Giotti, I.; Gensini, F.; Marozza, A.; Barilaro, A.; Porfirio, B.; et al. A novel PAX1 null homozygous mutation in autosomal recessive otofaciocervical syndrome associated with severe combined immunodeficiency. *Clin. Genet.* **2017**, *92*, 664–668. [CrossRef]
26. Yamazaki, Y.; Urrutia, R.; Franco, L.M.; Giliani, S.; Zhang, K.; Alazami, A.M.; Dobbs, A.K.; Masneri, S.; Joshi, A.; Otaizo-Carrasquero, F.; et al. PAX1 is essential for development and function of the human thymus. *Sci. Immunol.* **2020**, *5*, eaax1036. [CrossRef]
27. Tropeano, M.; Howley, D.; Gazzellone, M.J.; Wilson, C.E.; Ahn, J.W.; Stavropoulos, D.J.; Murphy, C.M.; Eis, P.S.; Hatchwell, E.; Dobson, R.J.; et al. Microduplications at the pseudoautosomal SHOX locus in autism spectrum disorders and related neurodevelopmental conditions. *J. Med. Genet.* **2016**, *53*, 536–547. [CrossRef]
28. Koolen, D.A.; Kramer, J.M.; Neveling, K.; Nillesen, W.M.; Moore-Barton, H.; Elmslie, F.V.; Toutain, A.; Amiel, J.; Malan, V.; Tsai, A.C.-H.; et al. Mutations in the chromatin modifier gene KANSL1 cause the 17q21.31 microdeletion syndrome. *Nat. Genet.* **2012**, *44*, 639–641. [CrossRef]
29. Smith, E.R.; Cayrou, C.; Huang, R.; Lane, W.S.; Cote, J.; Lucchesi, J.C. A human protein complex homologous to the Drosophila MSL complex is responsible for the majority of histone H4 acetylation at lysine 16. *Mol. Cell. Biol.* **2005**, *25*, 9175–9188. [CrossRef]
30. Sullivan, K.E. Chromosome 22q11.2 deletion syndrome and DiGeorge syndrome. *Immunol. Rev.* **2019**, *287*, 186–201. [CrossRef]
31. Piliero, L.M.; Sanford, A.N.; McDonald-McGinn, D.M.; Zackai, E.H.; Sullivan, K.E. T-cell homeostasis in humans with thymic hypoplasia due to chromosome 22q11.2 deletion syndrome. *Blood* **2004**, *103*, 1020–1025. [CrossRef] [PubMed]
32. Marcovecchio, G.E.; Bortolomai, I.; Ferrua, F.; Fontana, E.; Imberti, L.; Conforti, E.; Amodio, D.; Bergante, S.; Macchiarulo, G.; D'Oria, V.; et al. Thymic Epithelium Abnormalities in DiGeorge and Down Syndrome Patients Contribute to Dysregulation in T Cell Development. *Front. Immunol.* **2019**, *10*, 447. [CrossRef] [PubMed]
33. Montin, D.; Marolda, A.; Licciardi, F.; Robasto, F.; Di Cesare, S.; Ricotti, E.; Ferro, F.; Scaioli, G.; Giancotta, C.; Amodio, D.; et al. Immunophenotype Anomalies Predict the Development of Autoimmune Cytopenia in 22q11.2 Deletion Syndrome. *J. Allergy Clin. Immunol. Pract.* **2019**, *7*, 2369–2376. [CrossRef] [PubMed]
34. Singh, S.; Gupta, A.; Zech, M.; Sigafoos, A.N.; Clark, K.J.; Dincer, Y.; Wagner, M.; Humberson, J.B.; Green, S.; van Gassen, K.; et al. De novo variants of NR4A2 are associated with neurodevelopmental disorder and epilepsy. *Genet. Med.* **2020**, *22*, 1413–1417. [CrossRef]
35. Odagiu, L.; May, J.; Boulet, S.; Baldwin, T.A.; Labrecque, N. Role of the Orphan Nuclear Receptor NR4A Family in T-Cell Biology. *Front. Endocrinol. (Lausanne)* **2021**, *11*, 624122. [CrossRef]
36. Farley, A.M.; Morris, L.X.; Vroegindeweij, E.; Depreter, M.L.; Vaidya, H.; Stenhouse, F.H.; Tomlinson, S.R.; Anderson, R.A.; Cupedo, T.; Cornelissen, J.J.; et al. Dynamics of thymus organogenesis and colonization in early human development. *Development* **2013**, *140*, 2015–2026. [CrossRef]
37. Bhalla, P.; Wysocki, C.A.; van Oers, N.S.C. Molecular Insights into the Causes of Human Thymic Hypoplasia with Animal Models. *Front. Immunol.* **2020**, *11*, 830. [CrossRef]
38. Clement-Jones, M.; Schiller, S.; Rao, E.; Blaschke, R.J.; Zuniga, A.; Zeller, R.; Robson, S.C.; Binder, G.; Glass, I.; Strachan, T.; et al. The short stature homeobox gene SHOX is involved in skeletal abnormalities in Turner syndrome. *Hum. Mol. Genet.* **2000**, *9*, 695–702. [CrossRef]
39. Dingemans, A.J.M.; Stremmelaar, D.E.; van der Donk, R.; Vissers, L.E.L.M.; Koolen, D.A.; Rump, P.; Hehir-Kwa, J.Y.; de Vries, B.B.A. Quantitative facial phenotyping for Koolen-de Vries and 22q11.2 deletion syndrome. *Eur. J. Hum. Genet.* **2021**, *29*, 1418–1423. [CrossRef]
40. Bertini, V.; Azzarà, A.; Legitimo, A.; Milone, R.; Battini, R.; Consolini, R.; Valetto, A. Deletion Extents Are Not the Cause of Clinical Variability in 22q11.2 Deletion Syndrome: Does the Interaction between DGCR8 and miRNA-CNVs Play a Major Role? *Front. Genet.* **2017**, *8*, 47. [CrossRef]
41. León, L.E.; Benavides, F.; Espinoza, K.; Vial, C.; Alvarez, P.; Palomares, M.; Lay-Son, G.; Miranda, M.; Repetto, G.M. Partial microduplication in the histone acetyltransferase complex member KANSL1 is associated with congenital heart defects in 22q11.2 microdeletion syndrome patients. *Sci. Rep.* **2017**, *7*, 1795. [CrossRef] [PubMed]
42. Cordeddu, V.; Macke, E.L.; Radio, F.C.; Lo Cicero, S.; Pantaleoni, F.; Tatti, M.; Bellacchio, E.; Ciolfi, A.; Agolini, E.; Bruselles, A.; et al. Refinement of the clinical and mutational spectrum of UBE2A deficiency syndrome. *Clin. Genet.* **2020**, *98*, 172–178. [CrossRef] [PubMed]
43. Wolańska, E.; Pollak, A.; Rydzanicz, M.; Pesz, K.; Kłaniewska, M.; Rozensztrauch, A.; Skiba, P.; Stawiński, P.; Płoski, R.; Śmigiel, R. The Role of the Reanalysis of Genetic Test Results in the Diagnosis of Dysmorphic Syndrome Caused by Inherited xq24 Deletion including the UBE2A and CXorf56 Genes. *Genes* **2021**, *12*, 350. [CrossRef] [PubMed]

Journal of
Clinical Medicine

Epigenetic Alterations in Inborn Errors of Immunity

Roberta Romano [1,†], Francesca Cillo [1,†], Cristina Moracas [1], Laura Pignata [2], Chiara Nannola [1], Elisabetta Toriello [1], Antonio De Rosa [1], Emilia Cirillo [1], Emma Coppola [1], Giuliana Giardino [1], Nicola Brunetti-Pierri [1], Andrea Riccio [2,*] and Claudio Pignata [1,*]

[1]　Department of Translational Medical Sciences, Università degli Studi di Napoli "Federico II", 80125 Naples, Italy; roberta.romanomd@gmail.com (R.R.); francescacillo97@gmail.com (F.C.); cristinamoracas@gmail.com (C.M.); nannola.chiara@libero.it (C.N.); betty.toriello@gmail.com (E.T.); antonioderosa06@libero.it (A.D.R.); emiliacirillo83@gmail.com (E.C.); emmacopp24@gmail.com (E.C.); giuliana.giardino@unina.it (G.G.); brunetti@tigem.it (N.B.-P.)

[2]　Department of Environmental Biological and Pharmaceutical Sciences and Technologies, Università degli Studi della Campania "Luigi Vanvitelli", 81100 Caserta, Italy; laura.pignata@unicampania.it

*　Correspondence: andrea.riccio@unicampania.it (A.R.); pignata@unina.it (C.P.)

†　These authors contributed equally to this work.

Abstract: The epigenome bridges environmental factors and the genome, fine-tuning the process of gene transcription. Physiological programs, including the development, maturation and maintenance of cellular identity and function, are modulated by intricate epigenetic changes that encompass DNA methylation, chromatin remodeling, histone modifications and RNA processing. The collection of genome-wide DNA methylation data has recently shed new light into the potential contribution of epigenetics in pathophysiology, particularly in the field of immune system and host defense. The study of patients carrying mutations in genes encoding for molecules involved in the epigenetic machinery has allowed the identification and better characterization of environment-genome interactions via epigenetics as well as paving the way for the development of new potential therapeutic options. In this review, we summarize current knowledge of the role of epigenetic modifications in the immune system and outline their potential involvement in the pathogenesis of inborn errors of immunity.

Keywords: epigenetics; DNA methylation; inborn errors of immunity

Citation: Romano, R.; Cillo, F.; Moracas, C.; Pignata, L.; Nannola, C.; Toriello, E.; De Rosa, A.; Cirillo, E.; Coppola, E.; Giardino, G.; et al. Epigenetic Alterations in Inborn Errors of Immunity. *J. Clin. Med.* **2022**, *11*, 1261. https://doi.org/10.3390/jcm11051261

Academic Editors: Rita Consolini and Giorgio Costagliola

Received: 20 January 2022
Accepted: 24 February 2022
Published: 25 February 2022

Publisher's Note: MDPI stays neutral with regard to jurisdictional claims in published maps and institutional affiliations.

1. Introduction

Epigenetics is now emerging as an important tool of fine-tuning of gene transcription, and is thus directly implicated in cell maturation and functionality. The collection of genome-wide DNA methylation data has also recently shed new light onto the potential contribution of epigenetics in pathophysiology and, in particular, in the field of immune response. Here, we focus on the association between the pathogenesis of inborn errors of immunity and alterations of epigenetic modifications. In particular, we will summarize well-known disorders or newly identified syndromes in which disturbances of epigenetic machinery may help explain undefined cases and that may, in turn, clarify the contribution of the epigenome to immune system development.

2. Physiologic Roles of Epigenetics

During development, stable and heritable mechanisms, such as histone modifications and DNA methylation, are employed for the functional regulation of gene expression [1,2]. These processes are referred to as "epigenetics", indicating changes occurring without a direct alteration of the DNA sequence. Under this term are included all the changes exerted via the regulation of chromatin functions and states of activation that are critical for the control of DNA accessibility and transcription. Indeed, the three-dimensional folding of the nuclear genome is tightly linked to the functional DNA-dependent processes of replication and transcription. In particular, DNA replication is a complex and dynamic phenomenon based on the interplay among the epigenetic signature, the transcriptional activity and the structure of chromatin into which DNA is folded and condensed into the nucleus [3]. The double-stranded DNA is wrapped around a core of 8 histone proteins, including two copies of each histone, H2A, H2B, H3, and H4, forming superordinate biomolecular structures, namely nucleosomes, that build up chromatin fibers [4] (Figure 1A). Because of the flexibility of chromatin fibers, target DNA sequences can contact their regulatory elements, even though they are distantly located. Chromosomes segregate into two mutually exclusive types of chromatin, "A" and "B" compartments, including gene-rich active and repressive chromatin, respectively [5,6]. The A compartment is located centrally, whereas the B compartment is typically located peripherally in the nucleus [6]. Besides chromatin structure remodeling, DNA methylation, histone post-translational modifications and non-coding RNAs (ncRNAs) are the key epigenetic factors involved in the dynamics of transcriptional control.

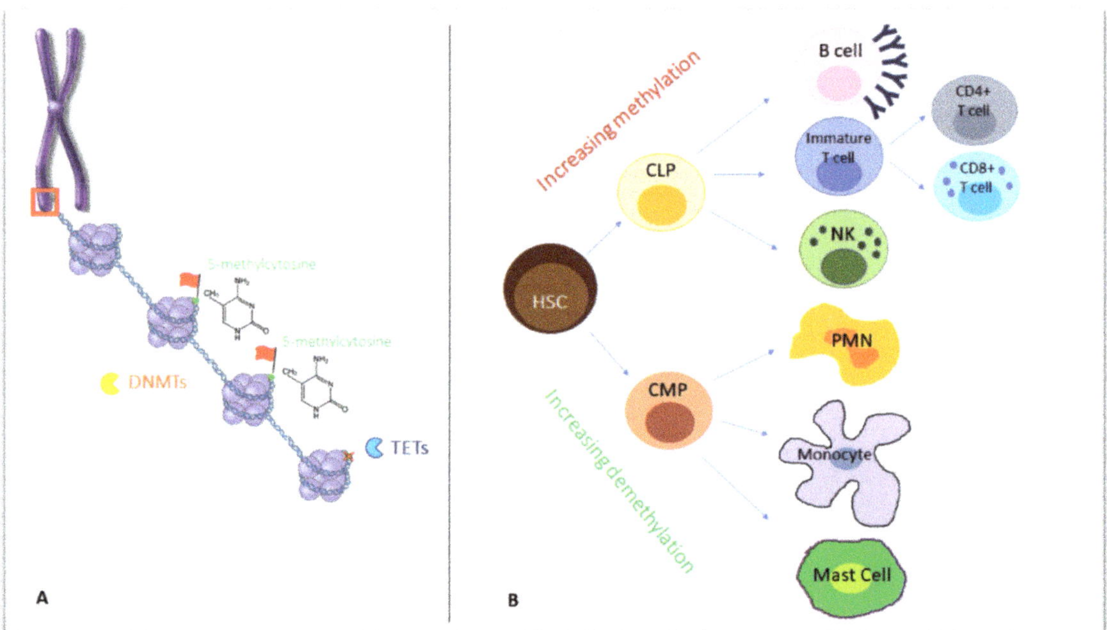

Figure 1. (**A**) Chromosomes formed by chromatin fibers organized into nucleosomes, in which they are wrapped around eight histone proteins (as shown in the red box, zooming in chromosome structure). On a deeper level, DNA methyltransferases (DNMTs) act on double-stranded DNA, adding a methyl group to the carbon 5 (5meC) of cytosine-followed-by-guanine dinucleotides, while Ten-Eleven Translocation (TET) enzymes are responsible for demethylation, removing a 5meC (red X in the figure). (**B**) The differentiation from a common progenitor, a hematopoietic stem cell, to lymphoid and myeloid lineage, is accompanied by a wave of increasing methylation or demethylation, respectively. DNMTs: DNA methyltransferases, TET: Ten-Eleven Translocation; HSC: hematopoietic stem cell, CLP: common lymphoid progenitor, CMP: common myeloid progenitor, PMN: polymorphonucleate.

For DNA methylation, a methyl group is added to the carbon 5 (5meC) of cytosine-followed-by-guanine dinucleotides (CG or CpG sites) by DNA methyltransferases (DN-MTs), a family of four enzymes. Unlike in mammals, bacterial methylation may also occur at N4 cytosine (4meC) and N6 adenine (6meA), the latter being the most prevalent in bacteria [7].

In mammals, DNMT3A, DNMT3B and DNMT3L catalyze de novo DNA methylation with differential kinetics and patterns during male and female gametogenesis and within cell lineage specification in post-implantation development [8,9], while DNMT1 secures the maintenance of DNA methylation following replication through cell division [10].

In the primordial germ cells and pre-implantation embryo, two waves of extensive erasure involving both passive and active mechanisms occur. Consecutive cell divisions may be followed by passive DNA demethylation, while enzyme members of the Ten-Eleven Translocation (TET) family mediate active demethylation through the oxidation of 5-methylcytosine (5mC) to 5-formylcytosine (5fC), 5-hydroxymethylcytosine (5hmC) and 5-carboxylcytosine (5caC), followed by the replication-dependent dilution of oxidized 5mC or the thymine DNA glycosylase (TDG)-mediated excision of 5fC and 5caC coupled with base excision repair [10].

In general, DNA methylation is high across gene bodies and inter-genic regions, and low at regulatory regions, such as promoters and enhancers. Once established on regulatory regions, methylation can be repressive for transcription because it either directly inhibits the binding of transcription factors or indirectly inhibits the activity of methyl-binding proteins and chromatin modifiers. The methylation of gene bodies is not a repressive mark, but it prevents spurious transcription initiation [11].

Although it is prevalent in mammals, it is worth mentioning that CpG methylation is not the only model of methylation: evidence of non-CpG methylation sites has been found in human embryonic stem cells, induced pluripotent stem cells and brain tissues [12,13]. Hence, novel research to ascertain its role in the maintenance of pluripotency, as well as in the pathophysiology of cancer and neurodegeneration, has developed significantly in recent years [14–16].

Histone marks, such as histone H3 modifications, correlate with gene expression. For example, histone 3 lysine 4 trimethylation (H3K4me3) and/or histone 3 lysine 27 acetylation (H3K27ac) are active marks, found at active promoters and/or enhancers. They correlate negatively with DNA methylation and positively with gene expression [17].

Proper DNA methylation is required for normal human development [18]. Methylation abnormalities may be associated with genetic defects involving *cis*-acting elements or *trans*-acting factors, but can also occur in the absence of obvious genetic changes as primary epimutations; these may represent stochastic or environment-driven errors in the establishment or maintenance of an epigenetic program [19]. Single-locus methylation defects can be a consequence of a variant occurring in *cis*, while, when multiple loci are involved, this may be due to a variant occurring in *trans*. In the latter case, variants in DNMTs or mutations in chromatin modifiers or transcription factors may alter genomic methylation [18].

In cancer, abnormal DNA methylation patterns have been frequently demonstrated, such as the hypermethylation of tumor suppressor gene promoters or the methylation changes of imprinted loci [20].

Recently, through the use of high-throughput screening platforms, an increasing number of disorders have been associated with specific "episignatures", indicating that DNA methylation analysis may represent a powerful tool for the more accurate classification of diseases with overlapping clinical signs and for categorizing cases with unclear genetic variants [21].

The so-called non-coding RNAs include microRNAs (mRNAs) and long-non-coding RNAs (lncRNAs), both of which are involved in the regulation of gene expression. The former are short molecules that bind to complementary sequences in the $3'$ UTR region of the mRNA, directly inhibiting its translation or inducing its degradation. The latter are longer than 200 nucleotides and act by binding to histone modifiers or transcription regulation proteins [22]

Since they play a crucial role in regulating processes such as proliferation, differentiation, development, and apoptosis, it is not surprising that the disruption of their function is also relevant for human diseases, as uncovered by evidence gained in the field of tumorigenesis [23].

3. Epigenetics in the Immune System

A growing body of evidence suggests that epigenetic mechanisms, including DNA methylation, play a key role in hematopoiesis, contributing to the differentiation of the hematopoietic stem cell (HSC) into different subsets of immune cells, namely towards the lymphoid and myeloid lineages. Indeed, each cell subset exhibits a unique methylation profile, with remarkable differences between the cells of the myeloid and lymphoid lineages [24].

DNA methylation is increased with lymphoid differentiation but reduced in myeloid differentiation [25] (Figure 1B). Interestingly, in humans, the inactivation of DNA demethylating enzymes TET has been associated with several myeloid malignancies, as myeloproliferation replaces cell differentiation [26].

During each step of their development, B cells undergo methylation changes in up to one third of all their genome CpGs. In the early phases of differentiation in the bone marrow, these changes are considered lineage-determining. Non-CpGs demethylation occurs upon B-cell commitment in pre-B2 cells, while CpG methylation changes in effector genes are detected in all other stages of B-cell maturation and activation after B-cell receptor stimulation by antigen binding in the spleen [27].

DNA methylation and histone acetylation are also involved in V(D)J recombination, a process that causes changes in chromatin structure and allows recombination steps through the activity of RAG1/2 enzyme, which recognizes specific signal sequences [28].

As for the T-cell compartment, when the lineage choice of T cells occurs, DNA methylation of the *Cd4* locus is required for its repression in CD8+ cells and its expression in CD4+ cells, as demonstrated in mouse models [29]. In the thymus, DNMT1 interaction with FOXP3 (Forkhead Box P3) transcription factor induces Tregs development. Tregs are a heterogeneous population of CD4-positive T cells characterized by a high expression of CD25 and a low expression of CD127 [30]. After T cell activation, active DNA demethylation is essential for interleukin-2 (IL2) synthesis and for lineage polarization into T helper-1 (Th1), Th2, and Th17 [31,32].

DNA methylation plays a critical role in CD4+ T-cell differentiation: DNMT1 loss leads to decreased peripheral T-cell proliferation and the increased expression of cytokines such as IL-2, IL-3, IL-4 and IFNγ, in activated CD4+ (and CD8+) T cells, suggesting a repressive function of DNMT1 towards cytokine production. Under TH2 polarizing conditions, DNMT1 dissociates from the IL4 locus, enabling the demethylation of the locus and the increased expression of IL-4 [33].

The shift to a memory-like phenotype induced in NK cells by some viral infections may also rely on changes in the methylome profiling of promoters of cytokines, including IL13, IL5, and IFN, which become demethylated, as observed in T-cell activation [33,34].

The role of DNA methylation machinery has also been described in the mononuclear-phagocyte system during monocyte differentiation into macrophages and their polarization to a "M1" state or an anti-inflammatory "M2" phenotype, as well as in keeping the neutrophil phenotype fully differentiated [35].

Extensive mRNA expression profiling has widely demonstrated how hematopoiesis and cell lineage commitment are also accompanied and orchestrated by changes in mRNA signatures [36]. For instance, relevant steps in both T- and B-cell lymphopoiesis rely on gene regulation by specific sets of miRNA [37]. Notably, hematopoiesis also undergoes regulation by lncRNAs that stimulate the proliferation and differentiation of erythroid progenitors by targeting GATA1, TAL1 and KLF1, as well as granulocyte differentiation, thanks to HOTAIRM1, that acts as a regulator of cell cycle [38–40].

4. Epigenetic Alterations in Inborn Errors of Immunity

Since the proper establishment of DNA methylation patterns is necessary for the differentiation of cells of the immune system, the impairment of DNA methylation machinery results in immune dysfunction and diseases. Historically known as primary immunodeficiencies, Mendelian disorders of the immune system are now referred to as Inborn Errors of Immunity (IEI), a more precise and wider definition that takes into account the traditionally known feature of increased susceptibility to infections along with remarkable immune dysregulation and/or hyperinflammation [41,42]. More than 400 genes have been included in the most recent classification of by the International Union of Immunological Sciences [43,44]. In the following sections, we review the potential involvement of epigenetic alterations in the pathogenesis of some inborn errors of immunity, whose features are summarized in Table 1.

Table 1. Representative gene defects causing epigenetic changes and immunological alterations within defined syndromes.

Humoral Immunity	Disorder	Altered Epigenetic Mechanism	Genes	Major Immunological Alteration
	CVID	DNA methylation	*PAX5, PIK3CD, BCL2L1, RPS6KB2, TCF3, KCNN4*	Agammaglobulinemia, impaired response to vaccines, autoimmunity, CLD, enteropathy
	ICF1	DNA methylation	*DNMT3B*	Agammaglobulinemia or hypogammaglobulinemia, recurrent infections
	ICF2	DNA methylation	*ZBTB24*	Agammaglobulinemia or hypogammaglobulinemia, recurrent infections
	ICF3	DNA methylation	*CDCA7*	Agammaglobulinemia or hypogammaglobulinemia, recurrent infections
	ICF4	DNA methylation	*HELLS*	Agammaglobulinemia or hypogammaglobulinemia, recurrent infections
	KS1	Histone modification	*KMT2D*	Hypogammaglobulinemia, autoimmune cytopenia
	KS2	Histone modification	*KDM6A*	Hypogammaglobulinemia, autoimmune cytopenia
Adaptive immunity				
	22q11.2 DS	DNA methylation Non-coding RNAs	*TBX1*	Lymphopenia, recurrent infections, autoimmunity
	Schimke immuno-osseous dysplasia	Chromatin remodeling	*SMARCAL1*	Lymphopenia, recurrent infections
Immune dys-regulation				
	TET2 loss-of-function	DNA methylation	*TET2*	Hepatosplenomegaly, lymphadenopathy, autoimmunity

4.1. Inborn Errors of Humoral Immunity

Common Variable Immunodeficiency (CVID) is a heterogeneous group of disorders characterized by hypogammaglobulinemia and impaired response to vaccinations. CVID is characterized by marked genetic and phenotypic heterogeneity and monogenic variants have been identified in no more than 10% of patients [45]. Thus, the majority of CVID patients lack a monogenic basis and a polygenic origin may be assumed in most cases. Since a genetic diagnosis of CVID can be achieved only in a small percentage of patients [46], epigenetic alterations, such as DNA methylation and histone modifications, may be theoretically envisioned as potential mechanisms implicated in genetically undefined cases, as a few studies, described below, seem to suggest.

In the early stages of B cell differentiation, during the transition from pro-B to pre-B cells, an alteration in DNA methylation occurs, especially in intragenic and intronic regions [47] closely associated with transcription factor sites related to B cell development, such as *EBF1*, *E2F*, and *PAX5* [48]. Tallmadge et al. analyzed the transcriptome sequencing of horses affected by CVID, revealing a significant down-regulation in *PAX5* expression. The suspicion of an epigenetic mechanism responsible for this down-regulation was confirmed by the analysis of the epigenomic profile, which revealed a hypermethylation of the *PAX5* enhancer in the bone marrow of CVID-affected horses [49].

However, the most important alterations in DNA methylation are observed in the transition from naïve B cells to germinal center memory and plasma cells. B-cell differentiation is associated with a gradual DNA demethylation [27], with a similar grade of DNA methylation in memory and plasma cells, although these two cell lines have different transcriptional profiles [50]. A study on CVID-discordant monozygotic twins revealed an increase in the DNA methylation of critical B lymphocyte genes, such as *PIK3CD*, *BCL2L1*, *RPS6KB2*, *TCF3* and *KCNN4* in the affected sibling, as compared to the healthy sibling. This hypermethylation, observed in both unswitched- and switched-memory B cells, led to a down-regulation of those genes and, consequently, to B cell dysfunction [51].

In another study, the DNA methylome of CVID patients was compared with that of healthy donors, underpinning the hypothesis that altered demethylation during B cell differentiation may contribute to the pathogenesis of CVID, with a reduction in memory B cells paralleling the degree of demethylation impairment [51].

Immunodeficiency with centromeric instability and facial anomalies syndrome (ICF) is a rare disease caused by biallelic mutations in DNA methyltransferases, characterized by instability of the pericentromeric heterochromatin of chromosomes 1, 9 and 16, peculiar facial anomalies and immune deficiency. The latter may have a variable degree of severity, ranging from complete agammaglobulinemia to decreased levels of single classes of immunoglobulins, lymphopenia, T-cell proliferative response [52–54] and, rarely, autoimmunity [52]. Recurrent respiratory and gastrointestinal infections are typical features.

ICFs are classified according to genetic defects in ICF1, ICF2, ICF3 and ICF4, due to mutations in the *DNMT3B*, *ZBTB24*, *CDCA7* and *HELLS* genes, respectively [55–57].

As for ICF1, most patients harbor mutations in the catalytic domain of DNMT3B and show hypomethylation of DNA at determined noncoding repetitive sequences and genes located in inactive heterochromatin, causing chromatin decondensation and chromosomal instability [58]. It has been hypothesized that dysregulated DNA methylation underlies an abnormal maturation of B cells and the generation of immunologic memory [59]. Indeed, lymphoblastoid B cell lines from patients with ICF1 show an impaired expression of the genes involved in critical processes such as lymphocyte signaling, maturation and migration. When compared with controls, almost half of these genes appear to be up-regulated. Additionally, the finding of an increased histone trimethylation at lysine-4, H3K4me3 supported this hypothesis [59].

ZBTB24 is another regulator of hematopoietic development and, being highly expressed in naïve B cells, has a paramount role in B-cell differentiation [60]. In ICF patients with *ZBTB24* mutations, a normal number of total B lymphocytes, as well as naïve and unswitched-memory B cells, has been described, associated with a decrease in switched-memory B cells [53].

Finally, the *HELLS* gene, causing ICF4, encodes a lymphoid-specific, ATP-dependent, chromatin-remodeling enzyme, which forms a complex with CDC7A protein, whose gene defect underlies ICF3. Together, they activate chromatin-remodeling activity and, presumably, as in mouse models, exert epigenetic control over B cell development [61].

Kabuki syndrome (KS), a rare, multisystemic genetic syndrome associated with an immune disorder, has an estimated prevalence of 1:30,000–1:40,000 individuals. It is characterized by typical facial features, mild-to-moderate developmental delay, cardiac, skeletal and/or renal malformations and immunological abnormalities [62]. Children with KS might share some immune system abnormalities overlapping with CVID, such as hypogammaglobulinemia, increased susceptibility to upper and lower respiratory tract infections and a higher risk of lymphoproliferation [63]. Autoimmune manifestations have also been reported, the most common being autoimmune thrombocytopenia, with or without hemolytic anemia, followed by thyroiditis, celiac disease and vitiligo [64,65]. Seventy percent of KS cases are caused by mutations in the histone methyltransferase *KMT2D* [66], whereas the remaining cases are due to mutations in the histone demethylase *KDM6A* [67]. Both genes contribute to gene expression during embryogenesis. In particular, KMT2D is a lysine H3K4 mono-methyltransferase belonging to the SET domain containing 1/Mixed-Lineage Leukemia (SET1/MLL) protein family, whereas KDM6A acts on H3K27-methylated lysine to remove a repressive mark [68]. Antibody deficiency, as well as a reduction in B cells, total-memory B cells and class-switched-memory B cells, have been detected in KS patients [69].

The immune defects described in KS patients may depend on a loss of H3K4 methylation occurring at crucial transcription factors, dysregulating T and B lymphocyte differentiation. KMT2D loss-of-function might also cause a direct alteration of the antibody maturation, reducing the efficiency of class-switch recombination, while autoimmunity may derive from B-cell tolerance breakage or defective Treg generation [70].

4.2. Inborn Errors of Adaptive Immunity

The 22q11.2 Deletion Syndrome (22q11.2 DS) is the most common chromosomal microdeletion disorder. It is characterized by a wide phenotypic spectrum and includes multi-organ defects with congenital heart disease, immunodeficiency, hypoparathyroidism, genitourinary problems, palatal abnormalities, developmental delay and psychiatric symptoms [71]. So far, no single gene has been identified to explain all the features of 22q11.2DS and epigenetic mechanisms have been proposed to explain the clinical variability [72]. The phenotype of 22q11.2 DS could be the sum of the haploinsufficiency of 22q11.2 genes, as well as histone and DNA methylation defects [73]. *TBX1* (T-box 1) is the main candidate gene to explain the disease manifestations and it is involved in chromatin accessibility and transcriptional regulation [74]. *TBX1* was found to co-localize with three H3K4 methyltransferases in ChIP–Western blot analyses of co-immunoprecipitation experiments. In mouse models, *Tbx1* haploinsufficiency is associated with a global reduction in H3K4me1 histone monomethylation levels, causing the differential expression of some protein-coding genes [75]. A genome-wide DNA methylation analysis conducted on 22q11.2DS patients by Rooney et al. [73] led to the identification of 160 differentially methylated CpG probes, retained for the epigenetic signature of the syndrome. Moreover, the DNA methylation profile described was different in patients carrying typical deletions as compared to patients with atypical distal deletions. Identifying the target genes and functional consequences of the histone and DNA methylation alteration in 22q11.2 DS will help to better understand the pathogenesis of the syndrome.

The dysregulation of miRNAs and lncRNAs due to microdeletion may also partially account for the heterogeneity of the immunological and clinical phenotypes of the syndrome.

Moreover, a reduced function of miR1857, among others, may contribute to a decreased expression of Bruton's tyrosine kinase (Btk) and marginal-zone B1 protein (Mzb1), thus explaining a subsequent reduction in memory B cells [72].

Schimke immuno-osseous dysplasia (OMIM 242900) is an autosomal recessive disorder, due to mutations in *SMARCAL1* gene encoding SWI/SNF-related, matrix-associated, actin-dependent regulator of chromatin, subfamily A like 1, a chromatin-remodeling enzyme. The function of SMARCAL1 is to regulate transcription through chromatin remodeling [76]. The clinical phenotype includes: dysmorphic features, short stature with skeletal abnormalities, such as spondyloepiphyseal dysplasia and exaggerated lumbar lordosis; and arteriopathy. Impaired kidney function and immune deficiency consisting in recurrent bacterial, viral, or fungal infections have also been reported [77]. Laboratory tests show lymphopenia, absent mitogen-induced proliferation response, reduced CD8 and CD3/CD4 T cells [78]. Loss-of-function mutations in *SMARCAL1* may lead to genome instability, since the enzyme recognizes transitions from single- to double-stranded DNA.

4.3. Inborn Errors of Innate Immunity

No studies of the methylation changes that could occur in this subgroup of disorders have been conducted. However, given that the signaling pathways affected in these diseases, such as that of TLR4, have been described in some cases to cause modifications in DNA methylation, it is conceivable to hypothesize that DNA methylation may potentially exert a mechanistic role in the pathogenesis of undefined disorders or, putatively, in modulating the natural history.

Mendelian susceptibility to Mycobacterial disease (MSMD) is a rare inherited condition characterized by selective predisposition to clinical disease caused by weakly virulent mycobacteria, such as bacillus Calmette–Guerin vaccines and non-tuberculous environmental mycobacteria [79] in otherwise healthy patients with no overt abnormalities in routine hematological and immunological functionality. MSMD patients are also at higher risk of tuberculosis, salmonellosis, candidiasis and, more rarely, to infections with other intra-macrophagic bacteria, fungi, or parasites [80]. Nine MSMD-causing genes, including seven autosomal (*IFNGR1*, *IFNGR2*, *STAT1*, *IL12B*, *IL12RB1*, *ISG15* and *IRF8*) and two X-linked (*NEMO* and *CYBB*) genes, have been described so far; all are involved in IFN-γ-dependent immunity [81]. Pacis et al. showed that the *Mycobacterium tuberculosis* infection of dendritic cells induces rapid loss of DNA methylation at distal enhancers that activate master immune transcription factors (including nuclear factor-kB and members of the Interferon Regulatory Factor family), suggesting an important role for DNA methylation in regulating innate immune responses [82,83].

4.4. Inborn Errors of Immunity with Immune Dysregulation

Bi-allelic loss-of-function variants in *TET2* in humans have been associated with immunodeficiency and autoimmune lymphoproliferative syndrome (ALPS)-like phenotypes with remarkable predisposition to lymphoma [84]. TET2 is a crucial epigenetic regulatory factor in hematopoietic cells, facilitating demethylation by oxidizing 5-methylcytosine (5mC) to 5-hydroxymethylcytosine (5hmC) and other oxidation products. Loss-of-function mutations in TET2 are responsible for DNA methylation increases in hematologic cells, thus accounting for the failure of the controlled development of B cells and the expansion of double-negative T cells [85]. In Tregs, TET is implicated in the stability of Foxp3 molecules. The haploinsufficiency of TET2 is related to hematological neoplasia. However, it should be mentioned that TET2 mutations also occur in healthy subjects with clonal hematopoiesis, implying that they are sufficient to induce cancer alone [86,87]. Extrinsic factors, namely infections through hyperinflammation, seem to be co-factors in carcinogenesis.

The activity of activation-induced cytidine deaminase (AID) is hampered in TET2-/- mice, leading to abnormal demethylation. Altogether, these changes impair the transcription of genes critical for germinal center exit, antigen presentation and the differentiation of germinal center B cells, concurring with the development of diffuse large B-cell lymphomas.

Therefore, it is conceivable to presume that TET2 has a crucial role in cell proliferation and differentiation [87].

TET proteins are also essential for specific points of B cell development, such as the transition from pro-B to pre-B and the differentiation of plasma cells [88].

5. Conclusions

Gene expression in the immune system is tightly regulated by epigenetic processes, including DNA methylation, chromatin remodeling and histone modifications, that orchestrate development, maturation and cell lineage commitment. In line with this, specific DNA methylation signatures and histone modification patterns can be detected for each cell population.

Next-generation sequencing technologies have enabled the identification of several new forms of IEI, surprisingly changing the scenario and expanding the knowledge of their molecular basis. Nonetheless, a genetic etiology still needs to be elucidated for many of them; hence, it is reasonable that alterations to the epigenetic mechanisms that control the transcription of genes involved in immune response may contribute to the pathogenesis of at least some of these disorders.

In addition, although most genetic IEIs are paradigmatic examples of monogenic disorders, a broad spectrum of severity and clinical phenotypes is widely recognized. Therefore, epigenetic signatures may be implicated in the regulation of disease expressivity and penetrance, possibly expanding the phenotype.

Author Contributions: R.R., G.G, A.R., N.B.-P. and C.P. conceived and designed the review; R.R., G.G., F.C., C.M., L.P., C.N., E.T., E.C. (Emma Coppola), A.D.R. and E.C. (Emilia Cirillo) wrote the paper; R.R. and F.C. created the figure; A.R. and C.P. supervised the Review. All authors have read and agreed to the published version of the manuscript.

Funding: This work was supported by the Ministero della Salute (RF 2016-02364303).

Institutional Review Board Statement: Not applicable.

Informed Consent Statement: Not applicable.

Data Availability Statement: Not applicable.

Conflicts of Interest: The authors declare no conflict of interest.

References

1. Ikegami, K.; Ohgane, J.; Tanaka, S.; Yagi, S.; Shiota, K. Interplay between DNA methylation, histone modification and chromatin remodeling in stem cells and during development. *Int. J. Dev. Biol.* **2009**, *53*, 203–214. [CrossRef] [PubMed]
2. Shiota, K. DNA methylation profiles of CpG islands for cellular differentiation and development in mammals. *Cytogenet. Genome Res.* **2004**, *105*, 325–334. [CrossRef] [PubMed]
3. Mishra, A.; Hawkins, R.D. Three-dimensional genome architecture and emerging technologies: Looping in disease. *Genome Med.* **2017**, *9*, 87. [CrossRef] [PubMed]
4. Luger, K.; Dechassa, M.L.; Tremethick, D.J. New insights into nucleosome and chromatin structure: An ordered state or a disordered affair? *Nat. Rev. Mol. Cell Biol.* **2012**, *13*, 436–447. [CrossRef]
5. Lieberman-Aiden, E.; van Berkum, N.L.; Williams, L.; Imakaev, M.; Ragoczy, T.; Telling, A.; Amit, I.; Lajoie, B.R.; Sabo, P.J.; Dorschner, M.O.; et al. Comprehensive mapping of long-range interactions reveals folding principles of the human genome. *Science* **2009**, *326*, 289–293. [CrossRef] [PubMed]
6. Rao, S.S.; Huntley, M.H.; Durand, N.C.; Stamenova, E.K.; Bochkov, I.D.; Robinson, J.T.; Sanborn, A.L.; Machol, I.; Omer, A.D.; Lander, E.S.; et al. A 3D map of the human genome at kilobase resolution reveals principles of chromatin looping. *Cell* **2014**, *159*, 1665–1680. [CrossRef]
7. Kim, J.K.; Samaranayake, M.; Pradhan, S. Epigenetic mechanisms in mammals. *Cell Mol. Life Sci.* **2009**, *66*, 596–612. [CrossRef]
8. Cerrato, F.; Sparago, A.; Ariani, F.; Brugnoletti, F.; Calzari, L.; Coppedè, F.; De Luca, A.; Gervasini, C.; Giardina, E.; Gurrieri, F.; et al. DNA Methylation in the Diagnosis of Monogenic Diseases. *Genes* **2020**, *11*, 355. [CrossRef]
9. Argelaguet, R.; Clark, S.J.; Mohammed, H.; Stapel, L.C.; Krueger, C.; Kapourani, C.A.; Imaz-Rosshandler, I.; Lohoff, T.; Xiang, Y.; Hanna, C.W.; et al. Multi-omics profiling of mouse gastrulation at single-cell resolution. *Nature* **2019**, *576*, 487–491. [CrossRef]
10. Jones, P.A.; Liang, G. Rethinking how DNA methylation patterns are maintained. *Nat. Rev. Genet.* **2009**, *10*, 805–811. [CrossRef]

11. Neri, F.; Rapelli, S.; Krepelova, A.; Incarnato, D.; Parlato, C.; Basile, G.; Maldotti, M.; Anselmi, F.; Oliviero, S. Intragenic DNA methylation prevents spurious transcription initiation. *Nature* **2017**, *543*, 72–77. [CrossRef] [PubMed]
12. Lister, R.; Pelizzola, M.; Dowen, R.H.; Hawkins, R.D.; Hon, G.; Tonti-Filippini, J.; Nery, J.R.; Lee, L.; Ye, Z.; Ngo, Q.M.; et al. Human DNA methylomes at base resolution show widespread epigenomic differences. *Nature* **2009**, *462*, 315–322. [CrossRef] [PubMed]
13. Xie, W.; Barr, C.L.; Kim, A.; Yue, F.; Lee, A.Y.; Eubanks, J.; Dempster, E.L.; Ren, B. Base-resolution analyses of sequence and parent-of-origin dependent DNA methylation in the mouse genome. *Cell* **2012**, *148*, 816–831. [CrossRef] [PubMed]
14. Fuso, A.; Lucarelli, M. CpG and Non-CpG Methylation in the Diet-Epigenetics-Neurodegeneration Connection. *Curr. Nutr. Rep.* **2019**, *8*, 74–82. [CrossRef] [PubMed]
15. Ramasamy, D.; Rao, A.K.D.M.; Rajkumar, T.; Mani, S. Non-CpG methylation-a key epigenetic modification in cancer. *Brief Funct. Genom.* **2021**, *20*, 304–311. [CrossRef]
16. Joe, S.; Nam, H. Prediction model construction of mouse stem cell pluripotency using CpG and non-CpG DNA methylation markers. *BMC Bioinform.* **2020**, *21*, 175. [CrossRef]
17. Smith, Z.D.; Meissner, A. DNA methylation: Roles in mammalian development. *Nat. Rev. Genet.* **2013**, *14*, 204–220. [CrossRef]
18. Velasco, G.; Francastel, C. Genetics meets DNA methylation in rare diseases. *Clin. Genet.* **2019**, *95*, 210–220. [CrossRef]
19. Monk, D.; Mackay, D.J.G.; Eggermann, T.; Maher, E.R.; Riccio, A. Genomic imprinting disorders: Lessons on how genome, epigenome and environment interact. *Nat. Rev. Genet.* **2019**, *20*, 235–248. [CrossRef]
20. Pignata, L.; Palumbo, O.; Cerrato, F.; Acurzio, B.; de Álava, E.; Roma, J.; Gallego, S.; Mora, J.; Carella, M.; Riccio, A.; et al. Both Epimutations and Chromosome Aberrations Affect Multiple Imprinted Loci in Aggressive Wilms Tumors. *Cancers* **2020**, *12*, 3411. [CrossRef]
21. Sadikovic, B.; Aref-Eshghi, E.; Levy, M.A.; Rodenhiser, D. DNA methylation signatures in mendelian developmental disorders as a diagnostic bridge between genotype and phenotype. *Epigenomics* **2019**, *11*, 563–575. [CrossRef] [PubMed]
22. Panni, S.; Lovering, R.C.; Porras, P.; Orchard, S. Non-coding RNA regulatory networks. *Biochim. Biophys. Acta Gene Regul. Mech.* **2020**, *1863*, 194417. [CrossRef] [PubMed]
23. Esteller, M. Non-coding RNAs in human disease. *Nat. Rev. Genet.* **2011**, *12*, 861–874. [CrossRef] [PubMed]
24. Zouali, M. DNA methylation signatures of autoimmune diseases in human B lymphocytes. *Clin. Immunol.* **2021**, *222*, 108622. [CrossRef]
25. Ji, H.; Ehrlich, L.I.; Seita, J.; Murakami, P.; Doi, A.; Lindau, P.; Lee, H.; Aryee, M.J.; Irizarry, R.A.; Kim, K.; et al. Comprehensive methylome map of lineage commitment from haematopoietic progenitors. *Nature* **2010**, *467*, 338–342. [CrossRef]
26. Moran-Crusio, K.; Reavie, L.; Shih, A.; Abdel-Wahab, O.; Ndiaye-Lobry, D.; Lobry, C.; Figueroa, M.E.; Vasanthakumar, A.; Patel, J.; Zhao, X.; et al. Tet2 loss leads to increased hematopoietic stem cell self-renewal and myeloid transformation. *Cancer Cell* **2011**, *20*, 11–24. [CrossRef]
27. Kulis, M.; Merkel, A.; Heath, S.; Queirós, A.C.; Schuyler, R.P.; Castellano, G.; Beekman, R.; Raineri, E.; Esteve, A.; Clot, G.; et al. Whole-genome fingerprint of the DNA methylome during human B cell differentiation. *Nat. Genet.* **2015**, *47*, 746–756. [CrossRef]
28. Selimyan, R.; Gerstein, R.M.; Ivanova, I.; Precht, P.; Subrahmanyam, R.; Perlot, T.; Alt, F.W.; Sen, R. Localized DNA demethylation at recombination intermediates during immunoglobulin heavy chain gene assembly. *PLoS Biol.* **2013**, *11*, e1001475. [CrossRef]
29. Sellars, M.; Huh, J.R.; Day, K.; Issuree, P.D.; Galan, C.; Gobeil, S.; Absher, D.; Green, M.R.; Littman, D.R. Regulation of DNA methylation dictates Cd4 expression during the development of helper and cytotoxic T cell lineages. *Nat. Immunol.* **2015**, *16*, 746–754. [CrossRef]
30. Piotrowska, M.; Gliwiński, M.; Trzonkowski, P.; Iwaszkiewicz-Grzes, D. Regulatory T Cells-Related Genes Are under DNA Methylation Influence. *Int. J. Mol. Sci.* **2021**, *22*, 7144. [CrossRef]
31. Gamper, C.J.; Agoston, A.T.; Nelson, W.G.; Powell, J.D. Identification of DNA methyltransferase 3a as a T cell receptor-induced regulator of Th1 and Th2 differentiation. *J. Immunol.* **2009**, *183*, 2267–2276. [CrossRef] [PubMed]
32. Feng, Y.; Arvey, A.; Chinen, T.; van der Veeken, J.; Gasteiger, G.; Rudensky, A.Y. Control of the inheritance of regulatory T cell identity by a cis element in the Foxp3 locus. *Cell* **2014**, *158*, 749–763. [CrossRef]
33. Correa, L.O.; Jordan, M.S.; Carty, S.A. DNA Methylation in T-Cell Development and Differentiation. *Crit. Rev. Immunol.* **2020**, *40*, 135–156. [CrossRef]
34. Luetke-Eversloh, M.; Cicek, B.B.; Siracusa, F.; Thom, J.T.; Hamann, A.; Frischbutter, S.; Baumgrass, R.; Chang, H.D.; Thiel, A.; Dong, J.; et al. NK cells gain higher IFN-γ competence during terminal differentiation. *Eur. J. Immunol.* **2014**, *44*, 2074–2084. [CrossRef] [PubMed]
35. Mills, C.D.; Kincaid, K.; Alt, J.M.; Heilman, M.J.; Hill, A.M. M-1/M-2 macrophages and the Th1/Th2 paradigm. *J. Immunol.* **2000**, *164*, 6166–6173. [CrossRef] [PubMed]
36. Petriv, O.I.; Kuchenbauer, F.; Delaney, A.D.; Lecault, V.; White, A.; Kent, D.; Marmolejo, L.; Heuser, M.; Berg, T.; Copley, M.; et al. Comprehensive microRNA expression profiling of the hematopoietic hierarchy. *Proc. Natl. Acad. Sci. USA* **2010**, *107*, 15443–15448. [CrossRef] [PubMed]
37. Johanson, T.M.; Skinner, J.P.; Kumar, A.; Zhan, Y.; Lew, A.M.; Chong, M.M. The role of microRNAs in lymphopoiesis. *Int. J. Hematol.* **2014**, *100*, 246–253. [CrossRef]
38. Tian, X.; Tian, J.; Tang, X.; Ma, J.; Wang, S. Long non-coding RNAs in the regulation of myeloid cells. *J. Hematol. Oncol.* **2016**, *9*, 99. [CrossRef]

39. Zhang, X.; Weissman, S.M.; Newburger, P.E. Long intergenic non-coding RNA HOTAIRM1 regulates cell cycle progression during myeloid maturation in NB4 human promyelocytic leukemia cells. *RNA Biol.* **2014**, *11*, 777–787. [CrossRef]
40. Dahariya, S.; Paddibhatla, I.; Kumar, S.; Raghuwanshi, S.; Pallepati, A.; Gutti, R.K. Long non-coding RNA: Classification, biogenesis and functions in blood cells. *Mol. Immunol.* **2019**, *112*, 82–92. [CrossRef]
41. Giardino, G.; Gallo, V.; Prencipe, R.; Gaudino, G.; Romano, R.; De Cataldis, M.; Lorello, P.; Palamaro, L.; Di Giacomo, C.; Capalbo, D.; et al. Unbalanced Immune System: Immunodeficiencies and Autoimmunity. *Front. Pediatr.* **2016**, *4*, 107. [CrossRef] [PubMed]
42. Delmonte, O.M.; Castagnoli, R.; Calzoni, E.; Notarangelo, L.D. Inborn Errors of Immunity with Immune Dysregulation: From Bench to Bedside. *Front. Pediatr.* **2019**, *7*, 353. [CrossRef] [PubMed]
43. Tangye, S.G.; Al-Herz, W.; Bousfiha, A.; Chatila, T.; Cunningham-Rundles, C.; Etzioni, A.; Franco, J.L.; Holland, S.M.; Klein, C.; Morio, T.; et al. Human Inborn Errors of Immunity: 2019 Update on the Classification from the International Union of Immunological Societies Expert Committee. *J. Clin. Immunol.* **2020**, *40*, 24–64. [CrossRef] [PubMed]
44. Bousfiha, A.; Jeddane, L.; Picard, C.; Al-Herz, W.; Ailal, F.; Chatila, T.; Cunningham-Rundles, C.; Etzioni, A.; Franco, J.L.; Holland, S.M.; et al. Human Inborn Errors of Immunity: 2019 Update of the IUIS Phenotypical Classification. *J. Clin. Immunol.* **2020**, *40*, 66–81. [CrossRef]
45. van Schouwenburg, P.A.; Davenport, E.E.; Kienzler, A.K.; Marwah, I.; Wright, B.; Lucas, M.; Malinauskas, T.; Martin, H.C.; Lockstone, H.E.; Cazier, J.B.; et al. Application of whole genome and RNA sequencing to investigate the genomic landscape of common variable immunodeficiency disorders. *Clin. Immunol.* **2015**, *160*, 301–314. [CrossRef] [PubMed]
46. Abolhassani, H.; Hammarström, L.; Cunningham-Rundles, C. Current genetic landscape in common variable immune deficiency. *Blood* **2020**, *135*, 656–667. [CrossRef] [PubMed]
47. Almamun, M.; Levinson, B.T.; Gater, S.T.; Schnabel, R.D.; Arthur, G.L.; Davis, J.W.; Taylor, K.H. Genome-wide DNA methylation analysis in precursor B-cells. *Epigenetics* **2014**, *9*, 1588–1595. [CrossRef]
48. Lee, S.T.; Xiao, Y.; Muench, M.O.; Xiao, J.; Fomin, M.E.; Wiencke, J.K.; Zheng, S.; Dou, X.; de Smith, A.; Chokkalingam, A.; et al. A global DNA methylation and gene expression analysis of early human B-cell development reveals a demethylation signature and transcription factor network. *Nucleic Acids Res.* **2012**, *40*, 11339–11351. [CrossRef]
49. Tallmadge, R.L.; Shen, L.; Tseng, C.T.; Miller, S.C.; Barry, J.; Felippe, M.J. Bone marrow transcriptome and epigenome profiles of equine common variable immunodeficiency patients unveil block of B lymphocyte differentiation. *Clin. Immunol.* **2015**, *160*, 261–276. [CrossRef]
50. Lai, A.Y.; Mav, D.; Shah, R.; Grimm, S.A.; Phadke, D.; Hatzi, K.; Melnick, A.; Geigerman, C.; Sobol, S.E.; Jaye, D.L.; et al. DNA methylation profiling in human B cells reveals immune regulatory elements and epigenetic plasticity at Alu elements during B-cell activation. *Genome Res.* **2013**, *23*, 2030–2041. [CrossRef]
51. Rodríguez-Cortez, V.C.; del Pino-Molina, L.; Rodríguez-Ubreva, J.; Ciudad, L.; Gómez-Cabrero, D.; Company, C.; Urquiza, J.M.; Tegnér, J.; Rodríguez-Gallego, C.; López-Granados, E.; et al. Monozygotic twins discordant for common variable immunodeficiency reveal impaired DNA demethylation during naïve-to-memory B-cell transition. *Nat. Commun.* **2015**, *6*, 7335. [CrossRef] [PubMed]
52. Sterlin, D.; Velasco, G.; Moshous, D.; Touzot, F.; Mahlaoui, N.; Fischer, A.; Suarez, F.; Francastel, C.; Picard, C. Genetic, Cellular and Clinical Features of ICF Syndrome: A French National Survey. *J. Clin. Immunol.* **2016**, *36*, 149–159. [CrossRef] [PubMed]
53. Lana, E.; Mégarbané, A.; Tourrière, H.; Sarda, P.; Lefranc, G.; Claustres, M.; De Sario, A. DNA replication is altered in Immunodeficiency Centromeric instability Facial anomalies (ICF) cells carrying DNMT3B mutations. *Eur. J. Hum. Genet.* **2012**, *20*, 1044–1050. [CrossRef] [PubMed]
54. Kamae, C.; Imai, K.; Kato, T.; Okano, T.; Honma, K.; Nakagawa, N.; Yeh, T.W.; Noguchi, E.; Ohara, A.; Shigemura, T.; et al. Clinical and Immunological Characterization of ICF Syndrome in Japan. *J. Clin. Immunol.* **2018**, *38*, 927–937. [CrossRef] [PubMed]
55. Chouery, E.; Abou-Ghoch, J.; Corbani, S.; El Ali, N.; Korban, R.; Salem, N.; Castro, C.; Klayme, S.; Rjeily, M.A.A.; Khoury-Matar, R.; et al. A novel deletion in ZBTB24 in a Lebanese family with immunodeficiency, centromeric instability, and facial anomalies syndrome type 2. *Clin. Genet.* **2012**, *82*, 489–493. [CrossRef] [PubMed]
56. Thijssen, P.E.; Ito, Y.; Grillo, G.; Wang, J.; Velasco, G.; Nitta, H.; Unoki, M.; Yoshihara, M.; Suyama, M.; Sun, Y.; et al. Mutations in CDCA7 and HELLS cause immunodeficiency-centromeric instability-facial anomalies syndrome. *Nat. Commun.* **2015**, *6*, 7870. [CrossRef]
57. Toubiana, S.; Velasco, G.; Chityat, A.; Kaindl, A.M.; Hershtig, N.; Tzur-Gilat, A.; Francastel, C.; Selig, S. Subtelomeric methylation distinguishes between subtypes of Immunodeficiency, Centromeric instability and Facial anomalies syndrome. *Hum. Mol. Genet.* **2018**, *27*, 3568–3581. [CrossRef]
58. Xu, G.L.; Bestor, T.H.; Bourc'his, D.; Hsieh, C.L.; Tommerup, N.; Bugge, M.; Hulten, M.; Qu, X.; Russo, J.J.; Viegas-Péquignot, E. Chromosome instability and immunodeficiency syndrome caused by mutations in a DNA methyltransferase gene. *Nature* **1999**, *402*, 187–191. [CrossRef]
59. Jin, B.; Tao, Q.; Peng, J.; Soo, H.M.; Wu, W.; Ying, J.; Fields, C.R.; Delmas, A.L.; Liu, X.; Qiu, J.; et al. DNA methyltransferase 3B (DNMT3B) mutations in ICF syndrome lead to altered epigenetic modifications and aberrant expression of genes regulating development, neurogenesis and immune function. *Hum. Mol. Genet.* **2008**, *17*, 690–709. [CrossRef]
60. de Greef, J.C.; Wang, J.; Balog, J.; den Dunnen, J.T.; Frants, R.R.; Straasheijm, K.R.; Aytekin, C.; van der Burg, M.; Duprez, L.; Ferster, A.; et al. Mutations in ZBTB24 are associated with immunodeficiency, centromeric instability, and facial anomalies syndrome type 2. *Am. J. Hum. Genet.* **2011**, *88*, 796–804. [CrossRef]

61. Jenness, C.; Giunta, S.; Müller, M.M.; Kimura, H.; Muir, T.W.; Funabiki, H. HELLS and CDCA7 comprise a bipartite nucleosome remodeling complex defective in ICF syndrome. *Proc. Natl. Acad. Sci. USA* **2018**, *115*, E876–E885. [CrossRef] [PubMed]

62. Stagi, S.; Gulino, A.V.; Lapi, E.; Rigante, D. Epigenetic control of the immune system: A lesson from Kabuki syndrome. *Immunol. Res.* **2016**, *64*, 345–359. [CrossRef] [PubMed]

63. Margot, H.; Boursier, G.; Duflos, C.; Sanchez, E.; Amiel, J.; Andrau, J.-C.; Arpin, S.; Brischoux-Boucher, E.; Boute, O.; Burglen, L.; et al. Immunopathological manifestations in Kabuki syndrome: A registry study of 177 individuals. *Genet. Med.* **2020**, *22*, 181–188. [CrossRef]

64. Ming, J.E.; Russell, K.L.; McDonald-McGinn, D.M.; Zackai, E.H. Autoimmune disorders in Kabuki syndrome. *Am. J. Med. Genet. A* **2005**, *132*, 260–262. [CrossRef] [PubMed]

65. Hoffman, J.D.; Ciprero, K.L.; Sullivan, K.E.; Kaplan, P.B.; McDonald-McGinn, D.M.; Zackai, E.H.; Ming, J.E. Immune abnormalities are a frequent manifestation of Kabuki syndrome. *Am. J. Med. Genet. A* **2005**, *135*, 278–281. [CrossRef]

66. Ng, S.B.; Bigham, A.W.; Buckingham, K.J.; Hannibal, M.C.; McMillin, M.J.; Gildersleeve, H.I.; Beck, A.E.; Tabor, H.K.; Cooper, G.M.; Mefford, H.C.; et al. Exome sequencing identifies MLL2 mutations as a cause of Kabuki syndrome. *Nat. Genet.* **2010**, *42*, 790–793. [CrossRef]

67. Lederer, D.; Grisart, B.; Digilio, M.C.; Benoit, V.; Crespin, M.; Ghariani, S.C.; Maystadt, I.; Dallapiccola, B.; Verellen-Dumoulin, C. Deletion of KDM6A, a histone demethylase interacting with MLL2, in three patients with Kabuki syndrome. *Am. J. Hum. Genet.* **2012**, *90*, 119–124. [CrossRef]

68. Van Laarhoven, P.M.; Neitzel, L.R.; Quintana, A.M.; Geiger, E.A.; Zackai, E.H.; Clouthier, D.E.; Artinger, K.B.; Ming, J.E.; Shaikh, T.H. Kabuki syndrome genes KMT2D and KDM6A: Functional analyses demonstrate critical roles in craniofacial, heart and brain development. *Hum. Mol. Genet.* **2015**, *24*, 4443–4453. [CrossRef]

69. Lindsley, A.W.; Saal, H.M.; Burrow, T.A.; Hopkin, R.J.; Shchelochkov, O.; Khandelwal, P.; Xie, C.; Bleesing, J.; Filipovich, L.; Risma, K.; et al. Defects of B-cell terminal differentiation in patients with type-1 Kabuki syndrome. *J. Allergy Clin. Immunol.* **2016**, *137*, 179–187.e10. [CrossRef]

70. Aref-Eshghi, E.; Rodenhiser, D.I.; Schenkel, L.C.; Lin, H.; Skinner, C.; Ainsworth, P.; Paré, G.; Hood, R.L.; Bulman, D.E.; Kernohan, K.D.; et al. Genomic DNA Methylation Signatures Enable Concurrent Diagnosis and Clinical Genetic Variant Classification in Neurodevelopmental Syndromes. *Am. J. Hum. Genet.* **2018**, *102*, 156–174. [CrossRef]

71. McDonald-McGinn, D.M.; Sullivan, K.E.; Marino, B.; Philip, N.; Swillen, A.; Vorstman, J.A.; Zackai, E.H.; Emanuel, B.S.; Vermeesch, J.R.; Morrow, B.E.; et al. 22q11.2 deletion syndrome. *Nat. Rev. Dis. Primers* **2015**, *1*, 15071. [CrossRef] [PubMed]

72. Du, Q.; de la Morena, M.T.; van Oers, N.S.C. The Genetics and Epigenetics of 22q11.2 Deletion Syndrome. *Front. Genet.* **2019**, *10*, 1365. [CrossRef] [PubMed]

73. Rooney, K.; Levy, M.A.; Haghshenas, S.; Kerkhof, J.; Rogaia, D.; Tedesco, M.G.; Imperatore, V.; Mencarelli, A.; Squeo, G.M.; Di Venere, E.; et al. Identification of a DNA Methylation Episignature in the 22q11.2 Deletion Syndrome. *Int. J. Mol. Sci.* **2021**, *22*, 8611. [CrossRef] [PubMed]

74. Baldini, A.; Fulcoli, F.G.; Illingworth, E. Tbx1: Transcriptional and Developmental Functions. *Curr. Top Dev. Biol.* **2017**, *122*, 223–243. [CrossRef]

75. Fulcoli, F.G.; Franzese, M.; Liu, X.; Zhang, Z.; Angelini, C.; Baldini, A. Rebalancing gene haploinsufficiency in vivo by targeting chromatin. *Nat. Commun.* **2016**, *7*, 11688. [CrossRef]

76. Boerkoel, C.F.; Takashima, H.; John, J.; Yan, J.; Stankiewicz, P.; Rosenbarker, L.; André, J.L.; Bogdanovic, R.; Burguet, A.; Cockfield, S.; et al. Mutant chromatin remodeling protein SMARCAL1 causes Schimke immuno-osseous dysplasia. *Nat. Genet.* **2002**, *30*, 215–220. [CrossRef]

77. Bertulli, C.; Marzollo, A.; Doria, M.; Di Cesare, S.; La Scola, C.; Mencarelli, F.; Pasini, A.; Affinita, M.C.; Vidal, E.; Magini, P.; et al. Expanding Phenotype of Schimke Immuno-Osseous Dysplasia: Congenital Anomalies of the Kidneys and of the Urinary Tract and Alteration of NK Cells. *Int. J. Mol. Sci.* **2020**, *21*, 8604. [CrossRef]

78. Saraiva, J.M.; Dinis, A.; Resende, C.; Faria, E.; Gomes, C.; Correia, A.J.; Gil, J.; da Fonseca, N. Schimke immuno-osseous dysplasia: Case report and review of 25 patients. *J. Med. Genet.* **1999**, *36*, 786–789. [CrossRef]

79. Indumathi, C.K.; Bustamante, J. Clinical and immunological profile of children with Mendelian Susceptibility to Mycobacterial Diseases (MSMD) from an Indian tertiary care hospital. *Indian J. Tuberc.* **2021**, *68*, 292–297. [CrossRef]

80. Bustamante, J.; Boisson-Dupuis, S.; Abel, L.; Casanova, J.L. Mendelian susceptibility to mycobacterial disease: Genetic, immunological, and clinical features of inborn errors of IFN-γ immunity. *Semin. Immunol.* **2014**, *26*, 454–470. [CrossRef]

81. Bustamante, J. Mendelian susceptibility to mycobacterial disease: Recent discoveries. *Hum. Genet.* **2020**, *139*, 993–1000. [CrossRef] [PubMed]

82. Pacis, A.; Mailhot-Léonard, F.; Tailleux, L.; Randolph, H.E.; Yotova, V.; Dumaine, A.; Grenier, J.C.; Barreiro, L.B. Gene activation precedes DNA demethylation in response to infection in human dendritic cells. *Proc. Natl. Acad. Sci. USA* **2019**, *116*, 6938–6943. [CrossRef] [PubMed]

83. Pacis, A.; Tailleux, L.; Morin, A.M.; Lambourne, J.; MacIsaac, J.L.; Yotova, V.; Dumaine, A.; Danckaert, A.; Luca, F.; Grenier, J.C.; et al. Bacterial infection remodels the DNA methylation landscape of human dendritic cells. *Genome Res.* **2015**, *25*, 1801–1811. [CrossRef] [PubMed]

84. Kaasinen, E.; Kuismin, O.; Rajamäki, K.; Ristolainen, H.; Aavikko, M.; Kondelin, J.; Saarinen, S.; Berta, D.G.; Katainen, R.; Hirvonen, E.A.M.; et al. Impact of constitutional TET2 haploinsufficiency on molecular and clinical phenotype in humans. *Nat. Commun.* **2019**, *10*, 1252. [CrossRef]

85. Spegarova, J.S.; Lawless, D.; Mohamad, S.M.B.; Engelhardt, K.R.; Doody, G.; Shrimpton, J.; Rensing-Ehl, A.; Ehl, S.; Rieux-Laucat, F.; Cargo, C.; et al. Germline TET2 loss of function causes childhood immunodeficiency and lymphoma. *Blood* **2020**, *136*, 1055–1066. [CrossRef]

86. Muto, H.; Sakata-Yanagimoto, M.; Nagae, G.; Shiozawa, Y.; Miyake, Y.; Yoshida, K.; Enami, T.; Kamada, Y.; Kato, T.; Uchida, K.; et al. Reduced TET2 function leads to T-cell lymphoma with follicular helper T-cell-like features in mice. *Blood Cancer J.* **2014**, *4*, e264. [CrossRef]

87. Rosikiewicz, W.; Chen, X.; Dominguez, P.M.; Ghamlouch, H.; Aoufouchi, S.; Bernard, O.A.; Melnick, A.; Li, S. TET2 deficiency reprograms the germinal center B cell epigenome and silences genes linked to lymphomagenesis. *Sci. Adv.* **2020**, *6*, eaay5872. [CrossRef]

88. Lio, C.W.; Zhang, J.; González-Avalos, E.; Hogan, P.G.; Chang, X.; Rao, A. Tet2 and Tet3 cooperate with B-lineage transcription factors to regulate DNA modification and chromatin accessibility. *Elife* **2016**, *5*, e18290. [CrossRef]

Journal of
Clinical Medicine

Review

The Importance of the Transcription Factor Foxp3 in the Development of Primary Immunodeficiencies

Paulina Mertowska, Sebastian Mertowski *, Martyna Podgajna and Ewelina Grywalska

Department of Experimental Immunology, Medical University of Lublin, Chodźki 4a St., 20-093 Lublin, Poland;
paulinamertowska@umlub.pl (P.M.); marpodgajna@gmail.com (M.P.); ewelina.grywalska@umlub.pl (E.G.)
* Correspondence: sebastianmertowski@umlub.pl

Abstract: Transcription factors are an extremely important group of proteins that are responsible for the process of selective activation or deactivation of other cellular proteins, usually at the last stage of signal transmission in the cell. An important family of transcription factors that regulate the body's response is the FOX family which plays an important role in regulating the expression of genes involved in cell growth, proliferation, and differentiation. The members of this family include the intracellular protein Foxp3, which regulates the process of differentiation of the T lymphocyte subpopulation, and more precisely, is responsible for the development of regulatory T lymphocytes. This protein influences several cellular processes both directly and indirectly. In the process of cytokine production regulation, the Foxp3 protein interacts with numerous proteins and transcription factors such as NFAT, nuclear factor kappa B, and Runx1/AML1 and is involved in the process of histone acetylation in condensed chromatin. Malfunctioning of transcription factor Foxp3 caused by the mutagenesis process affects the development of disorders of the immune response and autoimmune diseases. This applies to the impairment or inability of the immune system to fight infections due to a disruption of the mechanisms supporting immune homeostasis which in turn leads to the development of a special group of disorders called primary immunodeficiencies (PID). The aim of this review is to provide information on the role of the Foxp3 protein in the human body and its involvement in the development of two types of primary immunodeficiency diseases: IPEX (Immunodysregulation Polyendocrinopathy Enteropathy X-linked syndrome) and CVID (Common Variable Immunodeficiency).

Keywords: Foxp3; IPEX; CVID; primary immunodeficiencies

Citation: Mertowska, P.; Mertowski, S.; Podgajna, M.; Grywalska, E. The Importance of the Transcription Factor Foxp3 in the Development of Primary Immunodeficiencies. *J. Clin. Med.* **2022**, *11*, 947. https://doi.org/10.3390/jcm11040947

Academic Editors: Andreas Hecker, Rita Consolini and Giorgio Costagliola

Received: 31 December 2021
Accepted: 9 February 2022
Published: 11 February 2022

Publisher's Note: MDPI stays neutral with regard to jurisdictional claims in published maps and institutional affiliations.

1. Introduction

Transcription factors are a group of proteins that exhibit the ability to bind to genetic material (DNA). The site of attachment of transcription factors in the DNA region may be a promoter or an enhancer sequence in a specific site or region that regulates the transcription process. The action of transcription factors can be selectively activated in the cell or deactivated by other cellular proteins which usually takes place at the last stage of signal transmission in the cell [1]. One of the more important families among transcription factors is the FOX family, which plays an important role in regulating the expression of genes involved in cell growth, proliferation, and differentiation. This family includes the Foxp3 protein (Forkhead box protein P3) which is not only a transcription factor but also a key molecule involved in the development of regulatory T cells (Treg) [2,3]. It plays an important role in maintaining the homeostasis of the immune system, enables the complete stability of the Treg lineage, and directly modulates the expansion and function of conventional T-cells [3,4]. The Foxp3 protein can act both as a repressor and activator of the transcription process, while the type of its activity depends on its interaction with other transcription factors present in the cell [4]. Disturbances in the proper functioning of the Foxp3 protein in the human body (usually caused by mutations within the coding sequences) dysregulate the immune homeostasis and leads to the development of diseases called immunodeficiencies [5,6]. This is a special group of disorders in which the patient's immune system is characterized by a reduced ability or lack of defense against various pathogens. It is manifested by frequent, recurrent infections with microorganisms (bacteria, viruses, fungi) that are resistant to long-term therapy [7]. In the vast majority of cases it is caused by microbes commonly inhabiting the human body which, as a result of weakened immune reactions, cause the so-called opportunistic infections [8]. Apart from the tendency of recurrent infections, immunodeficiency is also accompanied by other health problems such as allergic phenomena [9], autoimmune phenomena [10,11], granulomas [12], tumors [13,14], endocrine disorders [15,16], and various cytopenias (most often thrombocytopenia [17] and neutropenia [18,19]) as well as diseases of the lungs [20] and gastrointestinal tract [21] (e.g., irritable bowel syndrome) [22,23]. Due to the underlying causes of immunodeficiencies, we can divide them into two groups: Primary Immunodeficiencies (PID), which are genetically determined and associated with mutations that are either inherited or appear de novo and Secondary Immunodeficiencies (SID), which are caused by external factors or the presence of comorbid disease [24,25].

The aim of this review is to present the role of the Foxp3 transcription factor in the human body in maintaining immune homeostasis and its significance during the development of immunodeficiency. Due to the Foxp3 protein's involvement in the regulation of transcription, we will try to explain its role in the development of primary immunodeficiency diseases that are strongly associated with genetic disorders.

2. Molecular Characterization of the Foxp3 Protein

The human *FOXP3* gene is located on the X chromosome, more specifically, by genomic sequence analysis, it was shown to be on the p arm at position Xp11.23 [26,27]. The gene is composed of 12 exons whose exon-intron boundaries are identical in the gene coding regions of both mice and humans [28]. The degree of nucleotide sequence identity between these organisms is 84.98% [26–28]. The *FOXP3* gene is responsible for the encoding of a protein product composed of 431 amino acids with a total molecular weight of 47.27 kDa and an isoelectric puncture of 8.62 (Table 1). The analysis of the protein sequence in the Foxp3 protein showed that its structure contains 47.33% hydrophobic amino acids and 52.67% hydrophilic amino acids (Table 1). According to the literature, the Foxp3 protein can also occur in the human body in the form of three other isoforms: isoforms 2, 3, and 4 [29]. They differ mainly in amino acid length and molecular weight which changes the amino acid composition, isoelectric point, and secondary structure of the protein. Our research team performed a detailed analysis of the similarity of the amino acid sequence of the Foxp3 protein and its three isoforms using the UniProt database. The differences made on this basis are shown in Table 1 and Figure 1.

Table 1. Characteristics of the basic properties of the Foxp3 protein and its isoforms.

| Name | Amino Acid Length | Mass [kDa] | Isoelectric Point | Amino Acid Composition | | Secondary Structure | | Protein ID |
				Hydrophobic Amino Acids [%]	Hydrophilic Amino Acids [%]	α-Helix	β-Starnad	
Foxp3	431	47.24	8.62	53.60	46.40	12 (26.15%)	10 (6.26%)	Q9BZS1
Isoform2	396	43.41	8.53	52.27	47.73	10 (25.25%)	9 (5.55%)	Q9BZS1-2
Isoform 3	456	49.84	8.00	53.51	46.49	14 (27.41%)	10 (6.56%)	Q9BZS1-3
Isoform 4	404	44.41	8.52	53.47	46.53	12 (24.50%)	11 (7.43%)	Q9BZS1-4

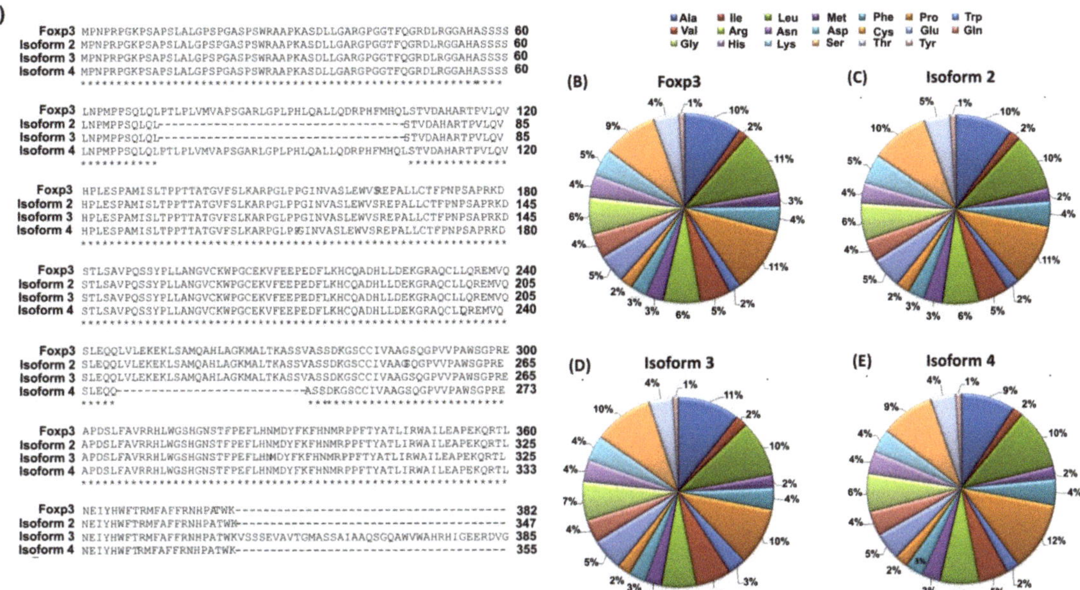

Figure 1. Characterization of the amino acid sequence of the Foxp3 protein and its isoforms (**A**) Comparison of the amino acid sequence of the Foxp3 protein and its isoforms; (**B–E**) Amino acid scald of Foxp3 protein and its isoforms [own elaboration]. Marks: (*) means the amino acids are identical in the sequence, while (-) means the lack of amino acids in the sequence.

For isoform 2, 34 amino acids are missing from the Foxp3 protein sequence (located at positions 72–106). The same is the case for isoform 3 where at position 382, a fragment consisting of 61 amino acids (-KVSSSEVAVTGMASSAIAAQSGQAWVWAHRHIGEERDVGCW WWLLASEVDAHLLPVPGLPQ-) is also added [29]. Additionally, between the Foxp3 protein and isoform 4, there is a difference of 26 amino acids located between amino acids 246–272 which determines the highest degree of sequence identity. The degree of identity of the amino acid sequence of the Foxp3 protein and its isoforms is quite diverse and ranges from 75.15% (between isoforms 3 and 4) to 93.73% (between the Foxp3 protein and isoform 4). Detailed information on the degree of amino acid sequence identity between individual proteins is provided in Table 2.

Table 2. The degree of identity of the amino acid sequence of the Foxp3 protein and its isoforms.

	Foxp3	Isoform 2	Isoform 3	Isoform 4
Foxp3	-	91.88%	80.65%	93.73%
Isoform 2	91.88%	-	86.84%	85.61%
Isoform 3	80.65%	86.84%	-	75.15%
Isoform 4	93.73%	85.61%	75.15%	-

Foxp3 isoforms differ not only in their structure and the degree of similarity of the amino acid sequence but also in their ability to interact with different proteins. The Foxp3 protein is able to interact with the IKZE3 protein (via the LXXLL motif), while as reported in the literature, isoform 2 does not [4,30]. It has been shown that it is able to interact with the ZFP90 protein and create complexes with the TRIM28 protein [31]. However, further research is needed to understand the differential function of Foxp3 isoforms.

The Foxp3 protein has several distinctive motifs in its structure. Two of them, located between amino acids 68–76 and amino acids 239–248, are nuclear export signals which are short peptides containing hydrophobic residues targeted for export from the cell nucleus into the cytoplasm through the nuclear pore complex (Figure 2) [32]. Another example is the LXXLL motif, located between amino acids 92–96, which is involved in many protein-protein interactions related to various aspects of transcription regulation (Figure 2) [33–35]. These motifs are present in many transcription factors and cofactors, mediating interactions that may activate or suppress transcription [36,37]. Several recently reported 3D structures of protein-LXXLL motif complexes have been associated with leukemia, further highlighting the diversity and regulatory importance of this seemingly simple motif [36,38,39].

Figure 2. Motifs and domains occurring in the amino acid sequence of the Foxp3 protein (own elaboration based on the Uniprot database [29]).

Another motif located between amino acids 414–417 is the nuclear localization signal which is an amino acid sequence that "tags" a protein for import into the cell nucleus via nuclear transport (Figure 2). Typically, this signal consists of one or more short sequences of positively charged lysines or arginines exposed on the protein surface [40]. Two extremely important domains within the sequence of this protein should also be noted. The first is a zinc finger domain of 26 amino acids located in the region between amino acids 197 and 222 (Figure 2). It is a structure found in DNA binding proteins and is directly involved in the binding of a nucleic acid molecule by a protein [41]. The presence of the zinc ion (Zn^{2+}) in the domain is crucial for the stability of the entire structure as its absence may result in structural and functional changes. The zinc atom in this structure is coordinated by residues C198, C203, H216, H221, and partially by D220 [42,43]. Near the zinc finger domain, there is another leucine zipper motif (239–260 amino acids) which is very important due to the functions performed by the Foxp3 protein [44,45]. This motif mediates intermolecular interactions, indicating the possible interaction in the dimerization process which is the essential function of the transcriptional regulator [45]. As indicated by research data, this motif is involved not only in homo-association but also in hetero-association with the Foxp1 protein [44,45]. Additionally, it has been shown that the presence of mutations within the leucine zipper motif can significantly reduce the binding affinity of the Foxp3 protein to the promoter regions in vitro [42]. The discoveries made by Mackey-Cushman et al., (2011) showed that the leucine zipper motif also mediates interactions between the Foxp3 protein and histones (the H1.5 histone, precisely) which suppress interleukin 2 (IL-2) transcription in T cells [46]. It should also be mentioned that the zinc finger motif is not directly involved in the dimerization process and the very mechanism of this process is not fully understood [42]. The second extremely important structure is the domain called "fork-head", which is often described as the transcription factor, whose purpose is to bind DNA. In the case of the Foxp3 protein, this domain is 87 amino acids long and is located between amino acids 337–423 in the sequence (Figure 2) [47,48]. It is involved in the protein dimerization process and its interaction with nuclear factor NFATC2 is responsible for activating T cells [49]. The NFATC2 protein is present in the cytosol and translocates only to the nucleus upon stimulation of the T-cell receptor (TCR) where it becomes a member of the nuclear factors of the activated T-cell transcription complex. This complex plays a key role in inducing gene transcription during the immune response [49,50].

3. The Occurrence, Interactions, and Importance of the Foxp3 Protein in the Human Body

3.1. The Occurrence of Foxp3 Protein on Cells of the Immune System and Its Functions in the Human Body

Treg cells belong to the subpopulation of lymphocytes responsible for suppressing an overly increased or autoreactive immune response, which may be specific or non-specific for a given antigen, without causing general immunodeficiency [51,52]. It is an extremely heterogeneous population with several subpopulations of cells with different levels of Foxp3 protein expression (Table 3) [53]. Treg cells expressing Foxp3+ can also be divided into two smaller subpopulations according to the origin of the cells. The first one is natural Treg lymphocytes (nTreg), which are formed in the thymus as a separate developmental line, while the second is induced (adoptive) Treg lymphocytes (iTreg or aTreg) [54,55]. The latter are Treg cells arising in peripheral tissues which initially do not express the Foxp3 factor but acquire it along with suppressive properties upon stimulation with an appropriate antigen. Natural nTreg lymphocytes express CTLA-4 (cytotoxic T cell antigen 4) [56], GITR (glucocorticoid-induced TNFR-related protein) [57], CCR4 (CC chemokine receptor type 4) [58,59], and CD62L (L-selectin) [60]. The formation of nTreg and iTreg lymphocytes requires the presence of interleukin 2 (IL-2) and transforming growth factor beta 1 (TGF-β) [61]. The similarities in the role of these cytokines in the maintenance and survival of both Treg lymphocyte populations are crucial. nTregs develop in response to contact with intrinsic antigens in the thymus and require high affinity between these antigens and MHC complexes [62,63]. This is probably because they arise from continuously proliferating precursor cells. They also require costimulation with CD28 (cluster of differentiation 28) which also plays an important role in the process of inducing the expression of the CTLA-4 molecule in T lymphocytes which can inhibit the CD28 signal and thus, is responsible for the negative feedback mechanism [64,65]. On the other hand, iTreg, through interactions with environmental antigens presented by dendritic cells in peripheral lymphoid organs and their conversion to iTreg CD25+Foxp3+, requires weaker, incomplete TCR stimulation [66]. Foxp3 protein is also known to convert naive T cells into Treg cells that are capable of suppression in vivo and in vitro, suggesting that Foxp3 can regulate the expression of suppression-mediating molecules [53]. Elucidation of Foxp3 gene targets may be crucial for understanding the Treg cell suppressive capacity [67–69].

Table 3. Treg lymphocyte subpopulations (based on [70]).

Name of the Subpopulation		Characteristics
Expressing Foxp3	T lymphocytes CD4+CD25+Foxp3+	The most widely studied and characterized sub-population of regulatory cells. Characterized by the expression of the Foxp3 transcription factor and accompanied by a high expression of the CD25 surface molecule.
	T lymphocytes CD8+CD25+Foxp3+	A subpopulation of CD8+ T cells that are much less well understood than CD4+Foxp3+ cells.
Not expressing Foxp3	Type 1 regulatory T cells (Tr1)	Cells with the CD4+Foxp3− phenotype secreting significant amounts of IL-10.
	Th3 lymphocytes	CD4+Foxp3− cells secreting significant amounts of TGF-β.
	CD8+CD28− lymphocytes	identified with pre-Ts lymphocytes that also do not express Foxp3.

Another group of immune system cells in which the Foxp3 protein is expressed is NKT cells [71]. This is a group of innate lymphocytes capable of producing cytokines characteristic for a Th1, Th2, or Th17 response [72,73]. They have also been confirmed to influence adaptive immunity by exacerbating or suppressing a variety of immune disorders such as autoimmunity and allergy. As mentioned before, Treg cells are characterized by strong immunosuppressive properties and expression of the Foxp3 transcription factor and constitute a key element in maintaining immune homeostasis in the human body. Studies from recent years have shown that NKT cells, like Treg cells, contribute to the maintenance of immune tolerance and are also capable of the Foxp3 protein expression. From animal model studies, scientists were able to establish that NKT cells in the lymph nodes of α-galactoceramide-stimulated mice increased the ability to express Foxp3 in response to TGF-β. However, further research is needed to use Foxp3+ NKT cells for therapeutic purposes in the treatment of immune response disorders [74,75].

In addition to T cells, expression of the Foxp3 protein is also observed on B cells. Regulatory B cells, which are characterized by the production of anti-inflammatory cytokines (e.g., IL-10 and IL-35), also contribute to the enhancement of the immune homeostasis in the human body [74]. Recent research has shown that some B cells are also capable of expressing the Foxp3 protein [76]. They are usually observed in patients diagnosed with multiple sclerosis [77] or in patients with systemic lupus erythema where their increased amount correlates with disease progression [78,79]. A study by Slobodin et al. (2010) shows that the percentage of Breg cells in the peripheral blood of SLE patients is higher than that of healthy controls and that Bregs have been shown to be functionally impaired. Additionally, they showed that with the expansion of CD25highIL-10highFoxP3high B regulator cells, increased disease activity occurs [80]. However, the current literature reports do not provide important information regarding the exact role of the Foxp3 protein on B lymphocytes and its role in the regulation of inflammation [70,81]. The subject of Foxp3 protein expression on cells of the immune system is quite debatable and requires a lot of research to fully understand the importance and role of this protein in the human body.

3.2. Interactions of Foxp3 Protein with Proteins and Transcription Factors

The Foxp3 protein plays an important role in the regulation of cytokine production through interaction with numerous proteins and transcription factors such as NFAT, nuclear factor kappa B (NF-κB, kappa-light-chain-enhancer of activated B cells) [82], and Runx1/AML1 (Runt-related transcription factor 1/acute myeloid leukemia 1 protein), also known as acute myeloid leukemia 1 protein (AML1) [83] or alpha-core binding factor 2 (CBFA2) subunit [84]. Transcription factors NFAT and Runx1/AML1 are necessary for the production of IL-2 following TCR receptor stimulation. The association of the Foxp3 protein with these factors inhibits the expression of IL-2, IL-4, (which are pro-inflammatory cytokines), and IFN-γ [85]. In vivo, the transcription factor NFAT, which is required to bind to the proximal region of the promoter for IL-2, participates in direct interactions with the forkhead domain of the Foxp3 protein [82]. These interactions are also necessary to stimulate the expression of some Treg lymphocyte antigens, such as CD25 or CTLA-4 [82,86,87]. CD25 is the alpha chain of the interleukin 2 receptor and is transiently expressed on activated T and B lymphocytes and constitutively present on Treg cells. The presence of this protein was also found on dendritic cells, fibroblasts, and endothelial cells [88,89]. In contrast, CTLA-4 is a protein receptor that acts as a checkpoint for immune responses. This protein is constitutively expressed in Treg lymphocytes, but when activated, it is only up-regulated in conventional T lymphocytes; this phenomenon is particularly evident in the case of neoplastic disorders [86,90]. The interaction of the Foxp3 protein with the Runx1/AML1 factor (association between the "fork-head" domain and the leucine zipper) consists of the attachment of this complex to the IL-2 promoter region. It should also be noted that the Foxp3 protein also participates in interactions with other members of the FOX family where it forms complexes capable of inhibiting the expression of selected genes [91]. The formation of either homodimeric or heterodimeric Foxp3 protein complexes occurs when a leucine zipper is used which associates with the IL-2 promoter in vivo. Scientific studies have also shown that the multimerization of the Foxp3 protein is extremely important for the proper functioning of Treg lymphocytes [92–95].

The N-terminus of the Foxp3 protein also has properties to inhibit the transcriptional activity of the NFAT factor and is necessary for the inhibition of IL-2 production by T cells. Studies have shown that the association of the Foxp3 protein with the IL-2 and IFN-γ promoters correlates with the process of hypoacetylation (low level of acetylation) of histones in condensed chromatin [4,96]. The binding of the Foxp3 protein to the CD25 and CTLA-4 promoter sequences, with the simultaneous hypoacetylation of condensed chromatin histones, suggests a direct mechanism of transcription activation [97–99].

In Treg cells, the Foxp3 protein is present as a part of a large complex that also includes histone acetyltransferases (HAT) and histone deacetylases (HDAC) [100,101]. The process of acetylation and deacetylation is one of the post-translational modifications of histones, involving lysine residues located at the N-terminus (protruding from the nucleosome core), which is an epigenetic mechanism for the control of gene expression. The enzymes catalyzing the reversible acetylation of histones are HAT and HDAC [102]. Research conducted in recent years has allowed us to establish that the former simultaneously act as coactivators of transcription, while the latter is its corepressors. Thus, it is possible to prove the relationship between the covalent modification of chromosomal proteins (acetylation of core histones) and gene expression. HAT and HDAC enzymes are responsible for the acetylation process of the Foxp3 protein and thus determine the functions of this transcription factor. Studies have shown that Foxp3 acetylation is related to its function in Treg cells [103–105]. Detailed analyzes have shown that the acetylation process of this protein is influenced by molecules such as KAT5 (TIP60), p300, HDAC6, HDAC7, HDAC9, HDAC10, and SIRT1 (Table 4).

Moreover, it has been shown that the stability of the Foxp3 protein is significantly higher in the presence of the enzyme p300 than in the presence of KAT5. After combining with the p300 protein, Foxp3 shows a higher affinity for DNA chromatin, while after interaction with KAT5, its affinity for the IL-2 promoter has significantly increased [106]. The importance and role of HDAC7 and HDAC9 in the acetylation of Foxp3 protein are poorly understood. The link is due to the fact that class II HDAC has no intrinsic functional catalytic activity and possibly, as the scientists suggest, works by recruiting class I HDAC into the complex. The mechanism of action of HDAC6, belonging to class II HDAC proteins, is based on the regulation of the acetylation process of many cellular proteins, among which α-tubulin and HSP90 deserve special attention [107,108]. Studies conducted by Zoeten et al., showed that the pharmacological inhibition or the use of a genetic knockout of the HDAC6 gene increased the level of acetylation of both Foxp3 and Hsp90 [107] which resulted in an increase in the immunosuppressive activity of Treg cells. According to the researchers, this process may be used in the future to suppress autoimmunity caused by Treg and be used in the prevention of transplant rejection [107,108]. The combination of KAT5 and HDAC7 is an essential mechanism in the inhibition of IL-2 transcription by the Foxp3 protein [100,109]. NAD-dependent sirtuin-1 deacetylase is considered a negative regulator of the acetylation process of the Foxp3 protein [55,110].

3.3. The Importance of Post-Transcriptional and Post-Translational Modifications in the Function of the Foxp3 Protein

The transcription factor Foxp3 is a major regulator of Treg cells' growth and suppression activity. As indicated in the literature, it is also subject to complex regulation by the participation of many posttranscriptional modifications as well as posttranslational modifications (PTM) which also indirectly affect Treg suppressor activity. The first modifications of the Foxp3 protein concern epigenetic changes related to the regulation of DNA methylation [111], histone modification [99], or nucleosome positioning [112]. In the literature, we can find reports on the regulation of the Foxp3 protein by conserved non-coding sequences 2 (CNS 2) [113,114] or the share of ubiquitin-specific peptidase 22

(USP22) [115,116]. Many factors also play an important role in regulating the conversion of Foxp3 precursor messenger RNA transcripts at the level of post-transcription modification. This concerns the participation of microRNAs (miRNAs), including miR-24, miR-31, and miR-210, the activity of which leads to the degradation of Foxp3 mRNA which consequently prevents the translation of Foxp3 [117,118]. PTM, which includes phosphorylation, O-GlcNAcylation, acetylation, ubiquitylation, or methylation, also significantly influences the activity of the Foxp3 protein [105,119]. It should be noted that most of the PTMs refer to enzymatic processes that are designed to change the protein after its synthesis. Moreover, all the induced modifications affect the characteristics of the protein, including its location and interactions. This is also the case with the Foxp3 protein, in which the occurring PTMs affect the very structure of the protein (its stabilization or degradation) and its interactions with other proteins as well as reducing or increasing the activity of suppressive Treg lymphocytes [105]. Detailed information on the positive and negative contribution of individual PTMs to the suppressor functions of Treg cells is presented in Figure 3.

Table 4. Effect of enzymes on the acetylation process of the Foxp3 transcription factor.

Name of the Enzyme	Abbreviation	Functions	Reference
Histone acetyltransferase KAT5	TIP60	Performs histone acetylation in the nucleosome which changes the binding to DNA. Acetylation neutralizes the positive charge on the histones, reducing the binding affinity of negatively charged DNA. This in turn reduces the steric hindrance of DNA and increases the interaction of transcription factors and other proteins. The three key functions of KAT5 are its ability to regulate transcription, DNA repair, and apoptosis.	[30,120]
Histone acetyltransferase	p300	Acts as a histone acetyltransferase that regulates transcription through chromatin remodeling and is important in cell proliferation and differentiation. It mediates the regulation of the cAMP gene, binding specifically to the phosphorylated CREB protein, and also contains a bromodomain which is involved in IL6 signaling.	[106,121]
Histone deacetylase 6	HDAC6	This enzyme is located in the cytoplasm where it is responsible for the regulation of acetylation of α-tubulin, HSP90, or glucocorticoid receptors. Upon activation of Treg cells, this enzyme migrates to the cell nucleus where it participates in the regulation of the acetylation level of the Foxp3 protein. Pharmacological inhibition or the use of a genetic knockout of the HDAC6 gene have been shown to increase the level of acetylation of both Foxp3 and Hsp90 proteins which results in an increase in the immunosuppressive activity of Treg cells.	[107,108,122]
Histone deacetylase 7	HDAC7	HDAC7 has been shown to have low intrinsic deacetylase activity and studies have demonstrated that HDAC7 may have a variety of alternative developmentary, proliferative, and inflammatory functions.	[101,102,123]
Histone deacetylase 9	HDAC9	Represses the activity of MEF2 by recruiting multi-component complexes containing CtBP and HDAC. May play a role in the process of hematopoiesis.	[108,124]
Histone deacetylase 10	HDAC10	From studies performed in a mouse model, HDAC10 deletion did not adversely affect the health of mice that retained normal CD4+ and CD8+ T cell function. However, HDAC10$^{-/-}$ Treg showed enhanced suppressive function both in vitro and in vivo. In addition, HDAC10$^{-/-}$ mice that received a heart transplant with a completely mismatched MHC became more tolerant and showed longer allograft survival.	[101]

Table 4. *Cont.*

Name of the Enzyme	Abbreviation	Functions	Reference
NAD-dependent sirtuin-1 deacetylase	SIRT1	SIRT1 deacetylates and thus inactivates the p53 protein. SIRT1 also stimulates autophagy by preventing the acetylation of proteins (via deacetylation) required for autophagy, as demonstrated in cultured cells and embryonic and neonatal tissues. This feature provides a link between sirtuin expression and the cellular response to nutrient constraints due to caloric restriction. SIRT1 inhibits NF-κB regulated gene expression by deacetylating the RelA/p65 subunit of NF-κB in lysine 310. SIRT1 plays a role in activating T17 helper cells that contribute to autoimmune disease.	[125,126]

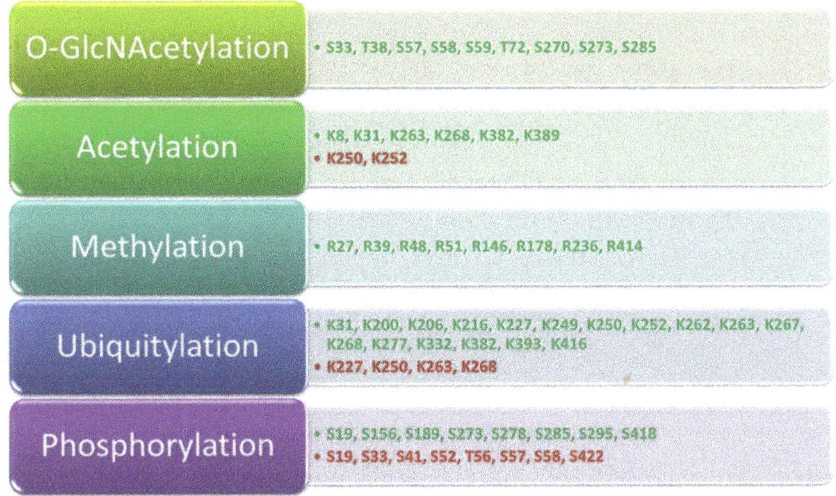

Figure 3. Contribution of PTM to the functioning of the Foxp3 protein and the suppressor functions of Treg lymphocytes. Modifications that positively affect the suppressor functions of Treg lymphocytes are marked in green while those that have a negative effect are in red; based on [105].

3.4. Importance of T Cell Metabolic Factors and the Level of Foxp3 Expression

Differentiation, proliferation, and suppressive function or survival of Treg cells are influenced by various factors of energy metabolism. Therefore, the role of the Foxp3 protein in the regulation of cellular metabolism is also an important issue [127]. Naive T cells have modest metabolic requirements which are mainly related to the oxidation of pyruvate and fatty acids in the tricarboxylic acid (TCA) cycle [127]. However, when activated, the energy requirements of these cells increase. This is possible due to significant metabolic changes induced by the TCR receptor and costimulatory molecules such as phosphoinositol 3-kinase (PI3K) as well as AKT and rapamycin 1 (mTOR1) complexes, the activation of which are responsible for the regulation of genes responsible for the uptake and breakdown of glucose and other energetic compounds, including acids [127,128]. Such changes provide not only the energy needed for proliferation but also the necessary biosynthetic raw materials. According to conducted studies, inhibition of glycolysis may direct the differentiation of T CD4+ lymphocytes towards an anergic state which is then accompanied by an increased expression of the Foxp3 protein [128,129]. As a result of genetic or chemical ablation of mTOR and elimination of glycolysis facilitators, it causes the generation of iTreg cells compared to effector lines [130]. Additionally, the forced activation of AMP-activated protein kinase (AMPK), a regulator of lipid metabolism involved in T cell differentiation, leads to increased Foxp3 protein expression and iTreg cell differentiation [131,132]. Moreover, the use of fatty acid inhibitors such as etomixir or the carnitine palmitoyltransferase 1A inhibitor reduces the degree of differentiation of iTreg cells [133–135]. From studies conducted in recent years it can be concluded that the process of induction of Foxp3 protein expression by iTreg cells is extremely sensitive to metabolic factors [136]. In addition, mutations leading to inappropriate dominance of the T-lymphocyte glycolytic pathway destabilize the Treg cell phenotype and result in the loss of the ability to express the Foxp3 protein which then leads to the inability of cells to suppress inflammation [127–129].

3.5. Regulation of Foxp3 Protein Expression as a Potential Therapeutic Strategy for Autoimmune and Neoplastic Diseases

Treg lymphocytes, due to their extremely important functions in the human body related to the suppression of immune reactions, have now become an important research target for many scientists. Particularly noteworthy are those studies that concern the possibility of their introduction as a treatment for autoimmune diseases and cancers [137]. The results of research conducted mainly on animal models show that the use of compounds as inhibitors or modulators of Foxp3 protein expression may not only confirm the correctness of the hypotheses put forward by scientists but also discover these molecules with high clinical potential [138].

The first group of such compounds are inhibitors of HDAC enzymes which, as shown by studies, can enhance not only the expression of *FOXP3* but also participate in the process of increasing the number of Treg cells and their functions. Currently, these compounds have found their application in animal models as therapeutics used to regulate the activity of Treg cells in autoimmune diseases (e.g., colitis, prostatitis) [124,139] and organ transplantation processes [107] as well as in the treatment of certain neoplastic diseases (e.g., breast cancer, lymphoma) [140,141]. The studies available in the literature show that the use of HDAC inhibitors influences the induction of FOXP3 acetylation in Treg cells, which prompt changes in markers related to the activity and functioning of Treg cells themselves, including receptors for TNF-α, CTLA-4 or PD-1 and IL-10 [142]. Obviously, a full understanding of the mechanisms of action of HDAC inhibitors still requires a lot of intensive research before such treatment strategies can be introduced into widespread clinical use.

Another group of compounds with therapeutic potential are histone acetyltransferase inhibitors [100]. The available studies on animal models show that the use of small-molecule allosteric compounds, which may interact with the cofactors of the Foxp3 protein, contributes to the regulation of the function of this protein [100]. The analyses conducted so far focus on two compounds, Tip60 and p300, for which allosteric modifiers were developed [143]. In the case of Tip60, these compounds reduce the process of histone acetylation and induce association with the Foxp3 protein which has been used to treat autoimmune diseases (mainly colitis and collagen-induced arthritis) [144–146]. There are also studies on the use of Tip60 and p300 inhibitors for antitumor therapy—mainly prostate cancer [137,147]. Tip60 inhibitors have been shown to be able to inhibit the cancer cells' growth by inducing the apoptosis process and to allow the reduction of Treg suppression without affecting the proliferation of T effector cells [148]. In the case of p300 inhibitors, their ability to suppress Treg function has been demonstrated [149].

In order to determine the best therapeutic strategy aimed at changing the level of Foxp3 protein expression many thorough studies should be carried out which will determine not only the mechanisms of action of selected allosteric molecules but also allow the evaluation of the effectiveness of such therapies in a clinical setting. However, therapies with the use of a targeted modification of the expression of the Foxp3 protein will certainly become a valuable tool in the future in the fight not only of autoimmune diseases but also of neoplastic diseases.

4. The Role of Foxp3 Protein in the Development of PIDs

PIDs are a group of genetically determined diseases that are characterized by the impairment of one or more mechanisms of innate or acquired immunity. Unlike secondary immunodeficiencies, the symptoms of PID occur throughout the patient's life. This disease has a heterogeneous course, characterized by a wide spectrum of symptoms with varying severity which may manifest themselves in childhood or even in adulthood [150]. PIDs are genetic disorders that may be inherited (most often autosomal recessive) or appear for the first time in a given patient. According to the literature data presented by the International Union of Immunological Societies (IUIS), by 2019 over 406 diseases called PIDs and ~430 genes (the damage of which may lead to their development) have been described [151]. Individual mutations underlie the malfunction or lack of one or more elements of the immune system, including B and T lymphocytes, NK cells, phagocytes, or components of the complement system [151,152]. According to general estimates, the incidence of PID is in 1:2000 to 1:3000 live births [153,154]. Due to the possible occurrence of symptoms later in life, it is very difficult to determine the prevalence of PID in adults. Several reports conducted in various countries around the world show that diseases classified as PID occur in the population ranging from 1:8500 to 1:100,000 [155]. In the current literature we find that PIDs are classified into nine classes which were proposed and updated by the IUIS in 2019 (Figure 4A) [156]. Currently, there is a view in the literature that some PIDs may have a multi-gene basis because only the presence of defects of several genes at the same time is clinically manifested. In addition, the presence of a PID phenocopy may also be the result of an autoimmune reaction against certain components of the immune system itself (e.g., against certain interferons) (Figure 4B) [157].

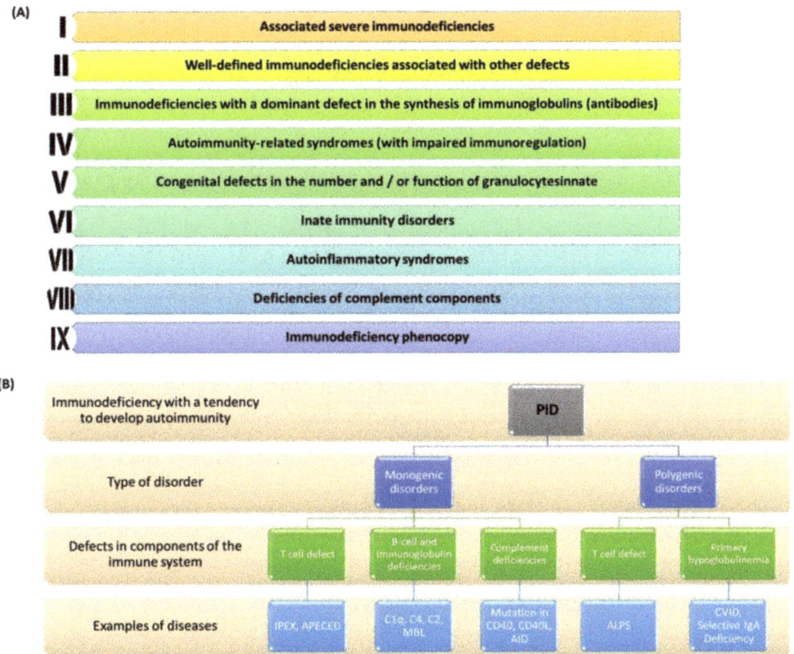

Figure 4. PID classification and division. (**A**) Updated PID classification by IUIS for 2019; (**B**) Division of PID based on the type of mono- and polygenic disorders with examples of diseases (prepared based on [156–158]). Abbreviations: APECED—Autoimmune polyendocrinopathy-candidiasis-ectodermal dystrophy; IPEX—immunodysregulation polyendocrinopathy enteropathy X-linked syndrome; C1q—complement component 1q; C4—Complement component 4; C2—Complement component 2; MBL—mannose-binding lectin; AID—Activation-induced cytidine deaminase; ALPS—Autoimmune Lymphoproliferative Syndrome; CVID—Common Variable Immunodeficiency; IgA—immunoglobulin A; CD40—cluster of differentiation 40; CD40L—cluster of differentiation 40 ligand.

The course of PIDs varies considerably which means that many clinical symptoms may be present from birth or may gradually worsen over time until the disease develops in childhood or adulthood. According to the literature, the mean time from the onset of symptoms to full PID diagnosis is on average ~5 years and largely depends not only on the type of deficiency and the patient's age but also on the patient's country of origin [159,160]. This is due to the lack of awareness and education in society as well as the lack of preparation in doctors to make accurate and quick diagnoses. Therefore, it is extremely important to search for diagnostic markers that allow for the shortening of time for accurate diagnosis as well as a comprehensive analysis of genetic disorders contributing to the development and progression of many diseases classified as PID [161,162] One such molecule is the Foxp3 protein in which mutation causes the immunodysregulation polyendocrinopathy enteropathy X-linked syndrome (IPEX) classified as a monogenic disorder [163]. Additionally, recent research sheds new light on the role of this protein in the development of CVID (common variable immunodeficiency) which is an example of a polygenic disorder. The *FOXP3* transcription factor is expressed on CD4+ Treg cells and is crucial for Treg function which is responsible for suppressing immune responses, especially at their early stages. The most important function of Foxp3 is its ability to confer suppressive activity on Treg cells, for example, by maintaining constitutive high expression of CTLA-4. It acts by preventing the activation and proliferation of B and T lymphocytes [164]. However, Foxp3 alone does not control all aspects of Treg biology and is not the initiating factor in Treg development [165,166]. The importance of Treg cells has been demonstrated in murine models—depletion of Foxp3+CD4+ Treg cells resulted in severe autoimmunity, allergy, and immunopathology (e.g., IPEX) in otherwise normal animals and those same diseases can be prevented by reconstituting Treg cells [167–169]. Moreover, it has also been demonstrated that Treg cells could prevent the progression of and even cure established autoimmune/inflammatory diseases [170,171]. They also play an important role in allergy prevention [172]. Foxp3+ Treg cells likewise depend on a lot of other suppressive molecules, such as IL-10, TGF-β, CD39, CD73, IL-35, and TIGIT, for their inhibitory function; most of which work by suppressing autoimmunity [173].

4.1. The Role of Mutations within the Foxp3 Protein in PID Development

The studies conducted so far indicate the presence of 63 identified mutations within the *FOXP3* gene which affect its proper functioning and contribute to the development of autoimmune diseases [174,175]. The vast majority of the discovered mutations concern the fork-head binding domain of the Foxp3 protein that influences the processes of nuclear import and DNA binding which are necessary for the suppressive activity of this protein. Some of the mutations were also found in the leucine zipper region which impaired the dimerization function of the Foxp3 protein; other mutations influenced the spatial change in the structure and position of the domains within the protein or led to a decrease in mRNA stability for the *FOXP3* gene [174–176].

4.1.1. The Role of Foxp3 Protein in IPEX Development

The occurrence of mutations in the *FOXP3* gene is associated with the development of the IPEX disease syndrome and was described for the first time in 1982 by Powell et al. as a rare immunodeficiency syndrome with a genetic predisposition [176]. This disease is characterized by the presence of three specific clinical symptoms such as enteropathy with chronic diarrhea (most often acute, watery, and bloody diarrhea), endocrinopathy (insulin-dependent diabetes type I), and dermatitis (Figure 5) [35]. IPEX is a recessive disorder related to the X chromosome; therefore, it occurs only in males (in the first six months of life) and causes T lymphocyte activation, accompanied by the overproduction of cytokines, and leads to autoimmune disorders with the presence of various autoantibodies [177]. This results in the development of many serious diseases such as type 1 diabetes and autoimmune hemolytic anemia as well as hypopituitarism or thyroid gland disorders and numerous skin lesions such as eczema. In many patients, exacerbation of the disease is also observed and caused by infections or food allergens. Subsequently, the main symptoms of the gastrointestinal tract or skin lesions are intensified or exacerbated as well as other disorders related to kidney, digestive, and immune system diseases (Figure 5) [163,178]. Due to such a wide range of symptoms, patients without appropriate treatment (immunosuppressants and bone marrow transplantation) die before reaching the age of 2 [176].

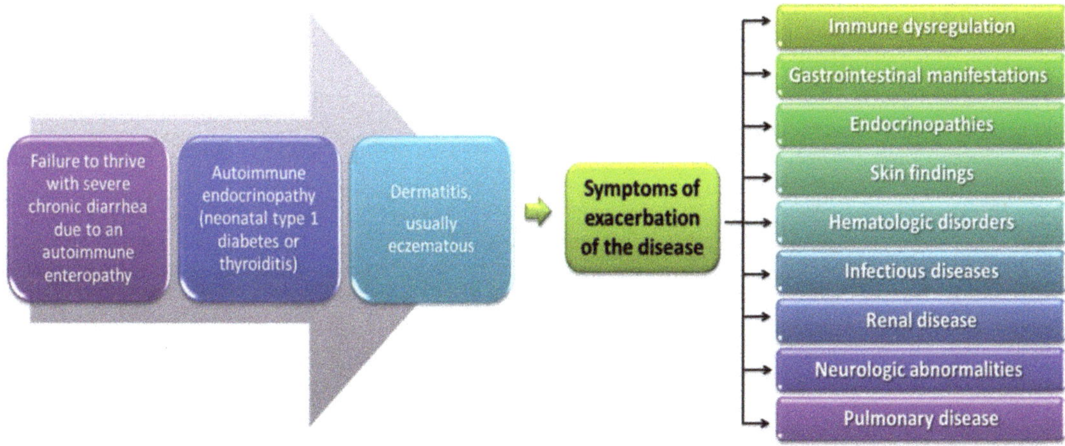

Figure 5. Symptoms of IPEX developed from [177].

The assessment of the presence of the IPEX syndrome requires the use of many extensive, basic, and specialized tests aimed at a correct diagnosis. Basic tests include a complete blood count with smear, determination of serum glucose concentration, thyroid function, and immunoglobulin levels as well as food hypersensitivity tests and the percentage of individual cells of the immune system, especially T and B lymphocytes (Table 5). Based on the obtained results of basic tests, advanced tests are ordered, including endoscopy with intestinal biopsy, skin biopsy, immunophenotyping of Treg lymphocytes, and sequencing of the *FOXP3* gene (Table 5) [177].

Table 5. Meanings of basic and specialized tests in the diagnosis of IPEX.

Type of Research	Type of Examination	The Importance of the Examination	Reference
Basic examination	Complete blood count with a smear	Presence of eosinophilia, neutropenia, anemia, or thrombocytopenia	[177]
	Serum glucose concentration	Glucose monitoring can help to detect the presence of type 1 diabetes	[177]
	Functioning of the thyroid gland	Elevated levels of anti-thyroid antibodies	[177]
	Concentration of immunoglobulins	Increase in IgE level in most patients, increase in IgA level in half of the patients, normal IgG and IgM	[178–182]
	Food hypersensitivity test	Presence of IgE-dependent food allergy	[181,183]
	Determination of the percentage of T and B lymphocytes	T and B cell subsets are usually normal	[177]

Table 5. *Cont.*

Type of Research	Type of Examination	The Importance of the Examination	Reference
Specialized research	Endoscopy with intestinal biopsy	Necessary to characterize the presence of enteropathy of the small intestine, combined with the performance of Foxp3 staining	[184]
	Skin biopsy	The presence of lymphocyte infiltrates in biopsy samples as an autoimmune process	[177]
	Treg lymphocyte immunophenotyping	Determines the amount of Treg and the expression level of Foxp3	[177]
	Sequencing of the *FOXP3* gene	Evaluates mutations within the *FOXP3* gene to confirm the clinical picture	[177]

Genetic conditions are related to the mutation within the transcription factor Foxp3 which was found in nearly 60% of patients with IPEX syndrome. These include missense or frame alteration mutations and insertion or deletion mutations at splicing sites which result in the loss of function of this protein. The occurring mutations also have consequences regarding the quantitative and functional disorders of Treg cells in the body deficiencies which cause autoimmune disease. We can find a description in the literature of most cases where mutations occurring within the Foxp3 protein are hereditary, although this does not exclude the occurrence of mutations in a sporadic manner [185–187]. These mutations take place in almost every region encoding the Foxp3 protein which clearly indicates that each domain of this protein is functional and necessary for its proper functioning. Although mutations in the non-coding regions are also known, their frequency is much lower compared to the mutations in the coding regions, however, their occurrence may also be the cause of the development of IPEX [28,188]. Nevertheless, scientists have not shown a clear correlation between genotype and phenotype [181,189,190].

An important diagnostic and cognitive tool turned out to be the use of flow cytometry which makes it possible to assess the expression of Foxp3 on cells of the immune system. Although high levels of *FOXP3* expression are commonly attributed to CD4+CD25$_{high}$ Treg cells, it is also induced by TCR stimulation in naive CD4+CD25− T cells and may persist for up to several days in activated CD4+ T cells [191]. The belief of many scientists that subsets of CD4+ T cells expressed in different tissues have different patterns of CD25 and other Treg cell markers further complicates the task of distinguishing Treg *FOXP3*+ from *FOXP3*+ found on other cells of the immune system. Therefore, it is necessary to search for new diagnostic markers to allow their safe differentiation. One such marker is the IL-7 receptor (CD127) promoter which is the target of *FOXP3* mediated transcriptional repression. Thanks to this discovery, scientists were able to identify the most suppressive population of human Treg cells, expressing the phenotype CD4+CD25+CD127$_{low}$ [192,193]. This has been used in the diagnostic methodology of the IPEX team where cells identified as CD4+CD25+CD127$_{low}$ correlate well with the population of CD4+CD25+*FOXP3*+ cells. Of course, the two populations should not be considered identical since conventional CD4+CD25-CD127+ T (Tcon) cells rapidly decrease CD127 expression following IL-7 signaling and TCR stimulation [194]. Studies have shown that cells with the CD4+CD25+CD127$_{low}$ phenotype may also be present in some IPEX patients who suffer from decreased FOXP3 expression due to the occurrence of hypomorphic mutations [55,195]. It has also been demonstrated that patients with IPEX syndrome who have missense mutations and deletions in splicing sites do not have Treg CD4+CD25+Foxp3+ lymphocytes and have a more severe form of the disease. Additionally, the absence of CD4+CD25+ cells confirms the diagnosis [35]. IPEX also has high levels of IgE and IgA immunoglobulins, as well as eosinophilia, which proves that the transcription factor Foxp3 is strongly associated with the human immune response [179–182].

Due to the special role of the Foxp3 protein in the development and progression of IPEX teams, many scientists are considering using this molecule as a therapeutic agent in the treatment of this disease. The studies available in the literature also show that in the absence of antigenic stimulation, *FOXP3* expression is promoted in CD4+CD25− T cells by signaling cytokines, which include STAT5, IL-2, IL-7, and IL-15, on peripheral blood cells [35,196,197]. Additionally, it has been shown that cytokine treatment does not induce *FOXP3* expression in lymph node derived CD4+ T cells [35,197]. The use of flow cytometry analysis has shown that the expression level of the Foxp3 protein on TCD4+ lymphocytes from peripheral blood can vary up to 40-fold between individual cells. This explains that the strength and duration of *FOXP3* induction may influence discrepancies in cytokine activation-induced T cell suppression [35,198–201].

Another therapy using the role of the Foxp3 protein proposed in the literature is the use of hematopoietic stem cell transplantation. Currently, this therapy is the only treatment option for sick patients with IPEX. According to the study by Passerini et al. in 2013, the conversion of CD4+ Treg cells after lentivirus-mediated transfer of the *FOXP3* gene results in a population of CD4+*FOXP3*+ T cells which exhibit a stable phenotype and preserved suppressive function. In addition, their studies have shown that CD4+*FOXP3*+ T cells are stable during inflammation not only in vitro but also in in vivo models. Therefore, based on the above observations, the researchers proposed the use of *FOXP3* gene transfer therapy in the IPEX syndrome in order to restore immune tolerance [202]. From the description of clinical cases and literature data, we can conclude that there are also patients with symptoms that resemble IPEX syndrome, including enteropathy, autoimmune endocrinopathy, and dermatitis, but the age and gender of these patients are more varied (these symptoms are also present in women) [203]. From the study by Ochs and Torgerson on 100 patients with a phenotype corresponding to IPEX syndrome, nearly half of them did not have a mutation in the *FOXP3* transcription factor gene [186]. Such diseases include CD25 or IL-2RA deficiency, mutations within STAT5b, STAT1 or STAT3, Dedicator of Cytokinesis 8 (DOCK8) deficiency as well as infantile or eosinophilic enteropathies and severe combined immunodeficiency (SCID) (Figure 6). The underlying genetic defect of these syndromes is unknown and requires further intensive research.

4.1.2. The Role of the Foxp3 Protein in the Development of CVID

Recent studies have also shown the role of the Foxp3 protein in the development of another PID disease, i.e., CVID. This disorder belongs to the humoral immunity deficiencies and is characterized by a relatively mild course. It is a type of polygenic disorder associated with the defective production of immunoglobulins, often accompanied by autoimmunity [204]. The epidemiology of this disorder is still difficult to define. Literature data show that the estimated prevalence of CVID in the population is 1 in 30,000 people. The diagnosis of CVID occurs in most patients between the age of 20 and 40 due to late symptoms [205,206]. Based on the analysis carried out by Grathman et al., in 2014, it was possible to establish the relationship between the clinical picture and the differences and effects of immunoglobulin treatment in 2212 patients from several European countries. The results of this review showed that patient survival depends on the time of diagnosis as well as the age when the first symptoms of the disease appeared [207]. The later the first symptoms appeared and the more delayed diagnosis of the disease, the greater the risk of death at any age (each year of delay in diagnosis increases the risk of death by about 4.5%) [207]. However, the etiology of CVID is not fully known, as only 20% of patients have the genetic cause identified. The most common form is sporadic cases with no family history of the disease (90%) [208]. They can be caused by a complex interaction of environmental and genetic components (multi-factor inheritance), but genes involved in the development and function of immune cells have now been shown to be the main cause [208,209]. As we know, the main role of the immune system is to defend against infections while protecting the body's own cells. Antibodies, also known as immunoglobulins, are proteins produced by B lymphocytes. In order for B lymphocytes to function effectively, they usually need the help of other immune cells such as T lymphocytes. Most people with CVID have a normal number of B cells, but this is characterized by a maturation disruption and thus a decrease in antibody synthesis. These disorders can be caused by a lack of needed help from T cells to develop a normal immune response [210–212]. As a result, people suffering from CVID will differ in their ability to elicit effective antibody responses due to the lowered levels of immunoglobulin. We can distinguish three types of disorders here; the first one concerns the three main types of immunoglobulins (IgG, IgA, and IgM), the second one concerns disorders within IgG and IgA, and the third group concerns those in which only IgG is lowered [213,214]. A diagnosis of CVID is usually confirmed by abnormal blood tests and medical history.

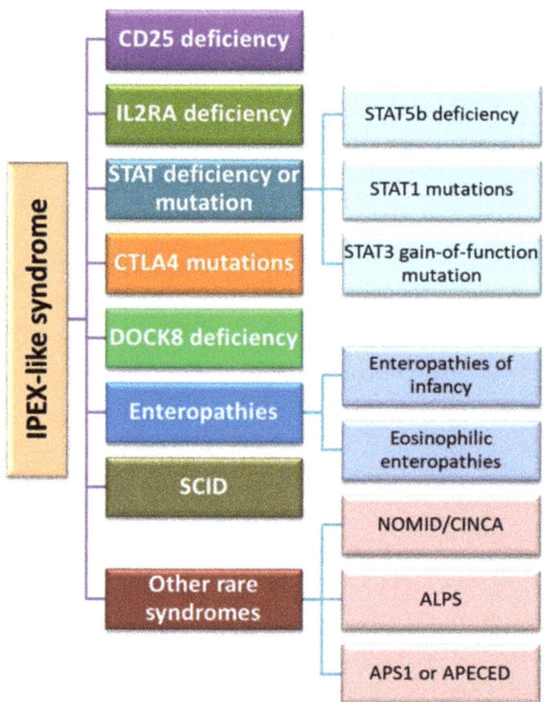

Figure 6. Diseases classified into IPEX-like syndromes (based on [177]). Abbreviations: IL2RA—Interleukin 2 Receptor Subunit Alpha; STAT—signal transducer and activator of transcription protein; STAT5—Signal Transducer And Activator Of Transcription 5; STAT1—Signal transducer and activator of transcription 1; STAT3—Signal Transducer And Activator Of Transcription 3; CTLA4—cytotoxic T cell antigen 4; DOCK8—Dedicator of cytokinesis 8; SCID—severe combined immunodeficiency; NOMID—neonatal onset multisystemic disease; CINCA—chronic infantile neurological, cutaneous, and articular syndrome; ALPS—Autoimmune Lymphoproliferative Syndrome; APS-1—Autoimmune Polyglandular Syndrome Type 1; APECED—Autoimmune polyendocrinopathy-candidiasis-ectodermal dystrophy.

Due to the important role of T lymphocytes in stimulating the synthesis of antibodies by B lymphocytes, some scientists have started research to determine whether the Foxp3 protein may be involved in this process [215]. The research conducted by Horn et al. in 2009 shows that this protein may indirectly influence the development of CVID. They analyzed the percentage of CD4+ Treg lymphocytes among patients diagnosed with CVID from different cities/countries: Freiburg, London, and Sydney, and correlated it with clinical symptoms. The percentage of Treg cells defined as CD25+Foxp3+ and CD25+CD127$_{low}$Foxp3+ or CD25+CD127$_{low}$CD4+ was analyzed and the results were compared with data from healthy patients. They found that, regardless of the phenotype used to define them, patients with CVID experienced a significant decrease in the percentage of Treg cells which correlated with the development of autoimmune disease. This provided evidence that a reduction in the number of Treg cells in CVIDs may play a role in the development and progression of clinical symptoms and may also contribute to understanding the pathogenesis of CVID complications [216]. Other research by Genere et al. built on the work presented by previous researchers; they showed that patients diagnosed with CVID and autoimmune disease had a significantly reduced frequency of CD4+CD25$_{HIGH}$Foxp3+ cells in the peripheral blood, accompanied by a reduced intensity of FOXP3 expression. Additionally, they found that although CVID patients with autoimmunity had a reduced frequency of CD4+CD25$_{HIGH}$Foxp3+ cells, FOXP3 expression levels did not differ from those of healthy controls. Thanks to the obtained results, the researchers showed that CD4+CD25$_{HIGH}$Foxp3+ cell homeostasis is disturbed in patients with CVID, especially in the presence of autoimmunity, which may indicate that Treg lymphocytes are involved in the pathological mechanisms of CVID [217]. The results of these studies were also confirmed by another research team (Arandi et al., 2013) which showed that the frequency of Treg was significantly lower in patients with CVID than in healthy subjects and that in patients with CVID, in whom autoimmunity was detected, the percentage of cells analyzed is significantly reduced compared to the cases without autoimmune diseases. There was also a significant difference in the expression level of the Foxp3 factor between patients with CVID and the control group [218]. The reduction of the expression level of the Foxp3 protein in patients with CVID was also analyzed and confirmed by Yu et al. [52]. They showed that the reduction in the levels of FoxP3, granzyme A, and pStat5 was significantly correlated with the degree of Treg dysfunction in CVID [52]. However, a full understanding of the role of Foxp3 in the development of CVID requires further extensive interdisciplinary research to understand its role in the pathogenesis of primary immunodeficiencies.

5. Materials and Methods

5.1. Search Strategy, Study Selection, and Data Extraction

The literature analysis was carried out on the PubMed database where the search for available articles was performed based on the following keywords: "Foxp3", "IPEX", "IPEX like", "Immunodysregulation, Polyendocrinopathy, and Enteropathy, X-Linked", "CVID", "Common Variable Immunodeficiency", and "Primary Immunodeficiency". The time range of the searched articles was established for the years 2000 to 2021 and filters related to the type of articles (clinical trials, review, systematic review) were used. Repetitions were rejected from the found articles. The suitability for the inclusion of each work into the publication was thoroughly assessed. Eventually, 223 articles were included in the review.

5.2. Biostatistical Analysis

For bioinformatic analysis, the amino acid sequences deposited in the UniProt database [219] were used. The identification numbers of the Foxp3 protein and its three isoforms, along with their amino acid sequences, are provided in Supplementary Materials Table S1. These sequences were used to carry out further bioinformatic analyses. The sequence length and molecular weight of individual proteins were pulled from the UniProt database. The determination of the isoelectric point of tested interleukins and their amino acid composition was carried out using the IPC isoelectric point calculator software available online [220]. The analysis of the second-order structure of interleukins was carried out using the NetSurfP-2.0 online program [221]. The amino acid sequences from the UniProt databases (Supplementary Materials Table S1) were used to analyze the identity of the amino acid sequences of the Foxp3 protein and its isoforms. The amino acid sequences of individual proteins were compared with each other using the Clustal Omega program available on their website [222]. The results of the analyses were presented as the percentage of identical amino acids in the analyzed amino acid sequences.

6. Conclusions

Numerous studies conducted in recent years have shown that Treg lymphocytes, which express Foxp3, appear in the human body immediately after birth and lead to the development of many inflammatory and autoimmune diseases after they are depleted. The Foxp3 protein has been shown to be necessary for lymphocytes in the thymus to differentiate into Treg lymphocytes. High expression of this transcription factor also guarantees their suppressive effect. The Foxp3 protein influences several cellular processes both directly and indirectly. In the process of cytokine production regulation, the Foxp3 protein interacts with numerous proteins and transcription factors such as NFAT, nuclear factor kappa B, and Runx1/AML1, and is involved in the process of histone acetylation in condensed chromatin. Thanks to their analyses and many experiments, scientists have shown that the similarity in the disturbance of the functioning of the FOXP3 gene in humans and mice is very similar. This allows for the conclusion that the process of dominant self-tolerance in these organisms is similar to each other. Scientists' persistence in researching the Foxp3 protein has led to including this factor in one of the most reliable molecular markers of natural Treg lymphocytes. In addition, studies on the dysfunction of the Foxp3 transcription factor caused by the mutagenesis process have shown that it significantly affects disorders of the immune response as well as the development and progression of primary immunodeficiencies or autoimmune diseases.

Supplementary Materials: The following are available online at https://www.mdpi.com/article/10.3390/jcm11040947/s1, Supplementary Materials Table S1: Amino acid sequences of the Foxp3 proprotein and its three isoforms from the UniProt database.

Author Contributions: Conceptualization, P.M., S.M., M.P. and E.G.; writing—original draft preparation, P.M., S.M. and M.P.; visualization, P.M.; methodology, P.M. and S.M.; software, P.M. and S.M.; formal analysis, P.M. and S.M.; investigation, P.M., S.M., M.P. and E.G.; writing—review and editing, E.G.; funding acquisition, E.G. All authors have read and agreed to the published version of the manuscript.

Funding: This work was supported by Research Grant No. DS640 of the Medical University of Lublin.

Institutional Review Board Statement: Not applicable.

Informed Consent Statement: Not applicable.

Data Availability Statement: Not applicable.

Conflicts of Interest: The authors declare no conflict of interest.

References

1. Mitsis, T.; Efthimiadou, A.; Bacopoulou, F.; Vlachakis, D.; Chrousos, G.; Eliopoulos, E. Transcription factors and evolution: An integral part of gene expression (Review). *World Acad. Sci. J.* **2020**, *2*, 3–8. [CrossRef]
2. Lu, L.; Barbi, J.; Pan, F. The regulation of immune tolerance by FOXP3. *Nat. Rev. Immunol.* **2017**, *17*, 703–717. [CrossRef] [PubMed]
3. Tai, X.; Erman, B.; Alag, A.; Mu, J.; Kimura, M.; Katz, G.; Guinter, T.; McCaughtry, T.; Etzensperger, R.; Feigenbaum, L.; et al. Foxp3 Transcription Factor Is Proapoptotic and Lethal to Developing Regulatory T Cells unless Counterbalanced by Cytokine Survival Signals. *Immunity* **2013**, *38*, 1116–1128. [CrossRef]
4. Colamatteo, A.; Carbone, F.; Bruzzaniti, S.; Galgani, M.; Fusco, C.; Maniscalco, G.T.; Di Rella, F.; De Candia, P.; De Rosa, V. Molecular Mechanisms Controlling Foxp3 Expression in Health and Autoimmunity: From Epigenetic to Post-translational Regulation. *Front. Immunol.* **2020**, *10*, 3136. [CrossRef] [PubMed]
5. Bluestone, J.A. FOXP3, the Transcription Factor at the Heart of the Rebirth of Immune Tolerance. *J. Immunol.* **2017**, *198*, 979–980. [CrossRef]
6. Chang, X.; Zheng, P.; Liu, Y. FoxP3: A genetic link between immunodeficiency and autoimmune diseases. *Autoimmun. Rev.* **2006**, *5*, 399–402. [CrossRef]
7. Justiz Vaillant, A.A.; Qurie, A. *Immunodeficiency*; StatPearls Publishing: Treasure Island, FL, USA, 2021.
8. Primary Immunodeficiency-Symptoms and Causes. Available online: https://www.mayoclinic.org/diseases-conditions/primary-immunodeficiency/symptoms-causes/syc-20376905 (accessed on 30 December 2021).
9. Sokol, K.; Milner, J.D. The overlap between allergy and immunodeficiency. *Curr. Opin. Pediatr.* **2018**, *30*, 848–854. [CrossRef]

10. Amaya-Uribe, L.; Rojas, M.; Azizi, G.; Anaya, J.-M.; Gershwin, M.E. Primary immunodeficiency and autoimmunity: A comprehensive review. *J. Autoimmun.* **2019**, *99*, 52–72. [CrossRef] [PubMed]

11. Todoric, K.; Koontz, J.B.; Mattox, D.; Tarrant, T.K. Autoimmunity in Immunodeficiency. *Curr. Allergy Asthma Rep.* **2013**, *13*, 361–370. [CrossRef]

12. Cowen, J.E.; Stevenson, J.; Paravasthu, M.; Darroch, J.; Jacob, A.; Tueger, S.; Gosney, J.R.; Simons, A.; Spencer, L.G.; Judge, E.P. Common variable immunodeficiency with granulomatous-lymphocytic interstitial lung disease and preceding neurological involvement: A case-report. *BMC Pulm. Med.* **2020**, *20*, 205. [CrossRef]

13. Mortaz, E.; Tabarsi, P.; Mansouri, D.; Khosravi, A.; Garssen, J.; Velayati, A.; Adcock, I.M. Cancers Related to Immunodeficiencies: Update and Perspectives. *Front. Immunol.* **2016**, *7*, 365. [CrossRef] [PubMed]

14. Haas, O.A. Primary Immunodeficiency and Cancer Predisposition Revisited: Embedding Two Closely Related Concepts into an Integrative Conceptual Framework. *Front. Immunol.* **2019**, *9*, 3136. [CrossRef] [PubMed]

15. Nozaki, T.; Takada, H.; Ishimura, M.; Ihara, K.; Imai, K.; Morio, T.; Kobayashi, M.; Nonoyama, S.; Hara, T. Endocrine complications in primary immunodeficiency diseases in Japan. *Clin. Endocrinol.* **2012**, *77*, 628–634. [CrossRef] [PubMed]

16. Coopmans, E.C.; Chunharojrith, P.; Neggers, S.J.C.M.M.; Van Der Ent, M.W.; Swagemakers, S.M.A.; Hollink, I.H.; Barendregt, B.H.; Van Der Spek, P.J.; Van Der Lely, A.-J.; Van Hagen, P.M.; et al. Endocrine Disorders Are Prominent Clinical Features in Patients with Primary Antibody Deficiencies. *Front. Immunol.* **2019**, *10*, 2079. [CrossRef] [PubMed]

17. Mohtashami, M.; Razavi, A.; Abolhassani, H.; Aghamohammadi, A.; Yazdani, R. Primary Immunodeficiency and Thrombocytopenia. *Int. Rev. Immunol.* **2021**, *40*, 1–43. [CrossRef] [PubMed]

18. Rezaei, N.; Moazzami, K.; Aghamohammadi, A.; Klein, C. Neutropenia and Primary Immunodeficiency Diseases. *Int. Rev. Immunol.* **2009**, *28*, 335–366. [CrossRef]

19. Sullivan, K.E. Neutropenia as a sign of immunodeficiency. *J. Allergy Clin. Immunol.* **2019**, *143*, 96–100. [CrossRef]

20. Soler-Palacín, P.; De Gracia, J.; Gonzalez-Granado, L.I.; Martín, C.; Rodríguez-Gallego, C.; Sánchez-Ramón, S. Lung ID-Signal Group Primary immunodeficiency diseases in lung disease: Warning signs, diagnosis and management. *Respir. Res.* **2018**, *19*, 219. [CrossRef]

21. Agarwal, S.; Mayer, L. Diagnosis and Treatment of Gastrointestinal Disorders in Patients with Primary Immunodeficiency. *Clin. Gastroenterol. Hepatol.* **2013**, *11*, 1050–1063. [CrossRef]

22. Lazaridis, N. Current insights into the innate immune system dysfunction in irritable bowel syndrome. *Ann. Gastroenterol.* **2018**, *31*, 171–187. [CrossRef]

23. Brede, K.K.; Wandel, M.; Wiig, I.; von der Lippe, C. Primary Immunodeficiency Diseases and Gastrointestinal Distress: Coping Strategies and Dietary Experiences to Relieve Symptoms. *Qual. Health Res.* **2020**, *31*, 361–372. [CrossRef] [PubMed]

24. Overview of Immunodeficiency Disorders-Immunology; Allergic Disorders. Available online: https://www.msdmanuals.com/en-kr/professional/immunology-allergic-disorders/immunodeficiency-disorders/overview-of-immunodeficiency-disorders (accessed on 30 December 2021).

25. Sánchez-Ramón, S.; Bermúdez, A.; González-Granado, L.I.; Rodríguez-Gallego, C.; Sastre, A.; Soler-Palacín, P.; the ID-Signal Onco-Haematology Group; Allende, L.; Alsina, L.; Bielsa, A.M.; et al. Primary and Secondary Immunodeficiency Diseases in Oncohaematology: Warning Signs, Diagnosis, and Management. *Front. Immunol.* **2019**, *10*, 586. [CrossRef] [PubMed]

26. Brunkow, M.E.; Jeffery, E.W.; Hjerrild, K.A.; Paeper, B.; Clark, L.B.; Yasayko, S.-A.; Wilkinson, J.E.; Galas, D.; Ziegler, S.F.; Ramsdell, F. Disruption of a new forkhead/winged-helix protein, scurfin, results in the fatal lymphoproliferative disorder of the scurfy mouse. *Nat. Genet.* **2001**, *27*, 68–73. [CrossRef] [PubMed]

27. Bennett, C.; Yoshioka, R.; Kiyosawa, H.; Barker, D.F.; Fain, P.R.; Shigeoka, A.O.; Chance, P.F. X-Linked Syndrome of Polyendocrinopathy, Immune Dysfunction, and Diarrhea Maps to Xp11.23-Xq13.3. *Am. J. Hum. Genet.* **2000**, *66*, 461–468. [CrossRef] [PubMed]

28. Georgiev, P.; Charbonnier, L.-M.; Chatila, T.A. Regulatory T Cells: The Many Faces of Foxp3. *J. Clin. Immunol.* **2019**, *39*, 623–640. [CrossRef]

29. FOXP3-Forkhead Box Protein P3-Homo Sapiens (Human)-FOXP3 Gene & Protein. Available online: https://www.uniprot.org/uniprot/Q9BZS1 (accessed on 30 December 2021).

30. Kwon, H.; Chen, H.-M.; Mathis, D.; Benoist, C. Different molecular complexes that mediate transcriptional induction and repression by FoxP3. *Nat. Immunol.* **2017**, *18*, 1238–1248. [CrossRef] [PubMed]

31. Huang, C.; Martin, S.; Pfleger, C.; Du, J.; Buckner, J.H.; Bluestone, J.A.; Riley, J.L.; Ziegler, S.F. Cutting Edge: A Novel, Human-Specific Interacting Protein Couples FOXP3 to a Chromatin-Remodeling Complex That Contains KAP1/TRIM28. *J. Immunol.* **2013**, *190*, 4470–4473. [CrossRef]

32. Magg, T.; Mannert, J.; Ellwart, J.W.; Schmid, I.; Albert, M.H. Subcellular localization of FOXP3 in human regulatory and nonregulatory T cells. *Eur. J. Immunol.* **2012**, *42*, 1627–1638. [CrossRef]

33. Du, J.; Huang, C.; Zhou, B.; Ziegler, S.F. Isoform-Specific Inhibition of RORα-Mediated Transcriptional Activation by Human FOXP3. *J. Immunol.* **2008**, *180*, 4785–4792. [CrossRef]

34. Mailer, R.K.W. Alternative Splicing of FOXP3—Virtue and Vice. *Front. Immunol.* **2018**, *9*, 530. [CrossRef]

35. Mailer, R.K. IPEX as a Consequence of Alternatively Spliced FOXP3. *Front. Pediatr.* **2020**, *8*, 594375. [CrossRef] [PubMed]

36. Kawajiri, K.; Ikuta, T.; Suzuki, T.; Kusaka, M.; Muramatsu, M.; Fujieda, K.; Tachibana, M.; Morohashi, K.-I. Role of the LXXLL-Motif and Activation Function 2 Domain in Subcellular Localization of Dax-1 (Dosage-Sensitive Sex Reversal-Adrenal Hypoplasia Congenita Critical Region on the X Chromosome, Gene 1). *Mol. Endocrinol.* **2003**, *17*, 994–1004. [CrossRef] [PubMed]
37. Nakae, J.; Cao, Y.; Daitoku, H.; Fukamizu, A.; Ogawa, W.; Yano, Y.; Hayashi, Y. The LXXLL motif of murine forkhead transcription factor FoxO1 mediates Sirt1-dependent transcriptional activity. *J. Clin. Investig.* **2006**, *116*, 2473–2483. [CrossRef] [PubMed]
38. Clerc, I.; Polakowski, N.; André-Arpin, C.; Cook, P.; Barbeau, B.; Mesnard, J.-M.; Lemasson, I. An Interaction between the Human T Cell Leukemia Virus Type 1 Basic Leucine Zipper Factor (HBZ) and the KIX Domain of p300/CBP Contributes to the Down-regulation of Tax-dependent Viral Transcription by HBZ. *J. Biol. Chem.* **2008**, *283*, 23903–23913. [CrossRef] [PubMed]
39. Garber, P.M.; Vidanes, G.M.; Toczyski, D.P. Damage in transition. *Trends Biochem. Sci.* **2005**, *30*, 63–66. [CrossRef]
40. Ikuta, T.; Watanabe, J.; Kawajiri, K. Characterization of the LxxLL Motif in the Aryl Hydrocarbon Receptor: Effects on Subcellular Localization and Transcriptional Activity. *J. Biochem.* **2002**, *131*, 79–85. [CrossRef]
41. Cassandri, M.; Smirnov, A.; Novelli, F.; Pitolli, C.; Agostini, M.; Malewicz, M.; Melino, G.; Raschellà, G. Zinc-finger proteins in health and disease. *Cell Death Discov.* **2017**, *3*, 17071. [CrossRef]
42. Song, X.; Li, B.; Xiao, Y.; Chen, C.; Wang, Q.; Liu, Y.; Berezov, A.; Xu, C.; Gao, Y.; Li, Z.; et al. Structural and Biological Features of FOXP3 Dimerization Relevant to Regulatory T Cell Function. *Cell Rep.* **2012**, *1*, 665–675. [CrossRef]
43. Li, J.; Jiang, L.; Liang, X.; Qu, L.; Wu, D.; Chen, X.; Guo, M.; Chen, Z.; Chen, L.; Chen, Y. DNA-binding properties of FOXP3 transcription factor. *Acta Biochim. Biophys. Sin.* **2017**, *49*, 792–799. [CrossRef]
44. Chae, W.-J.; Henegariu, O.; Lee, S.-K.; Bothwell, A.L.M. The mutant leucine-zipper domain impairs both dimerization and suppressive function of Foxp3 in T cells. *Proc. Natl. Acad. Sci. USA* **2006**, *103*, 9631–9636. [CrossRef]
45. Lopes, J.E.; Torgerson, T.R.; Schubert, L.A.; Anover, S.D.; Ocheltree, E.L.; Ochs, H.D.; Ziegler, S.F. Analysis of FOXP3 Reveals Multiple Domains Required for Its Function as a Transcriptional Repressor. *J. Immunol.* **2006**, *177*, 3133–3142. [CrossRef] [PubMed]
46. Mackey-Cushman, S.L.; Gao, J.; Holmes, A.D.; Nunoya, J.-I.; Wang, R.; Unutmaz, D.; Su, L. FoxP3 interacts with linker histone H1.5 to modulate gene expression and program Treg cell activity. *Genes Immun.* **2011**, *12*, 559–567. [CrossRef] [PubMed]
47. Lozano, T.; Casares, N.; Lasarte, J.J. Searching for the Achilles Heel of FOXP3. *Front. Oncol.* **2013**, *3*, 294. [CrossRef] [PubMed]
48. Liu, Y.; Wang, L.; Han, R.; Beier, U.H.; Hancock, W.W. Two Lysines in the Forkhead Domain of Foxp3 Are Key to T Regulatory Cell Function. *PLoS ONE* **2012**, *7*, e29035. [CrossRef]
49. Wu, Y.; Borde, M.; Heissmeyer, V.; Feuerer, M.; Lapan, A.D.; Stroud, J.; Bates, D.L.; Guo, L.; Han, A.; Ziegler, S.F.; et al. FOXP3 Controls Regulatory T Cell Function through Cooperation with NFAT. *Cell* **2006**, *126*, 375–387. [CrossRef]
50. Torgerson, T.R.; Genin, A.; Chen, C.; Zhang, M.; Zhou, B.; Añover-Sombke, S.; Frank, M.B.; Dozmorov, I.; Ocheltree, E.; Kulmala, P.; et al. FOXP3 Inhibits Activation-Induced NFAT2 Expression in T Cells Thereby Limiting Effector Cytokine Expression. *J. Immunol.* **2009**, *183*, 907–915. [CrossRef]
51. Romano, M.; Fanelli, G.; Albany, C.J.; Giganti, G.; Lombardi, G. Past, Present, and Future of Regulatory T Cell Therapy in Transplantation and Autoimmunity. *Front. Immunol.* **2019**, *10*, 43. [CrossRef]
52. Yu, G.; Berquist, S.; Huang, J.; Martinez, J.; Hoyte, E.; Vanichsarn, C.; Nadeau, K. Natural Regulatory T Cell Dysfunction in Subjects with Common Variable Immunodeficiency Complicated by Autoimmune Disease is associated with activation of the AkT-mTOR pathway. *J. Allergy Clin. Immunol.* **2009**, *123*, S264. [CrossRef]
53. Rudensky, A.Y. Regulatory T cells and Foxp3. *Immunol. Rev.* **2011**, *241*, 260–268. [CrossRef]
54. Bacchetta, R.; Gambineri, E.; Roncarolo, M.-G. Role of regulatory T cells and FOXP3 in human diseases. *J. Allergy Clin. Immunol.* **2007**, *120*, 227–235. [CrossRef]
55. Li, Z.; Li, D.; Tsun, A.; Li, B. FOXP3+ regulatory T cells and their functional regulation. *Cell. Mol. Immunol.* **2015**, *12*, 558–565. [CrossRef] [PubMed]
56. Barnes, M.J.; Griseri, T.; Johnson, A.M.F.; Young, W.; Powrie, F.; Izcue, A. CTLA-4 promotes Foxp3 induction and regulatory T cell accumulation in the intestinal lamina propria. *Mucosal Immunol.* **2012**, *6*, 324–334. [CrossRef]
57. Xiao, X.; Shi, X.; Fan, Y.; Zhang, X.; Wu, M.; Lan, P.; Minze, L.; Fu, Y.-X.; Ghobrial, R.M.; Liu, W.; et al. GITR subverts Foxp3+ Tregs to boost Th9 immunity through regulation of histone acetylation. *Nat. Commun.* **2015**, *6*, 8266. [CrossRef]
58. Bertolini, T.B.; Piñeros, A.R.; Prado, R.Q.; Gembre, A.F.; Ramalho, L.N.Z.; Alves-Filho, J.C.; Bonato, V.L.D. CCR4-dependent reduction in the number and suppressor function of CD4+Foxp3+ cells augments IFN-γ-mediated pulmonary inflammation and aggravates tuberculosis pathogenesis. *Cell Death Dis.* **2018**, *10*, 11. [CrossRef] [PubMed]
59. Klapa, S.; Mueller, A.; Csernok, E.; Fagin, U.; Klenerman, P.; Holl-Ulrich, K.; Gross, W.L.; Lamprecht, P. Lower numbers of FoxP3 and CCR4 co-expressing cells in an elevated subpopulation of CD4+CD25high regulatory T cells from Wegener's granulomatosis. *Clin. Exp. Rheumatol.* **2010**, *28*, 72–80.
60. Pelletier, B.; Mondoulet, L.; Ligouis, M.; Dhelft, V.; Plaquet, C.; Dupont, C.M.; Benhamou, P.-H.; Sampson, H.A. Foxp3+CD62L+ Tregs induced by EPIT have the potency to suppress effector T cells proliferation in specific and bystander conditions. *J. Allergy Clin. Immunol.* **2017**, *139*, AB255. [CrossRef]
61. Xu, L.; Kitani, A.; Strober, W. Molecular mechanisms regulating TGF-β-induced Foxp3 expression. *Mucosal Immunol.* **2010**, *3*, 230–238. [CrossRef]
62. Lin, X.; Chen, M.; Liu, Y.; Guo, Z.; He, X.; Brand, D.; Zheng, S.G. Advances in distinguishing natural from induced Foxp3(+) regulatory T cells. *Int. J. Clin. Exp. Pathol.* **2013**, *6*, 116–123.

63. Cosovanu, C.; Neumann, C. The Many Functions of Foxp3+ Regulatory T Cells in the Intestine. *Front. Immunol.* **2020**, *11*, 600973. [CrossRef]

64. Lio, C.-W.J.; Dodson, L.F.; Deppong, C.M.; Hsieh, C.-S.; Green, J.M. CD28 Facilitates the Generation of Foxp3− Cytokine Responsive Regulatory T Cell Precursors. *J. Immunol.* **2010**, *184*, 6007–6013. [CrossRef]

65. He, X.; Smeets, R.L.; Van Rijssen, E.; Boots, A.M.H.; Joosten, I.; Koenen, H.J.P.M. Single CD28 stimulation induces stable and polyclonal expansion of human regulatory T cells. *Sci. Rep.* **2017**, *7*, 43003. [CrossRef]

66. Freudenberg, K.; Lindner, N.; Dohnke, S.; Garbe, A.I.; Schallenberg, S.; Kretschmer, K. Critical Role of TGF-β and IL-2 Receptor Signaling in Foxp3 Induction by an Inhibitor of DNA Methylation. *Front. Immunol.* **2018**, *9*, 125. [CrossRef]

67. Schuster, M.; Plaza-Sirvent, C.; Visekruna, A.; Huehn, J.; Schmitz, I. Generation of Foxp3+CD25− Regulatory T-Cell Precursors Requires c-Rel and IκBNS. *Front. Immunol.* **2019**, *10*, 1583. [CrossRef]

68. Won, H.Y.; Shin, J.H.; Oh, S.; Jeong, H.; Hwang, E.S. Enhanced CD25+Foxp3+ regulatory T cell development by amodiaquine through activation of nuclear receptor 4A. *Sci. Rep.* **2017**, *7*, 16946. [CrossRef]

69. Shevach, E.M. Foxp3+ T Regulatory Cells: Still Many Unanswered Questions—A Perspective after 20 Years of Study. *Front. Immunol.* **2018**, *9*, 1048. [CrossRef]

70. Chien, C.-H.; Chiang, B.-L. Regulatory T cells induced by B cells: A novel subpopulation of regulatory T cells. *J. Biomed. Sci.* **2017**, *24*, 86. [CrossRef]

71. Engelmann, P.; Farkas, K.; Kis, J.; Richman, G.; Zhang, Z.; Liew, C.W.; Borowiec, M.; Niewczas, M.A.; Jalahej, H.; Orbán, T. Characterization of human invariant natural killer T cells expressing FoxP3. *Int. Immunol.* **2011**, *23*, 473–484. [CrossRef] [PubMed]

72. Niemeyer, M.; Darmoise, A.; Mollenkopf, H.-J.; Hahnke, K.; Hurwitz, R.; Besra, G.S.; Schaible, U.E.; Kaufmann, S.H.E. Natural killer T-cell characterization through gene expression profiling: An account of versatility bridging T helper type 1 (Th1), Th2 and Th17 immune responses. *Immunology* **2007**, *123*, 45–56. [CrossRef]

73. Koguchi, Y.; Buenafe, A.C.; Thauland, T.J.; Gardell, J.L.; Bivins-Smith, E.R.; Jacoby, D.B.; Slifka, M.K.; Parker, D.C. Preformed CD40L Is Stored in Th1, Th2, Th17, and T Follicular Helper Cells as Well as CD4+8− Thymocytes and Invariant NKT Cells but Not in Treg Cells. *PLoS ONE* **2012**, *7*, e31296. [CrossRef] [PubMed]

74. Monteiro, M.; Almeida, C.F.; Caridade, M.; Ribot, J.; Duarte, J.; Agua-Doce, A.; Wollenberg, I.; Silva-Santos, B.; Graca, L. Identification of Regulatory Foxp3+ Invariant NKT Cells Induced by TGF-β. *J. Immunol.* **2010**, *185*, 2157–2163. [CrossRef] [PubMed]

75. Lu, Y.; Guo, Y.; Xu, L.; Li, Y.; Cao, L. Foxp3 regulates ratio of Treg and NKT cells in a mouse model of asthma. *Mol. Cell. Biochem.* **2015**, *403*, 25–31. [CrossRef] [PubMed]

76. Noh, J.; Choi, W.S.; Noh, G.; Lee, J.H. Presence ofFoxp3-expressing CD19(+)CD5(+) B Cells in Human Peripheral Blood Mononuclear Cells: Human CD19(+)CD5(+)Foxp3(+) Regulatory B Cell (Breg). *Immune Netw.* **2010**, *10*, 247–249. [CrossRef]

77. Mohajeri, M.; Farazmand, A.; Bonab, M.M.; Nikbin, B.; Minagar, A. FOXP3 gene expression in multiple sclerosis patients pre- and post mesenchymal stem cell therapy. *Iran. J. Allergy Asthma Immunol.* **2011**, *10*, 155–161.

78. Abbass, A.A.; Mohamed, A.N.; Abdel-Rehim, A.S.M. Association of FOXP3 regulatory gene expression with systemic lupus erythematosus disease activity among Egyptian patients. *Egypt. J. Immunol.* **2013**, *20*, 21–28. [PubMed]

79. Bonelli, M.; Göschl, L.; Blüml, S.; Rath, E.; Smolen, J.S.; Scheinecker, C. CD4+CD25-Foxp3+ T cells are increased in systemic lupus erythematosus patients with active glomerulonephritis. *Ann. Rheum. Dis.* **2011**, *70*, A47–A48. [CrossRef]

80. Slobodin, G.; Sheikh-Ahmad, M.; Rosner, I.; Peri, R.; Rozenbaum, M.; Kessel, A.; Toubi, E.; Odeh, M. Regulatory T cells (CD4+CD25brightFoxP3+) expansion in systemic sclerosis correlates with disease activity and severity. *Cell. Immunol.* **2010**, *261*, 77–80. [CrossRef]

81. Vadasz, Z.; Toubi, E. FoxP3 Expression in Macrophages, Cancer, and B Cells—Is It Real? *Clin. Rev. Allergy Immunol.* **2016**, *52*, 364–372. [CrossRef] [PubMed]

82. Lozano, T.; Villanueva, L.; Durantez, M.; Gorraiz, M.; Ruiz, M.; Belsue, V.; Riezu-Boj, J.-I.; Hervás-Stubbs, S.; Oyarzabal, J.; Bandukwala, H.; et al. Inhibition of FOXP3/NFAT Interaction Enhances T Cell Function after TCR Stimulation. *J. Immunol.* **2015**, *195*, 3180–3189. [CrossRef]

83. Ono, M.; Yaguchi, H.; Ohkura, N.; Kitabayashi, I.; Nagamura, Y.; Nomura, T.; Miyachi, Y.; Tsukada, T.; Sakaguchi, S. Foxp3 controls regulatory T-cell function by interacting with AML1/Runx1. *Nature* **2007**, *446*, 685–689. [CrossRef] [PubMed]

84. Maeda, M.; Matsuzaki, H.; Yamamoto, S.; Lee, S.; Kumagai-Takei, N.; Yoshitome, K.; Min, Y.; Sada, N.; Nishimura, Y.; Otsuki, T. Aberrant expression of FoxP3 in a human T cell line possessing regulatory T cell-like function and exposed continuously to asbestos fibers. *Oncol. Rep.* **2018**, *40*, 748–758. [CrossRef] [PubMed]

85. Kitoh, A.; Ono, M.; Naoe, Y.; Ohkura, N.; Yamaguchi, T.; Yaguchi, H.; Kitabayashi, I.; Tsukada, T.; Nomura, T.; Miyachi, Y.; et al. Indispensable Role of the Runx1-Cbfβ Transcription Complex for In Vivo-Suppressive Function of FoxP3+ Regulatory T Cells. *Immunity* **2009**, *31*, 609–620. [CrossRef]

86. Walker, L.S. Treg and CTLA-4: Two intertwining pathways to immune tolerance. *J. Autoimmun.* **2013**, *45*, 49–57. [CrossRef]

87. Fan, M.Y.; Low, J.S.; Tanimine, N.; Finn, K.K.; Priyadharshini, B.; Germana, S.K.; Kaech, S.M.; Turka, L.A. Differential Roles of IL-2 Signaling in Developing versus Mature Tregs. *Cell Rep.* **2018**, *25*, 1204–1213.e4. [CrossRef]

88. Ferreira, R.C.; Simons, H.Z.; Thompson, W.S.; Rainbow, D.B.; Yang, X.; Cutler, A.J.; Oliveira, J.; Dopico, X.C.; Smyth, D.; Savinykh, N.; et al. Cells with Treg-specific FOXP3 demethylation but low CD25 are prevalent in autoimmunity. *J. Autoimmun.* **2017**, *84*, 75–86. [CrossRef] [PubMed]

89. Driesen, J.; Popov, A.; Schultze, J.L. CD25 as an immune regulatory molecule expressed on myeloid dendritic cells. *Immunobiology* **2008**, *213*, 849–858. [CrossRef] [PubMed]
90. Brunner-Weinzierl, M.; Rudd, C.E. CTLA-4 and PD-1 Control of T-Cell Motility and Migration: Implications for Tumor Immunotherapy. *Front. Immunol.* **2018**, *9*, 2737. [CrossRef] [PubMed]
91. Kim, J.-H.; Hwang, J.; Jung, J.H.; Lee, H.-J.; Lee, D.Y.; Kim, S.-H. Molecular networks of FOXP family: Dual biologic functions, interplay with other molecules and clinical implications in cancer progression. *Mol. Cancer* **2019**, *18*, 180. [CrossRef]
92. Zorn, E.; Nelson, E.A.; Mohseni, M.; Porcheray, F.; Kim, H.; Litsa, D.; Bellucci, R.; Raderschall, E.; Canning, C.; Soiffer, R.J.; et al. IL-2 regulates FOXP3 expression in human CD4+CD25+ regulatory T cells through a STAT-dependent mechanism and induces the expansion of these cells in vivo. *Blood* **2006**, *108*, 1571–1579. [CrossRef]
93. Hench, V.K.; Su, L. Regulation of IL-2 gene expression by Siva and FOXP3 in human T cells. *BMC Immunol.* **2011**, *12*, 54. [CrossRef]
94. Bendfeldt, H.; Benary, M.; Scheel, T.; Steinbrink, K.; Radbruch, A.; Herzel, H.; Baumgrass, R. IL-2 Expression in Activated Human Memory FOXP3+ Cells Critically Depends on the Cellular Levels of FOXP3 as Well as of Four Transcription Factors of T Cell Activation. *Front. Immunol.* **2012**, *3*, 264. [CrossRef]
95. Dikiy, S.; Li, J.; Bai, L.; Jiang, M.; Janke, L.; Zong, X.; Hao, X.; Hoyos, B.; Wang, Z.-M.; Xu, B.; et al. A distal Foxp3 enhancer enables interleukin-2 dependent thymic Treg cell lineage commitment for robust immune tolerance. *Immunity* **2021**, *54*, 931–946.e11. [CrossRef] [PubMed]
96. Konopacki, C.; Pritykin, Y.; Rubtsov, Y.; Leslie, C.S.; Rudensky, A.Y. Transcription factor Foxp1 regulates Foxp3 chromatin binding and coordinates regulatory T cell function. *Nat. Immunol.* **2019**, *20*, 232–242. [CrossRef] [PubMed]
97. Chen, C.; Rowell, E.A.; Thomas, R.M.; Hancock, W.W.; Wells, A.D. Transcriptional Regulation by Foxp3 Is Associated with Direct Promoter Occupancy and Modulation of Histone Acetylation. *J. Biol. Chem.* **2006**, *281*, 36828–36834. [CrossRef]
98. Ohkura, N.; Sakaguchi, S. Transcriptional and epigenetic basis of Treg cell development and function: Its genetic anomalies or variations in autoimmune diseases. *Cell Res.* **2020**, *30*, 465–474. [CrossRef] [PubMed]
99. He, H.; Ni, B.; Tian, Y.; Tian, Z.; Chen, Y.; Liu, Z.; Yang, X.; Lv, Y.; Zhang, Y. Histone methylation mediates plasticity of human FOXP3+regulatory T cells by modulating signature gene expressions. *Immunology* **2013**, *141*, 362–376. [CrossRef] [PubMed]
100. Li, B.; Samanta, A.; Song, X.; Iacono, K.T.; Bembas, K.; Tao, R.; Basu, S.; Riley, J.; Hancock, W.W.; Shen, Y.; et al. FOXP3 interactions with histone acetyltransferase and class II histone deacetylases are required for repression. *Proc. Natl. Acad. Sci. USA* **2007**, *104*, 4571–4576. [CrossRef]
101. Dahiya, S.; Beier, U.H.; Wang, L.; Han, R.; Jiao, J.; Akimova, T.; Angelin, A.; Wallace, D.C.; Hancock, W.W. HDAC10 deletion promotes Foxp3+ T-regulatory cell function. *Sci. Rep.* **2020**, *10*, 424. [CrossRef] [PubMed]
102. Beier, U.; Akimova, T.; Liu, Y.; Wang, L.; Hancock, W.W. Histone/protein deacetylases control Foxp3 expression and the heat shock response of T-regulatory cells. *Curr. Opin. Immunol.* **2011**, *23*, 670–678. [CrossRef]
103. Zhang, H.; Xiao, Y.; Zhu, Z.; Li, B.; Greene, I.M. Immune regulation by histone deacetylases: A focus on the alteration of FOXP3 activity. *Immunol. Cell Biol.* **2011**, *90*, 95–100. [CrossRef]
104. van Loosdregt, J.; Vercoulen, Y.; Guichelaar, T.; Gent, Y.Y.J.; Beekman, J.M.; Van Beekum, O.; Brenkman, A.B.; Hijnen, D.; Mutis, T.; Kalkhoven, E.; et al. Regulation of Treg functionality by acetylation-mediated Foxp3 protein stabilization. *Blood* **2010**, *115*, 965–974. [CrossRef]
105. Deng, G.; Song, X.; Fujimoto, S.; Piccirillo, C.A.; Nagai, Y.; Greene, M.I. Foxp3 Post-translational Modifications and Treg Suppressive Activity. *Front. Immunol.* **2019**, *10*, 2486. [CrossRef] [PubMed]
106. Xiao, Y.; Nagai, Y.; Deng, G.; Ohtani, T.; Zhu, Z.; Zhou, Z.; Zhang, H.; Ji, M.Q.; Lough, J.W.; Samanta, A.; et al. Dynamic Interactions between TIP60 and p300 Regulate FOXP3 Function through a Structural Switch Defined by a Single Lysine on TIP60. *Cell Rep.* **2014**, *7*, 1471–1480. [CrossRef] [PubMed]
107. De Zoeten, E.; Wang, L.; Butler, K.V.; Beier, U.; Akimova, T.; Sai, H.; Bradner, J.E.; Mazitschek, R.; Kozikowski, A.P.; Matthias, P.; et al. Histone Deacetylase 6 and Heat Shock Protein 90 Control the Functions of Foxp3+ T-Regulatory Cells. *Mol. Cell. Biol.* **2011**, *31*, 2066–2078. [CrossRef] [PubMed]
108. Mollapour, M.; Neckers, L. Post-translational modifications of Hsp90 and their contributions to chaperone regulation. *Biochim. Biophys. Acta* **2012**, *1823*, 648–655. [CrossRef]
109. Nagai, Y.; Lam, L.; Greene, M.I.; Zhang, H. FOXP3 and Its Cofactors as Targets of Immunotherapies. *Engineering* **2019**, *5*, 115–121. [CrossRef]
110. Van Loosdregt, J.; Brunen, D.; Fleskens, V.; Pals, C.E.G.M.; Lam, E.W.F.; Coffer, P.J. Rapid Temporal Control of Foxp3 Protein Degradation by Sirtuin-1. *PLoS ONE* **2011**, *6*, e19047. [CrossRef] [PubMed]
111. Polansky, J.K.; Kretschmer, K.; Freyer, J.; Floess, S.; Garbe, A.; Baron, U.; Olek, S.; Hamann, A.; von Boehmer, H.; Huehn, J. DNA methylation controls Foxp3 gene expression. *Eur. J. Immunol.* **2008**, *38*, 1654–1663. [CrossRef] [PubMed]
112. Morikawa, H.; Ohkura, N.; Vandenbon, A.; Itoh, M.; Nagao-Sato, S.; Kawaji, H.; Lassmann, T.; Carninci, P.; Hayashizaki, Y.; Forrest, A.; et al. Differential roles of epigenetic changes and Foxp3 expression in regulatory T cell-specific transcriptional regulation. *Proc. Natl. Acad. Sci. USA* **2014**, *111*, 5289–5294. [CrossRef]
113. Maruyama, T.; Konkel, J.E.; Zamarron, B.F.; Chen, W. The molecular mechanisms of Foxp3 gene regulation. *Semin. Immunol.* **2011**, *23*, 418–423. [CrossRef]
114. Zheng, Y.; Josefowicz, S.; Chaudhry, A.; Peng, X.P.; Forbush, K.; Rudensky, A.Y. Role of conserved non-coding DNA elements in the Foxp3 gene in regulatory T-cell fate. *Nature* **2010**, *463*, 808–812. [CrossRef]

115. Montauti, E.; Zhang, Y.; Cortez, J.; Marson, A.; Fang, D. USP22 Stabilizes FoxP3 to Maintain Regulatory T Cell Function through a Dual Mechanism Resulting in Tumor Immune Evasion. *J. Immunol.* **2020**, *204*, 244.1.

116. Montauti, E.; Zhang, Y.; Fang, D. USP22 Maintains Regulatory T Cells by Stabilizing Foxp3 Protein Level and Impairs Anti-Tumor Immune Response. *J. Immunol.* **2018**, *200*, 57.10.

117. Zhang, Y.; Liu, W.; Chen, Y.; Liu, J.; Wu, K.; Su, L.; Zhang, W.; Jiang, Y.; Zhang, X.; Zhang, Y.; et al. A Cellular MicroRNA Facilitates Regulatory T Lymphocyte Development by Targeting the FOXP3 Promoter TATA-Box Motif. *J. Immunol.* **2017**, *200*, 1053–1063. [CrossRef]

118. Hippen, K.L.; Loschi, M.; Nicholls, J.; MacDonald, K.P.A.; Blazar, B.R. Effects of MicroRNA on Regulatory T Cells and Implications for Adoptive Cellular Therapy to Ameliorate Graft-versus-Host Disease. *Front. Immunol.* **2018**, *9*, 57. [CrossRef]

119. Dong, Y.; Yang, C.; Pan, F. Post-Translational Regulations of Foxp3 in Treg Cells and Their Therapeutic Applications. *Front. Immunol.* **2021**, *12*, 6172. [CrossRef] [PubMed]

120. Bin Dhuban, K.; D'Hennezel, E.; Nagai, Y.; Xiao, Y.; Shao, S.; Istomine, R.; Alvarez, F.; Ben-Shoshan, M.; Ochs, H.; Mazer, B.; et al. Suppression by human FOXP3+ regulatory T cells requires FOXP3-TIP60 interactions. *Sci. Immunol.* **2017**, *2*, eaai9297. [CrossRef]

121. Liu, Y.; Wang, L.; Han, R.; Beier, U.; Akimova, T.; Bhatti, T.; Xiao, H.; Cole, P.A.; Brindle, P.; Hancock, W.W. Two Histone/Protein Acetyltransferases, CBP and p300, Are Indispensable for Foxp3+ T-Regulatory Cell Development and Function. *Mol. Cell. Biol.* **2014**, *34*, 3993–4007. [CrossRef]

122. Wang, L.; Beier, U.; Akimova, T.; Dahiya, S.; Han, R.; Samanta, A.; Levine, M.H.; Hancock, W.W. Histone/protein deacetylase inhibitor therapy for enhancement of Foxp3+ T-regulatory cell function posttransplantation. *Am. J. Transplant.* **2018**, *18*, 1596–1603. [CrossRef]

123. Wang, L.; Liu, Y.; Han, R.; Beier, U.H.; Bhatti, T.R.; Akimova, T.; Greene, M.I.; Hiebert, S.W.; Hancock, W.W. FOXP3+ regulatory T cell development and function require histone/protein deacetylase 3. *J. Clin. Investig.* **2015**, *125*, 1111–1123. [CrossRef]

124. De Zoeten, E.; Wang, L.; Sai, H.; Dillmann, W.H.; Hancock, W.W. Inhibition of HDAC9 Increases T Regulatory Cell Function and Prevents Colitis in Mice. *Gastroenterology* **2010**, *138*, 583–594. [CrossRef]

125. Yang, X.; Lun, Y.; Jiang, H.; Liu, X.; Duan, Z.; Xin, S.; Zhang, J. SIRT1-Regulated Abnormal Acetylation of FOXP3 Induces Regulatory T-Cell Function Defect in Hashimoto's Thyroiditis. *Thyroid* **2018**, *28*, 246–256. [CrossRef]

126. Beier, U.H.; Wang, L.; Bhatti, T.R.; Liu, Y.; Han, R.; Ge, G.; Hancock, W.W. Sirtuin-1 Targeting Promotes Foxp3+ T-Regulatory Cell Function and Prolongs Allograft Survival. *Mol. Cell. Biol.* **2011**, *31*, 1022–1029. [CrossRef]

127. Kempkes, R.; Joosten, I.; Koenen, H.J.P.M.; He, X. Metabolic Pathways Involved in Regulatory T Cell Functionality. *Front. Immunol.* **2019**, *10*, 2839. [CrossRef]

128. De Rosa, V.; Galgani, M.; Porcellini, A.; Colamatteo, A.; Santopaolo, M.; Zuchegna, C.; Romano, A.; De Simone, S.; Procaccini, C.; La Rocca, C.; et al. Glycolysis controls the induction of human regulatory T cells by modulating the expression of FOXP3 exon 2 splicing variants. *Nat. Immunol.* **2015**, *16*, 1174–1184. [CrossRef] [PubMed]

129. Shi, H.; Chi, H. Metabolic Control of Treg Cell Stability, Plasticity, and Tissue-Specific Heterogeneity. *Front. Immunol.* **2019**, *10*, 2716. [CrossRef] [PubMed]

130. Chapman, N.M.; Chi, H. mTOR signaling, Tregs and immune modulation. *Immunotherapy* **2014**, *6*, 1295–1311. [CrossRef] [PubMed]

131. An, J.; Ding, Y.; Yu, C.; Li, J.; You, S.; Liu, Z.; Song, P.; Zou, M.-H. AMP-activated protein kinase alpha1 promotes tumor development via FOXP3 elevation in tumor-infiltrating Treg cells. *iScience* **2022**, *25*, 103570. [CrossRef]

132. He, N.; Fan, W.; Henriquez, B.; Yu, R.T.; Atkins, A.R.; Liddle, C.; Zheng, Y.; Downes, M.; Evans, R.M. Metabolic control of regulatory T cell (Treg) survival and function by Lkb1. *Proc. Natl. Acad. Sci. USA* **2017**, *114*, 12542–12547. [CrossRef]

133. Raud, B.; Roy, D.G.; Divakaruni, A.S.; Tarasenko, T.N.; Franke, R.; Ma, E.H.; Samborska, B.; Hsieh, W.Y.; Wong, A.H.; Stüve, P.; et al. Etomoxir Actions on Regulatory and Memory T Cells Are Independent of Cpt1a-Mediated Fatty Acid Oxidation. *Cell Metab.* **2018**, *28*, 504–515.e7. [CrossRef]

134. Zhu, X.; Ji, W.; Guo, S.; Zhu, D.; Yang, Y.; Liu, X. Glycolytic and lipid oxidative metabolic programs are essential for freshly-isolated regulatory T cells in mice with sepsis. *RSC Adv.* **2020**, *10*, 21000–21008. [CrossRef]

135. Brown, Z.J.; Fu, Q.; Ma, C.; Kruhlak, M.; Zhang, H.; Luo, J.; Heinrich, B.; Yu, S.J.; Zhang, Q.; Wilson, A.; et al. Carnitine palmitoyltransferase gene upregulation by linoleic acid induces CD4+ T cell apoptosis promoting HCC development. *Cell Death Dis.* **2018**, *9*, 1–14. [CrossRef] [PubMed]

136. Atif, M.; Mohr, A.; Conti, F.; Scatton, O.; Gorochov, G.; Miyara, M. Metabolic Optimisation of Regulatory T Cells in Transplantation. *Front. Immunol.* **2020**, *11*, 2005. [CrossRef] [PubMed]

137. Witt, D.R.; Kellogg, R.A.; Snyder, M.P.; Dunn, J. Windows into human health through wearables data analytics. *Curr. Opin. Biomed. Eng.* **2019**, *9*, 28–46. [CrossRef]

138. Yuan, Z.; Peng, L.; Radhakrishnan, R.; Seto, E. Histone Deacetylase 9 (HDAC9) Regulates the Functions of the ATDC (TRIM29) Protein. *J. Biol. Chem.* **2010**, *285*, 39329–39338. [CrossRef] [PubMed]

139. Shen, L.; Pili, R. Class I histone deacetylase inhibition is a novel mechanism to target regulatory T cells in immunotherapy. *OncoImmunology* **2012**, *1*, 948–950. [CrossRef]

140. Terranova-Barberio, M.; Thomas, S.; Ali, N.; Pawlowska, N.; Park, J.; Krings, G.; Rosenblum, M.D.; Budillon, A.; Munster, P.N. HDAC inhibition potentiates immunotherapy in triple negative breast cancer. *Oncotarget* **2017**, *8*, 114156–114172. [CrossRef]

141. Chen, I.-C.; Sethy, B.; Liou, J.-P. Recent Update of HDAC Inhibitors in Lymphoma. *Front. Cell Dev. Biol.* **2020**, *8*, 906. [CrossRef]

142. Akimova, T.; Ge, G.; Golovina, T.; Mikheeva, T.; Wang, L.; Riley, J.L.; Hancock, W.W. Histone/protein deacetylase inhibitors increase suppressive functions of human FOXP3+ Tregs. *Clin. Immunol.* **2010**, *136*, 348–363. [CrossRef]

143. Xiao, Y.; Li, B.; Zhou, Z.; Hancock, W.W.; Zhang, H.; I Greene, M. Histone acetyltransferase mediated regulation of FOXP3 acetylation and Treg function. *Curr. Opin. Immunol.* **2010**, *22*, 583–591. [CrossRef]

144. Bettini, M.L.; Pan, F.; Bettini, M.; Finkelstein, D.; Rehg, J.E.; Floess, S.; Bell, B.D.; Ziegler, S.F.; Huehn, J.; Pardoll, D.M.; et al. Loss of Epigenetic Modification Driven by the Foxp3 Transcription Factor Leads to Regulatory T Cell Insufficiency. *Immunity* **2012**, *36*, 717–730. [CrossRef]

145. Wang, L.; de Zoeten, E.; Greene, M.I.; Hancock, W.W. Immunomodulatory effects of deacetylase inhibitors: Therapeutic targeting of FOXP3+ regulatory T cells. *Nat. Rev. Drug Discov.* **2009**, *8*, 969–981. [CrossRef] [PubMed]

146. Saouaf, S.J.; Li, B.; Zhang, G.; Shen, Y.; Furuuchi, N.; Hancock, W.W.; Greene, M.I. Deacetylase inhibition increases regulatory T cell function and decreases incidence and severity of collagen-induced arthritis. *Exp. Mol. Pathol.* **2009**, *87*, 99–104. [CrossRef] [PubMed]

147. Lu, J.; He, X.; Zhang, L.; Zhang, R.; Li, W. Acetylation in Tumor Immune Evasion Regulation. *Front. Pharmacol.* **2021**, *12*, 1558. [CrossRef] [PubMed]

148. Grover, P.; Goel, P.N.; Piccirillo, C.A.; Greene, M.I. FOXP3 and Tip60 Structural Interactions Relevant to IPEX Development Lead to Potential Therapeutics to Increase FOXP3 Dependent Suppressor T Cell Functions. *Front. Pediatr.* **2021**, *9*, 7292. [CrossRef] [PubMed]

149. Santer, F.R.; Höschele, P.P.; Oh, S.J.; Erb, H.H.; Bouchal, J.; Cavarretta, I.T.; Parson, W.; Meyers, D.J.; Cole, P.A.; Culig, Z. Inhibition of the Acetyltransferases p300 and CBP Reveals a Targetable Function for p300 in the Survival and Invasion Pathways of Prostate Cancer Cell Lines. *Mol. Cancer Ther.* **2011**, *10*, 1644–1655. [CrossRef] [PubMed]

150. Hamid, I.J.A.; Azman, N.A.; Gennery, A.R.; Mangantig, E.; Hashim, I.F.; Zainudeen, Z.T. Systematic Review of Primary Immunodeficiency Diseases in Malaysia: 1979–2020. *Front. Immunol.* **2020**, *11*, 1923. [CrossRef] [PubMed]

151. Gupta, S.; Sehgal, S. Editorial: Advances in Primary Immunodeficiencies in India. *Front. Immunol.* **2021**, *12*, 1335. [CrossRef] [PubMed]

152. Al-Herz, W.; Bousfiha, A.; Casanova, J.-L.; Chatila, T.; Conley, M.E.; Cunningham-Rundles, C.; Etzioni, A.; Franco, J.L.; Gaspar, H.B.; Holland, S.M.; et al. Primary Immunodeficiency Diseases: An Update on the Classification from the International Union of Immunological Societies Expert Committee for Primary Immunodeficiency. *Front. Immunol.* **2014**, *5*, 162. [CrossRef]

153. Picard, C.; Bobby Gaspar, H.; Al-Herz, W.; Bousfiha, A.; Casanova, J.L.; Chatila, T.; Crow, Y.J.; Cunningham-Rundles, C.; Etzioni, A.; Franco, J.L.; et al. International Union of Immunological Societies: 2017 Primary Immunodeficiency Diseases Committee Report on Inborn Errors of Immunity. *J. Clin. Immunol.* **2018**, *38*, 96–128. [CrossRef]

154. Tangye, S.G.; Al-Herz, W.; Bousfiha, A.; Chatila, T.; Cunningham-Rundles, C.; Etzioni, A.; Franco, J.L.; Holland, S.M.; Klein, C.; Morio, T.; et al. Human Inborn Errors of Immunity: 2019 Update on the Classification from the International Union of Immunological Societies Expert Committee. *J. Clin. Immunol.* **2020**, *40*, 24–64. [CrossRef]

155. Abolhassani, H.; Azizi, G.; Sharifi, L.; Yazdani, R.; Mohsenzadegan, M.; Delavari, S.; Sohani, M.; Shirmast, P.; Chavoshzadeh, Z.; Mahdaviani, S.A.; et al. Global systematic review of primary immunodeficiency registries. *Expert Rev. Clin. Immunol.* **2020**, *16*, 717–732. [CrossRef]

156. Bousfiha, A.; Jeddane, L.; Picard, C.; Al-Herz, W.; Ailal, F.; Chatila, T.; Cunningham-Rundles, C.; Etzioni, A.; Franco, J.; Holland, S.M.; et al. Human Inborn Errors of Immunity: 2019 Update of the IUIS Phenotypical Classification. *J. Clin. Immunol.* **2020**, *40*, 66–81. [CrossRef] [PubMed]

157. Costagliola, G.; Cappelli, S.; Consolini, R. Autoimmunity in Primary Immunodeficiency Disorders: An Updated Review on Pathogenic and Clinical Implications. *J. Clin. Med.* **2021**, *10*, 4729. [CrossRef] [PubMed]

158. McCusker, C.; Upton, J.; Warrington, R. Primary immunodeficiency. *Allergy Asthma Clin. Immunol.* **2018**, *14*, 61. [CrossRef] [PubMed]

159. Lin, L.-J.; Wang, Y.-C.; Liu, X.-M. Clinical and Immunological Features of Common Variable Immunodeficiency in China. *Chin. Med. J.* **2015**, *128*, 310–315. [CrossRef]

160. Wu, J.; Zhong, W.; Yin, Y.; Zhang, H. Primary immunodeficiency disease: A retrospective study of 112 Chinese children in a single tertiary care center. *BMC Pediatr.* **2019**, *19*, 410–417. [CrossRef]

161. Oliveira, J.B.; Fleisher, T.A. Laboratory evaluation of primary immunodeficiencies. *J. Allergy Clin. Immunol.* **2010**, *125*, S297–S305. [CrossRef]

162. Heimall, J.R.; Hagin, D.; Hajjar, J.; Henrickson, S.; Hernandez-Trujillo, H.S.; Tan, Y.; Kobrynski, L.; Paris, K.; Torgerson, T.R.; Verbsky, J.W.; et al. Use of Genetic Testing for Primary Immunodeficiency Patients. *J. Clin. Immunol.* **2018**, *38*, 320–329. [CrossRef]

163. Bacchetta, R.; Barzaghi, F.; Roncarolo, M.-G. From IPEX syndrome to FOXP3 mutation: A lesson on immune dysregulation. *Ann. N. Y. Acad. Sci.* **2016**, *1417*, 5–22. [CrossRef]

164. Sakaguchi, S.; Yamaguchi, T.; Nomura, T.; Ono, M. Regulatory T Cells and Immune Tolerance. *Cell* **2008**, *133*, 775–787. [CrossRef]

165. Gavin, M.A.; Rasmussen, J.; Fontenot, J.D.; Vasta, V.; Manganiello, V.C.; Beavo, J.A.; Rudensky, A.Y. Foxp3-dependent programme of regulatory T-cell differentiation. *Nature* **2007**, *445*, 771–775. [CrossRef]

166. Lin, Y.-L.; Shieh, C.-C.; Wang, J.-Y. The functional insufficiency of human CD4+CD25high T-regulatory cells in allergic asthma is subjected to TNF-α modulation. *Allergy* **2007**, *63*, 67–74. [CrossRef]

167. Fontenot, J.D.; Gavin, M.A.; Rudensky, A.Y. Foxp3 programs the development and function of CD4+CD25+ regulatory T cells. *Nat. Immunol.* **2003**, *4*, 330–336. [CrossRef]

168. Hori, S.; Nomura, T.; Sakaguchi, S. Control of Regulatory T Cell Development by the Transcription Factor Foxp3. *Science* **2003**, *299*, 1057–1061. [CrossRef] [PubMed]

169. Khattri, R.; Cox, T.; Yasayko, S.-A.; Ramsdell, F. An essential role for Scurfin in CD4+CD25+ T regulatory cells. *Nat. Immunol.* **2003**, *4*, 337–342. [CrossRef] [PubMed]

170. Liu, H.; Hu, B.; Xu, D.; Liew, F.Y. CD4+CD25+ Regulatory T Cells Cure Murine Colitis: The Role of IL-10, TGF-β, and CTLA4. *J. Immunol.* **2003**, *171*, 5012–5017. [CrossRef] [PubMed]

171. Nguyen, M.-T.J.; Fryml, E.; Sahakian, S.K.; Liu, S.; Michel, R.P.; Lipman, M.L.; Mucsi, I.; Cantarovich, M.; Tchervenkov, J.I.; Paraskevas, S. Pretransplantation Recipient Regulatory T cell Suppressive Function Predicts Delayed and Slow Graft Function after Kidney Transplantation. *Transplantation* **2014**, *98*, 745–753. [CrossRef]

172. Palomares, O.; Yaman, G.; Azkur, A.K.; Akkoc, T.; Akdis, M.; Akdis, C.A. Role of Treg in immune regulation of allergic diseases. *Eur. J. Immunol.* **2010**, *40*, 1232–1240. [CrossRef]

173. Yamaguchi, T.; Wing, J.B.; Sakaguchi, S. Two modes of immune suppression by Foxp3+ regulatory T cells under inflammatory or non-inflammatory conditions. *Semin. Immunol.* **2011**, *23*, 424–430. [CrossRef]

174. Kwon, H.; Chen, H.-M.; Mathis, D.; Benoist, C. FoxP3 scanning mutagenesis reveals functional variegation and mild mutations with atypical autoimmune phenotypes. *Proc. Natl. Acad. Sci. USA* **2017**, *115*, E253–E262. [CrossRef]

175. Van Gool, F.; Nguyen, M.L.; Mumbach, M.R.; Satpathy, A.T.; Rosenthal, W.L.; Giacometti, S.; Le, D.T.; Liu, W.; Brusko, T.M.; Anderson, M.S.; et al. A Mutation in the Transcription Factor Foxp3 Drives T Helper 2 Effector Function in Regulatory T Cells. *Immunity* **2019**, *50*, 362–377.e6. [CrossRef]

176. Van Der Vliet, H.J.J.; Nieuwenhuis, E.E. IPEX as a Result of Mutations in FOXP3. *Clin. Dev. Immunol.* **2007**, *2007*, e89017. [CrossRef]

177. IPEX: Immune Dysregulation, Polyendocrinopathy, Enteropathy, X-Linked-UpToDate. Available online: https://www.uptodate.com/contents/ipex-immune-dysregulation-polyendocrinopathy-enteropathy-x-linked (accessed on 30 December 2021).

178. Hines, B.; Wright, B.L.; Wadera, S.; Cac, N.; Miller, H.K.; Abraham, R.S.; Bauer, C.S. IPEX Syndrome in Siblings with a Novel Variant in FOXP3. *J. Allergy Clin. Immunol.* **2019**, *143*, AB118. [CrossRef]

179. Ozcan, E.; Notarangelo, L.D.; Geha, R.S. Primary immune deficiencies with aberrant IgE production. *J. Allergy Clin. Immunol.* **2008**, *122*, 1054–1062. [CrossRef] [PubMed]

180. Torgerson, T.R.; Ochs, H.D. Immune dysregulation, polyendocrinopathy, enteropathy, X-linked: Forkhead box protein 3 mutations and lack of regulatory T cells. *J. Allergy Clin. Immunol.* **2007**, *120*, 744–750. [CrossRef]

181. Torgerson, T.R.; Linane, A.; Moes, N.; Anover, S.; Mateo, V.; Rieux-Laucat, F.; Hermine, O.; Vijay, S.; Gambineri, E.; Cerf-Bensussan, N.; et al. Severe Food Allergy as a Variant of IPEX Syndrome Caused by a Deletion in a Noncoding Region of the FOXP3 Gene. *Gastroenterology* **2007**, *132*, 1705–1717. [CrossRef] [PubMed]

182. Lucas, K.G.; Ungar, D.; Comito, M.; Bayerl, M.; Groh, B. Submyeloablative cord blood transplantation corrects clinical defects seen in IPEX syndrome. *Bone Marrow Transpl.* **2006**, *39*, 55–56. [CrossRef] [PubMed]

183. Chatila, T.A.; Blaeser, F.; Ho, N.; Lederman, H.M.; Voulgaropoulos, C.; Helms, C.; Bowcock, A. JM2, encoding a fork head–related protein, is mutated in X-linked autoimmunity–allergic disregulation syndrome. *J. Clin. Investig.* **2000**, *106*, R75–R81. [CrossRef]

184. Heltzer, M.L.; Choi, J.K.; Ochs, H.D.; Sullivan, K.E.; Torgerson, T.R.; Ernst, L. A Potential Screening Tool for IPEX Syndrome. *Pediatr. Dev. Pathol.* **2007**, *10*, 98–105. [CrossRef] [PubMed]

185. Owen, C.J.; Jennings, C.E.; Imrie, H.; Lachaux, A.; Bridges, N.A.; Cheetham, T.D.; Pearce, S. Mutational Analysis of the FOXP3 Gene and Evidence for Genetic Heterogeneity in the Immunodysregulation, Polyendocrinopathy, Enteropathy Syndrome. *J. Clin. Endocrinol. Metab.* **2003**, *88*, 6034–6039. [CrossRef] [PubMed]

186. Ochs, H.D.; Torgerson, T.R. Immune dysregulation, polyendocrinopathy, enteropathy, X-linked inheritance: Model for autoaggression. *Adv. Exp. Med. Biol.* **2007**, *601*, 27–36. [CrossRef]

187. Gambineri, E.; Perroni, L.; Passerini, L.; Bianchi, L.; Doglioni, C.; Meschi, F.; Bonfanti, R.; Sznajer, Y.; Tommasini, A.; Lawitschka, A.; et al. Clinical and molecular profile of a new series of patients with immune dysregulation, polyendocrinopathy, enteropathy, X-linked syndrome: Inconsistent correlation between forkhead box protein 3 expression and disease severity. *J. Allergy Clin. Immunol.* **2008**, *122*, 1105–1112.e1. [CrossRef] [PubMed]

188. Barzaghi, F.; Passerini, L.; Bacchetta, R. Immune Dysregulation, Polyendocrinopathy, Enteropathy, X-Linked Syndrome: A Paradigm of Immunodeficiency with Autoimmunity. *Front. Immunol.* **2012**, *3*, 211. [CrossRef] [PubMed]

189. Bacchetta, R.; Passerini, L.; Gambineri, E.; Dai, M.; Allan, S.E.; Perroni, L.; Dagna-Bricarelli, F.; Sartirana, C.; Matthes-Martin, S.; Lawitschka, A.; et al. Defective regulatory and effector T cell functions in patients with FOXP3 mutations. *J. Clin. Investig.* **2006**, *116*, 1713–1722. [CrossRef] [PubMed]

190. Fuchizawa, T.; Adachi, Y.; Ito, Y.; Higashiyama, H.; Kanegane, H.; Futatani, T.; Kobayashi, I.; Kamachi, Y.; Sakamoto, T.; Tsuge, I.; et al. Developmental changes of FOXP3-expressing CD4+CD25+ regulatory T cells and their impairment in patients with FOXP3 gene mutations. *Clin. Immunol.* **2007**, *125*, 237–246. [CrossRef]

191. Walker, M.R.; Kasprowicz, D.J.; Gersuk, V.H.; Bénard, A.; Van Landeghen, M.; Buckner, J.H.; Ziegler, S.F. Induction of FoxP3 and acquisition of T regulatory activity by stimulated human CD4+CD25− T cells. *J. Clin. Investig.* **2003**, *112*, 1437–1443. [CrossRef]

192. Liu, W.; Putnam, A.L.; Xu-Yu, Z.; Szot, G.L.; Lee, M.R.; Zhu, S.; Gottlieb, P.A.; Kapranov, P.; Gingeras, T.R.; de St Groth, B.F.; et al. CD127 expression inversely correlates with FoxP3 and suppressive function of human CD4+ T reg cells. *J. Exp. Med.* **2006**, *203*, 1701–1711. [CrossRef]

193. Seddiki, N.; Santner-Nanan, B.; Martinson, J.; Zaunders, J.; Sasson, S.; Landay, A.; Solomon, M.; Selby, W.; Alexander, S.I.; Nanan, R.; et al. Expression of interleukin (IL)-2 and IL-7 receptors discriminates between human regulatory and activated T cells. *J. Exp. Med.* **2006**, *203*, 1693–1700. [CrossRef]

194. Alves, N.L.; Van Leeuwen, E.M.M.; Derks, I.A.M.; van Lier, R. Differential Regulation of Human IL-7 Receptor α Expression by IL-7 and TCR Signaling. *J. Immunol.* **2008**, *180*, 5201–5210. [CrossRef]

195. Otsubo, K.; Kanegane, H.; Kamachi, Y.; Kobayashi, I.; Tsuge, I.; Imaizumi, M.; Sasahara, Y.; Hayakawa, A.; Nozu, K.; Iijima, K.; et al. Identification of FOXP3-negative regulatory T-like (CD4+CD25+CD127low) cells in patients with immune dysregulation, polyendocrinopathy, enteropathy, X-linked syndrome. *Clin. Immunol.* **2011**, *141*, 111–120. [CrossRef]

196. Imamichi, H.; Sereti, I.; Lane, H.C. IL-15 acts as a potent inducer of CD4+CD25hi cells expressing FOXP3. *Eur. J. Immunol.* **2008**, *38*, 1621–1630. [CrossRef]

197. Tabares, P.; Berr, S.; Langenhorst, D.; Sawitzki, B.; Berge, I.T.; Tony, H.-P.; Hünig, T. Short-term cytokine stimulation reveals regulatory T cells with down-regulated Foxp3 expression in human peripheral blood. *Eur. J. Immunol.* **2017**, *48*, 366–379. [CrossRef] [PubMed]

198. Allan, S.E.; Passerini, L.; Bacchetta, R.; Crellin, N.; Dai, M.; Orban, P.C.; Ziegler, S.F.; Roncarolo, M.G.; Levings, M.K. The role of 2 FOXP3 isoforms in the generation of human CD4+ Tregs. *J. Clin. Investig.* **2005**, *115*, 3276–3284. [CrossRef] [PubMed]

199. Aarts-Riemens, T.; Emmelot, M.E.; Verdonck, L.F.; Mutis, T. Forced overexpression of either of the two common human Foxp3 isoforms can induce regulatory T cells from CD4+CD25− cells. *Eur. J. Immunol.* **2008**, *38*, 1381–1390. [CrossRef] [PubMed]

200. Allan, S.E.; Crome, S.; Crellin, N.K.; Passerini, L.; Steiner, T.; Bacchetta, R.; Roncarolo, M.G.; Levings, M. Activation-induced FOXP3 in human T effector cells does not suppress proliferation or cytokine production. *Int. Immunol.* **2007**, *19*, 345–354. [CrossRef] [PubMed]

201. Wang, J.; Ioan-Facsinay, A.; van der Voort, E.I.H.; Huizinga, T.W.J.; Toes, R.E.M. Transient expression of FOXP3 in human activated nonregulatory CD4+ T cells. *Eur. J. Immunol.* **2006**, *37*, 129–138. [CrossRef]

202. Passerini, L.; Mel, E.R.; Sartirana, C.; Fousteri, G.; Bondanza, A.; Naldini, L.; Roncarolo, M.G.; Bacchetta, R. CD4+ T Cells from IPEX Patients Convert into Functional and Stable Regulatory T Cells by FOXP3 Gene Transfer. *Sci. Transl. Med.* **2013**, *5*, 215ra174. [CrossRef]

203. Jamee, M.; Zaki-Dizaji, M.; Lo, B.; Abolhassani, H.; Aghamahdi, F.; Mosavian, M.; Nademi, Z.; Mohammadi, H.; Jadidi-Niaragh, F.; Rojas, M.; et al. Clinical, Immunological, and Genetic Features in Patients with Immune Dysregulation, Polyendocrinopathy, Enteropathy, X-linked (IPEX) and IPEX-like Syndrome. *J. Allergy Clin. Immunol. Pract.* **2020**, *8*, 2747–2760.e7. [CrossRef]

204. Varricchi, G.; Poto, R.; Ianiro, G.; Punziano, A.; Marone, G.; Gasbarrini, A.; Spadaro, G. Gut Microbiome and Common Variable Immunodeficiency: Few Certainties and Many Outstanding Questions. *Front. Immunol.* **2021**, *12*, 2915. [CrossRef]

205. NORD (National Organization for Rare Disorders). Common Variable Immune Deficiency. Available online: https://rarediseases.org/rare-diseases/common-variable-immune-deficiency (accessed on 30 December 2021).

206. Yakaboski, E.; Fuleihan, R.L.; Sullivan, K.E.; Cunningham-Rundles, C.; Feuille, E. Lymphoproliferative Disease in CVID: A Report of Types and Frequencies from a US Patient Registry. *J. Clin. Immunol.* **2020**, *40*, 524–530. [CrossRef]

207. Gathmann, B.; Mahlaoui, N.; Gérard, L.; Oksenhendler, E.; Warnatz, K.; Schulze, I.; Kindle, G.; Kuijpers, T.W.; van Beem, R.T.; Guzman, D.; et al. Clinical picture and treatment of 2212 patients with common variable immunodeficiency. *J. Allergy Clin. Immunol.* **2014**, *134*, 116–126.e11. [CrossRef]

208. Pathogenesis of Common Variable Immunodeficiency-UpToDate. Available online: https://www.uptodate.com/contents/pathogenesis-of-common-variable-immunodeficiency (accessed on 30 December 2021).

209. Vořechovský, I.; Cullen, M.; Carrington, M.; Hammarström, L.; Webster, A.D.B. Fine Mapping of IGAD1 in IgA Deficiency and Common Variable Immunodeficiency: Identification and Characterization of Haplotypes Shared by Affected Members of 101 Multiple-Case Families. *J. Immunol.* **2000**, *164*, 4408–4416. [CrossRef]

210. Patuzzo, G.; Mazzi, F.; Vella, A.; Ortolani, R.; Barbieri, A.; Tinazzi, E.; Marchi, G.; Codella, O.; Beri, R.; Puccetti, A.; et al. Immunophenotypic Analysis of B Lymphocytes in Patients with Common Variable Immunodeficiency: Identification of CD23 as a Useful Marker in the Definition of the Disease. *ISRN Immunol.* **2013**, *2013*, 512527. [CrossRef]

211. Matson, E.M.; Abyazi, M.L.; Bell, K.A.; Hayes, K.M.; Maglione, P.J. B Cell Dysregulation in Common Variable Immunodeficiency Interstitial Lung Disease. *Front. Immunol.* **2021**, *11*, 3842. [CrossRef]

212. López-Herrera, G.; Segura-Méndez, N.H.; O'Farril-Romanillos, P.; Nez, M.E.N.-N.; Zarate-Hernández, M.C.; Mogica-Martínez, D.; Yamazaki-Nakashimada, M.A.; Staines-Boone, A.T.; Santos-Argumedo, L.; Berrón-Ruiz, L. Low Percentages of Regulatory T Cells in Common Variable Immunodeficiency (CVID) Patients with Autoimmune Diseases and Its Association with Increased Numbers of CD4+CD45RO+ T and CD21 Low B Cells. *Allergol. Immunopathol.* **2019**, *47*, 457–466. [CrossRef]

213. Ghafoor, A.; Joseph, S.M. Making a Diagnosis of Common Variable Immunodeficiency: A Review. *Cureus* **2020**, *12*, e6711. [CrossRef]

214. Salzer, U.; Warnatz, K.; Peter, H.H. Common variable immunodeficiency-an update. *Arthritis Res. Ther.* **2012**, *14*, 223. [CrossRef] [PubMed]

215. Kutukculer, N.; Azarsiz, E.; Aksu, G.; Karaca, N.E. CD4+CD25+Foxp3+ T regulatory cells, Th1 (CCR5, IL-2, IFN-γ) and Th2 (CCR4, IL-4, Il-13) type chemokine receptors and intracellular cytokines in children with common variable immunodeficiency. *Int. J. Immunopathol. Pharmacol.* **2015**, *29*, 241–251. [CrossRef] [PubMed]

216. Horn, J.; Manguiat, A.; Berglund, L.J.; Knerr, V.; Tahami, F.; Grimbacher, B.; Fulcher, D.A. Decrease in phenotypic regulatory T cells in subsets of patients with common variable immunodeficiency. *Clin. Exp. Immunol.* **2009**, *156*, 446–454. [CrossRef] [PubMed]

217. Genre, J.; Errante, P.R.; Kokron, C.M.; Toledo-Barros, M.; Câmara, N.; Rizzo, L.V. Reduced frequency of CD4+CD25HIGHFOXP3+ cells and diminished FOXP3 expression in paO.S.tients with Common Variable Immunodeficiency: A link to autoimmunity? *Clin. Immunol.* **2009**, *132*, 215–221. [CrossRef]

218. Arandi, N.; Mirshafiey, A.; Abolhassani, H.; Jeddi-Tehrani, M.; Edalat, R.; Sadeghi, B.; Shaghaghi, M.; Aghamohammadi, A. Frequency and Expression of Inhibitory Markers of CD4+CD25+FOXP3+Regulatory T Cells in Patients with Common Variable Immunodeficiency. *Scand. J. Immunol.* **2013**, *77*, 405–412. [CrossRef]

219. UniProtKB. Available online: http://www.uniprot.org (accessed on 30 December 2021).

220. Kozlowski, L.P. IPC–Isoelectric Point Calculator. *Biol. Direct* **2016**, *11*, 55. [CrossRef] [PubMed]

221. Services. Available online: https://services.healthtech.dtu.dk (accessed on 30 December 2021).

222. Align. Available online: https://www.uniprot.org/align (accessed on 30 December 2021).

Journal of
Clinical Medicine

Article

Assessment of COVID-19 Incidence and the Ability to Synthesise Anti-SARS-CoV-2 Antibodies of Paediatric Patients with Primary Immunodeficiency

Karolina Pieniawska-Śmiech [1,2,*], Anna Kuraszewicz [2], Joanna Sado [2], Karol Śmiech [3] and Aleksandra Lewandowicz-Uszyńska [2,4,*]

1   Department of Clinical Immunology, Wroclaw Medical University, 50-368 Wroclaw, Poland
2   Department of Clinical Immunology and Paediatrics, Provincial Hospital J. Gromkowski, 51-149 Wroclaw, Poland; akuraszewicz@gmail.com (A.K.); asia.sado@gmail.com (J.S.)
3   Department of Cardiology, Research and Development Center, Regional Specialist Hospital, 51-124 Wroclaw, Poland; smiech.karol@gmail.com
4   3rd Department and Clinic of Paediatrics, Immunology and Rheumatology of Developmental Age, Wroclaw Medical University, 50-367 Wroclaw, Poland
*   Correspondence: karolina.pieniawska-smiech@student.umw.edu.pl (K.P.-Ś.); aleksandra.lewandowicz-uszynska@umw.edu.pl (A.L.-U.); Tel.: +48-7139-253-97 (K.P.-Ś. & A.L.-U.)

Citation: Pieniawska-Śmiech, K.; Kuraszewicz, A.; Sado, J.; Śmiech, K.; Lewandowicz-Uszyńska, A. Assessment of COVID-19 Incidence and the Ability to Synthesise Anti-SARS-CoV-2 Antibodies of Paediatric Patients with Primary Immunodeficiency. *J. Clin. Med.* **2021**, *10*, 5111. https://doi.org/10.3390/jcm10215111

Academic Editors: Rita Consolini and Giorgio Costagliola

Received: 3 October 2021
Accepted: 28 October 2021
Published: 30 October 2021

Abstract: Background: Data regarding the course of SARS-CoV-2 infection in children with primary immunodeficiency (PID) is insufficient. The purpose of the study was to evaluate the morbidity and clinical course of COVID-19 and the ability to produce anti-SARS-CoV-2 IgG antibodies in children with PID. Methods: In this retrospective study, medical records of 99 patients aged 0–18 were evaluated. The patients were divided into three groups: PID group (68.69%), control group (19.19%) and patients with ongoing or previous paediatric inflammatory multisystem syndrome (12.12%). Data such as morbidity, clinical outcome, and IgG anti-SARS-CoV-2 antibody titres were assessed. Results: A confirmed diagnosis of SARS-CoV-2 infection has been established in 26.47% of patients with PID. Among patients with PID infected with SARS-CoV-2, only three cases were hospitalised. Mortality in the PID group was 0%. Throughout an observation period of 1 year, 47.06% of patients with PID were tested positive for the anti-SARS-CoV-2 antibody. Conclusions: In the study group, in most cases the disease had a mild and self-limiting course. Remarkably, even though IgG deficiency was the most prevalent form of PID in the study group, the patients were able to respond satisfactorily to the infection in terms of anti-SARS-CoV-2 IgG.

Keywords: anti-SARS-CoV-2 antibodies; COVID-19; primary immunodeficiency; SARS-CoV-2; PIMS-TS

1. Introduction

At the end of 2019, a new strain of pneumonia-causing coronavirus was identified in Wuhan, China [1]. Its rapid spread resulted in an outbreak of an epidemic that started in China and gradually expanded worldwide [2,3]. The causative virus, initially called 2019-nCoV, was named SARS-CoV-2, and the disease associated with it–COVID-19 [4].

Primary immunodeficiency (PID) manifests mainly as recurrent and/or severe infections, and patients affected by PID constitute a unique population [5]. If possible, medical interventions should focus on correcting the immune defect in the first place. Another important goal is the prevention and treatment of infections that are still an important cause of mortality in this patient group.

The Center for Disease Control and Prevention (CDC) recognises PID as a risk factor of severe clinical course of COVID-19 [6,7]. Data regarding the course of SARS-CoV-2 infection in children, including children with PID, are insufficient. Other areas that need further research are the duration time of immunity to reinfection and the applicability of serological methods in confirming previous infections [8–12]. Following appropriate validation, serologic tests detecting anti-SARS-CoV-2 antibodies might help identify patients who were infected with the new coronavirus in the past [13–16]. To increase the predictive value of serological methods, it has been suggested that only the tests with high specificity (>99.5%) were used only on individuals with a high clinical probability of a previous infection. The main disadvantages of serologic tests are limited use in the diagnosis of the acute phase of the infection, variable sensitivity and specificity, depending on the assay, and relatively high costs and absence of antibody synthesis in response to the infection in some patients [17]. It is obvious that such a phenomenon may be observed in patients with immune deficiency, both congenital and acquired. The purpose of the study was to evaluate the morbidity and clinical course of COVID-19 and the ability to produce the anti-SARS-CoV-2 IgG antibodies in children with PID. At the same time, the applicability of serological methods in the diagnosis of SARS-CoV-2 in this group of patients was assessed.

2. Materials and Methods

In this retrospective study medical records of 99 patients aged 0–18 who were admitted to the Department of Clinical Immunology and Paediatrics of J. Gromkowski Provincial Hospital in Wrocław from June 2020 to June 2021 were assessed. Testing for coronavirus infection (antigen/polymerase chain reaction-PCR tests) and serologic tests for IgM and IgG anti-SARS-CoV-2 antibodies titres were performed on all of the included patients. Samples were collected from June 2020 to June 2021, during hospitalisation. The patients were divided into three groups: patients already diagnosed with PID according to IUIS criteria and classification (study group) accounted for 68.69% ($n = 68$), patients without an established diagnosis of PID (control group) constituted 19.19% ($n = 19$), and patients with ongoing or previous paediatric inflammatory multisystem syndrome (PIMS-TS/MIS-C) constituted 12.12% ($n = 12$). The control group consisted of patients with recurrent respiratory tract infections diagnosed in the Department of Immunology and Paediatrics who did not show abnormalities in immunological tests and did not meet IUIS criteria for inborn errors of immunity (IEI).

A total of 63.64% ($n = 63$) of the patients were male, and 36.36% ($n = 36$) were female. The mean age of the patients was 7.3 years. All the patients in the study group had been managed in the department for their PID before—55.88% ($n = 38$) of them were treated with immunoglobulin substitution therapy, and the remaining 44.12% ($n = 30$) did not receive such treatment. The most common form of PID was antibody deficiency ($n = 44$) (Table 1).

Table 1. Types of PID and its prevalence in the study group.

Primary Immunodeficiency	Number of Patients; Percent
Combined immunodeficiencies with associated or syndromic features	$n = 20$; 29.14%
IgG subclass deficiency	$n = 18$; 26.47%
Hypogammaglobulinemia IgG	$n = 14$; 20.59%
Other hypogammaglobulinemias *	$n = 8$; 11.76%
Common variable immunodeficiency (CVID)	$n = 3$; 4.11%
Severe combined immunodeficiency (SCID)	$n = 2$; 2.94%
Other, unclassified **	$n = 2$; 2.94%
X-linked agammaglobulinemia	$n = 1$; 1.01%

Abbreviations: * IgG subclass deficiency with IgA deficiency/selective IgM deficiency/transient hypogammaglobulinemia of infancy/IgM and IgG subclass deficiency; ** isolated congenital asplenia/severe lymphocyte T deficiency during diagnostics.

IgM and IgG anti-SARS-CoV-2 antibody titres were measured quantitatively using chemiluminescence. All of the tests were performed in the same laboratory. IgG antibodies against S1/S2 antigens of SARS-CoV-2 were measured from June 2020 to March 2021 and the anti-trimeric spike glycoprotein of SARS-CoV-2 IgG antibodies were measured from March 2021 to June 2021.

Statistical analysis of data was conducted using the spreadsheet of Microsoft Office Excel (Microsoft Corp., Redmond, WA, USA) and Statistica v. 13–non-parametric Mann-Whitney U test. The significance level was defined as $\alpha = 0.05$. A *p*-value less than 0.05 was considered statistically significant.

Consent for the study was granted by the Bioethics Committee of the Wroclaw Medical University.

3. Results

3.1. Morbidity and Disease Course

Throughout an observation period of 1 year, a diagnosis of SARS-CoV-2 infection was confirmed (by means of a PCR or antigen laboratory test) in 18 out of 68 patients with PID (26.47%). Three cases were diagnosed incidentally during tests before non-COVID-19-related hospital admission (Figure 1). Signs of COVID-19 and/or high probability of the infection (e.g., positive result of a SARS-CoV-2 test in a close family member) were identified in 13 patients (19.12%).

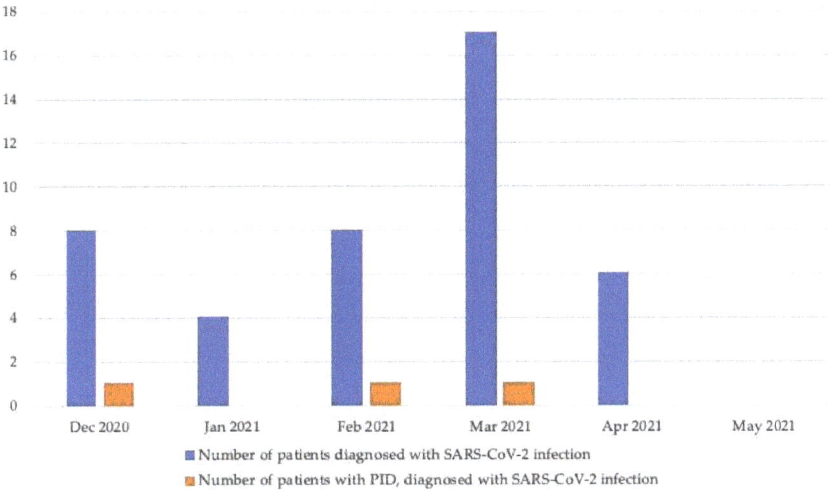

Figure 1. Number of cases of acute SARS-CoV-2 infection identified as a result of the routine testing performed before hospital admission.

Most of the confirmed and/or highly probable cases of COVID-19 ($n = 19$; 61.29%) were noted during the so-called 'second wave' of the pandemic (September 2020–January 2021, and none during 'first wave' (March 2020–August 2020). The predominant variant of SARS-CoV-2 in Poland at the end of 2020 and the beginning of 2021 was 20A but when 20I (also known as B.1.1.7) variant emerged in late December 2020, it quickly became the one responsible for the largest number of infections and started the so-called 'third wave'.

The most common symptom of the infection was elevated body temperature: fever ($n = 12$) or low-grade fever (subfebrile temperature; $n = 6$) (Table S1). Among patients with PID infected with SARS-CoV-2, hospital admission was necessary in only three cases (4.41%)—each of those children suffered from humoral immune disorders, one of them also had a diagnosis of Rubinstein-Taybi syndrome. One of the hospitalised patients required oxygen therapy and was treated with convalescent plasma, none of them required management in an intensive care unit. As for June 2021, mortality in the PID group was 0%.

3.2. Evaluation of Anti-SARS-CoV-2 Antibody Synthesis

Throughout an observation period of 1 year, 32 out of 68 patients with PID tested positive for anti-SARS-CoV2 antibodies (47.06%). By far the majority of these cases ($n = 31$;

96.88%) were associated with confirmed or highly probable (close contact with an infected individual, e.g., a parent, and/or symptoms characteristic of COVID-19) SARS-CoV-2 infection. Furthermore, among 18 patients with a positive test result for SARS-CoV-2 genetic material, only 2 (11.11%) did not produce antibodies directed against it—this included a 12 month-old child with IgG and IgA deficiency and a patient with Rubinstein-Taybi syndrome and IgA, IgM and IgG deficiency, who was also treated with convalescent plasma during the acute phase of the disease.

Among patients with PID who were infected or were most likely infected with COVID-19, there were 10 children (14.70% of all patients with PID) who have been treated with immunoglobulin substitution therapy at that time and only one of these patients required hospital admission. It is also noteworthy that a patient with a history of severe combined immune deficiency (SCID) and hematopoietic stem cell transplantation, which was performed a few years before, developed anti-SARS-CoV-2 antibodies following a symptomatic infection with the virus. Moreover, there was no need for hospital admission in this case.

As for the levels of IgG anti-SARS-CoV-2 antibodies in individuals with positive test results, there were no statistically significant differences when compared with the control group ($n = 19$) ($p > 0.05$) (Figure 2), as well as between the PID group and patients with ongoing or previous PIMS-TS and between the control group and the PIMS-TS group.

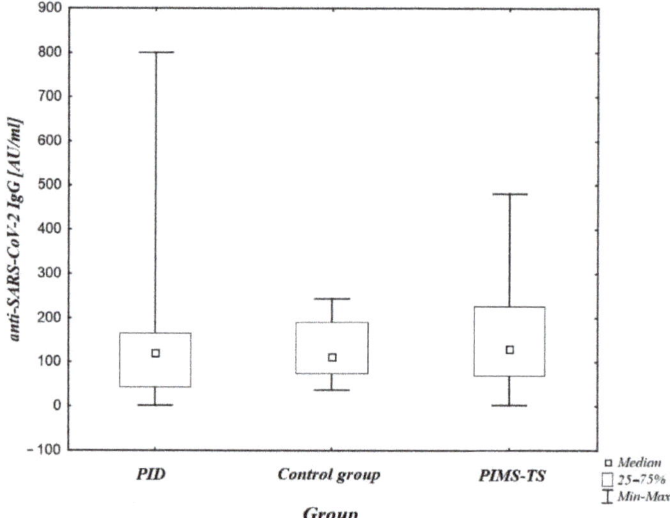

Figure 2. A comparison of anti-SARS-CoV-2 IgG antibody levels in patients with confirmed or highly probable COVID-19 in each patient group ($p > 0.05$). PID-primary immunodeficiency; PIMS-TS-paediatric inflammatory multisystem syndrome.

However, a statistically significant difference ($p = 0.0001$) between patients with PID receiving immunoglobulin substitution therapy and patients with PID without such treatment was noted (Figure 3).

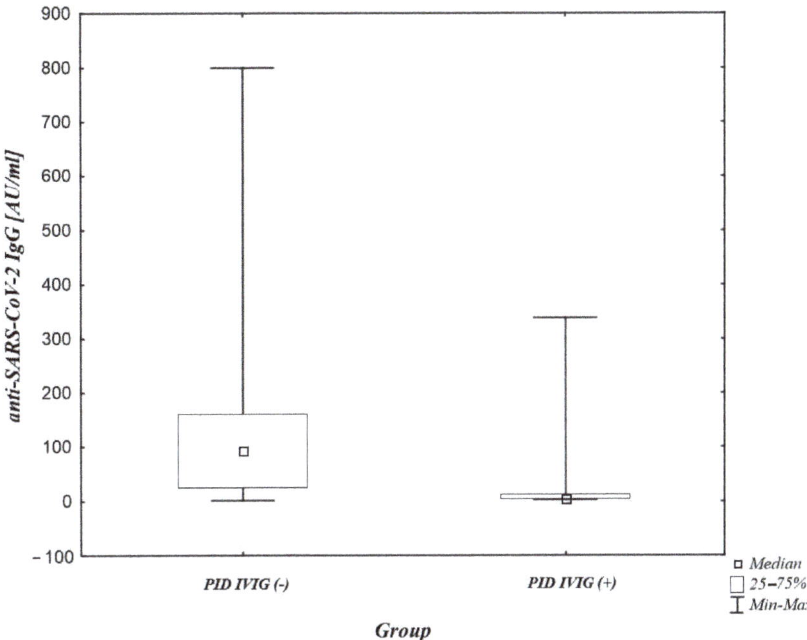

Figure 3. A comparison of anti-SARS-CoV2 IgG antibody levels between patients with PID receiving immunoglobulin substitution therapy (PID IVIG (+)) and patients with PID without such treatment (PID IVIG (−)) ($p = 0.0001$).

4. Discussion

In the absence of more extensive and/or thorough data, it remains unclear whether PID is a predisposing or, paradoxically, a protective factor for SARS-CoV-2 infection [18]. To answer this question, more data regarding COVID-19 morbidity, clinical course and mortality in patients with PID is necessary. Our study represents the experiences of one clinical centre and as such should be regarded as a single opinion in a broader discussion. It is worth considering if immunoglobulin replacement therapy is the protective factor for a severe course of COVID-19, even if immunoglobulins available on the market during the study (June 2020–June 2021) probably did not contain significant level of anti-SARS-CoV-2 IgG antibodies. This influence may be associated with the modulatory effect of immunoglobulins on the immune system, which is used in therapy of e.g., Kawasaki disease, Guillain-Barre syndrome [19].

In an observational study conducted in Israel [20], which was published in January 2021, amongst patients with PID aged 4 months to 6 years, a total number of 20 SARS-CoV-2 infections was recorded. The majority of these cases (95%) were reported during the second wave of the pandemic, which was consistent with our results. Moreover, children receiving immunoglobulin substitution therapy constituted the majority of the infected patient population. There were no cases of severe COVID-19, none of the infected patients required hospital admission and 35% of the affected children remained asymptomatic during the course of the disease. The authors implied that the COVID-19 pandemic had little impact on patients with PID.

The observations made by researchers in Iran [21] were different. In a prospective study, based on data acquired from the national registry, it was concluded that with only 1.23-fold higher incidence of infections, patients with PID, mainly those with combined immunodeficiency and immune dysregulation, present a 10-fold higher mortality rate compared to the general population. The study included 19 children with PID in whom

SARS-CoV-2 infection was confirmed using an RT-PCR test. Exposure to the virus from an unknown source or a source outside the patient's family accounted for 84.2% of the total number of cases. The results of our study were quite different, and contact with close relatives was the source of infection for many of the infected patients ($n = 11$).

The Iranian researchers showed that combined immunodeficiency ($n = 10$, all without hematopoietic stem cell transplantation or HSCT, 47.0%) was the major PID entity amongst COVID-19 positive cases, followed by humoral immunodeficiencies ($n = 4$), phagocytic defects ($n = 2$), immune dysregulation ($n = 2$), and autoinflammatory disorders ($n = 1$) [21], ergo studied population was different than ours. The discrepancies in the incidence of certain forms of PID might be a result of the relatively low number of patients in both study groups and different characteristics of the populations managed in each hospital department. It is worth mentioning that the difference between Iranian and Polish patients with PID is related to the high prevalence of consanguinity in Iran compared to Poland, and the consequent high prevalence of autosomal recessive immunodeficiency.

In an international study conducted by Meyts et al. [22] published in February 2021, 32 cases of COVID-19 were recorded in children with PID, nine of them required management in an intensive care unit (ICU) and two of them died. Among patients treated in the ICU settings there were patients with a diagnosis of chronic granulomatous disease ($n = 1$), trisomy 21 ($n = 1$), Wiskott-Aldrich syndrome ($n = 1$), nuclear factor κB mutation (NFKB2) ($n = 1$) and X-linked inhibitor of apoptosis protein (XIAP) deficiency ($n = 1$). Due to numerous comorbidities, the authors defined the connection between SARS-CoV-2 infection and the death of both patients as 'unclear'.

Throughout an observation period of 1 year, none of the patients managed by our department for PID required treatment in ICU while infected with SARS-CoV-2.

At present, the data regarding IgG anti-SARS-CoV-2 antibody synthesis in individuals with PID are insufficient.

However, the research conducted on the immunocompetent population presents some interesting information. A key factor in determining the appropriate time window for the use of serological tests is the occurrence of seroconversion. Recent publications indicate that the median IgG detection occurs 9 to 14 days after disease onset [23,24]. Peterson et al. reported that approximately 1 in 16 people lacked IgG antibodies following infection. Race/ethnicity, weight status, immunosuppressive therapy and illness severity were independent predictors of IgG antibody presence after SARS-CoV-2 infection [25].

In a study conducted by Venkatamaran et al. in India, the authors evaluated humoral immune response associated with anti-SARS-CoV-2 antibody synthesis in hospitalised patients by comparing antibody titres between children with and without PIMS-TS [26]. Almost half of seropositive children had PIMS-TS. Antibody levels may be helpful in the diagnosis and disease stratification of PIMS-TS. Nearly one-fifth of the hospitalised children tested serology positive over four months. Antibody levels in children with PIMS-TS were significantly higher in comparison to the other two groups (acute COVID-19 infection and children without PIMS-TS).

The main purpose of our study was to evaluate antibody synthesis in patients with primary immune deficiency, and so an additional comparison of antibody synthesis between this unique population and children with (both ongoing and previous) PIMS-TS was made. No difference in levels of the antibody was recorded (Figure 2). This observation requires further verification on a larger group of patients, including meticulous evaluation of synthesis and perseverance of the antibody during the acute phase of the disease and after its resolution. It is noteworthy that in one patient with a history of PIMS-TS, IgG subclass deficiency was detected twice. There was no reference, however, since no immunological studies were performed in this patient before PIMS occurred.

It is also important to note that amongst children receiving immunoglobulin substitution therapy, positive anti-SARS-CoV-2 IgG test was recorded only in individuals with infection confirmed with a PCR/antigen test or with a high probability of infection, which meant the presence of characteristic clinical symptoms and/or close contact with an infected person. Moreover, the mean level of anti-SARS-CoV-2 IgG in this group (IVIG +) was lower than in children not receiving such treatment and proved to be statistically significant (p = 0.0001). This leads to the conclusion that the IgG anti-CoV antibody test results were legitimate (throughout the observation period). Another important issue is that a significant percent of the study group with relatively mild immunodefciency (= not required immunoglobulin replacement therapy at the time of the study, i.e., isolated IgG subclass deficiency) may influence the results of the study.

Furthermore, it is worth mentioning, that the control group was not a classic control group and included patients with (mostly) mild recurrent upper respiratory tract infections, without abnormalities in immunological tests, who did not meet IUIS criteria for IEI.

The most important limitation of the study was the retrospective nature of the analysis which was based on collected medical records. More precise, prospective studies, evaluating the duration of antibody response in patients with PID and a history of COVID-19 are needed.

5. Conclusions

COVID-19 might be regarded as one of the main challenges for healthcare in the twenty-first century. However, based on the ongoing collection of data, it will be possible to identify the risk group of severe COVID-19 amongst individuals with PID in the future. Patients diagnosed with PID constitute a unique population. Usually, they are provided with high-quality medical care and were well isolated throughout the pandemic. Moreover, the caregivers responsible for them are fully aware of the danger and abide by all of the hygiene standards. As a result, throughout the first wave of the pandemic, the number of infections detected in patients with PID was much smaller than in immunocompetent patients. The second and the third wave were associated with an increase in the number of infections both in adults and in children. Many of the patients managed by the department for their PID became infected at that time. However, in most cases, the disease had a mild and self-limiting course. Study results indicate that COVID-19 is not only a less severe disease in children than in adults, but also is not as severe as one might expect in children with dysfunctional immune systems. Nevertheless, this observation should not affect the sanitary regime and safety regulations concerning the management of PID patients, especially in the context of the new B.1.617.2 (delta) variant.

Remarkably, even though the most prevalent form of PID in the study group was IgG deficiency, the patients were able to respond satisfactorily to the infection in terms of anti-SARS-CoV-2 IgG. Thus, some of PID may be a group with a significance in limitation of transmission of SARS-CoV-2 viral infection after COVID-19 vaccination. According to Polish consensus by group of experts, vaccination against COVID-19 should be recommended [27].

It seems that the main factor influencing the course of COVID-19 in both immunocompetent patients and the patients with PID is comorbidity.

Supplementary Materials: The following are available online at https://www.mdpi.com/article/10.3390/jcm10215111/s1, Table S1: Clinical characteristics of patients with PID with confirmed or suspected SARS-CoV-2 infection.

Author Contributions: Conceptualization, K.P.-Ś. and A.L.-U.; Data curation, K.P.-Ś.; Formal analysis, K.Ś.; Investigation, K.P.-Ś., A.K. and J.S.; Methodology, K.P.-Ś., K.Ś. and A.L.-U.; Project administration, K.P.-Ś.; Supervision, A.L.-U.; Visualization, K.Ś.; Writing—original draft, K.P.-Ś., J.S. and A.L.-U.; Writing—review & editing, K.P.-Ś., A.K. and A.L.-U. All authors have read and agreed to the published version of the manuscript.

Funding: Funding within the grant A240.210.084 by Wroclaw Medical University.

Institutional Review Board Statement: The study was conducted according to the guidelines of the Declaration of Helsinki, and approved by the Institutional Review Board (or Ethics Committee) of Wroclaw Medical University (protocol code 472/2021; date of approval 28 May 2021).

Informed Consent Statement: Not applicable.

Data Availability Statement: The authors confirm that the data supporting the findings of this study are available within the article and its supplementary materials.

Acknowledgments: The authors would like to thank Przemyslaw Poltorak for linguistic help during manuscript writing.

Conflicts of Interest: The authors declare no conflict of interest.

References

1. Phelan, A.L.; Katz, R.; Gostin, L.O. The Novel Coronavirus Originating in Wuhan, China: Challenges for Global Health Governance. *JAMA* **2020**, *323*, 709–710. [CrossRef] [PubMed]
2. Li, X.; Wang, W.; Zhao, X.; Zai, J.; Zhao, Q.; Li, Y.; Chaillon, A. Transmission dynamics and evolutionary history of 2019-nCoV. *J. Med. Virol.* **2020**, *92*, 501–511. [CrossRef] [PubMed]
3. Liu, Y.C.; Kuo, R.L.; Shih, S.R. COVID-19: The first documented coronavirus pandemic in history. *Biomed. J.* **2020**, *43*, 328–333. [CrossRef] [PubMed]
4. Habas, K.; Nganwuchu, C.; Shahzad, F.; Gopalan, R.; Haque, M.; Rahman, S.; Majumder, A.A.; Nasim, T. Resolution of coronavirus disease 2019 (COVID-19). *Expert Rev. Anti-Infect. Ther.* **2020**, *18*, 1201–1211. [CrossRef]
5. Lewandowicz-Uszyńska, A.; Pasternak, G.; Świerkot, J.; Bogunia-Kubik, K. Primary Immunodeficiencies: Diseases of Children and Adults-A Review. *Adv. Exp. Med. Biol.* **2021**, *1289*, 37–54. [CrossRef]
6. Available online: https://www.cdc.gov/coronavirus/2019-ncov/need-extra-precautions/people-with-medical-conditions.html (accessed on 13 July 2021).
7. Available online: https://www.cdc.gov/coronavirus/2019-ncov/hcp/clinical-care/underlyingconditions.html (accessed on 13 July 2021).
8. Gudbjartsson, D.F.; Norddahl, G.L.; Melsted, P.; Gunnarsdottir, K.; Holm, H.; Eythorsson, E.; Arnthorsson, A.O.; Helgason, D.; Bjarnadottir, K.; Ingvarsson, R.F.; et al. Humoral Immune Response to SARS-CoV-2 in Iceland. *N. Engl. J. Med.* **2020**, *383*, 1724. [CrossRef]
9. Isho, B.; Abe, K.T.; Zuo, M.; Jamal, A.J.; Rathod, B.; Wang, J.H.; Li, Z.; Chao, G.; Rojas, O.L.; Bang, Y.M.; et al. Persistence of serum and saliva antibody responses to SARS-CoV-2 spike antigens in COVID-19 patients. *Sci. Immunol.* **2020**, *5*, eabe5511. [CrossRef]
10. Iyer, A.S.; Jones, F.K.; Nodoushani, A.; Kelly, M.; Becker, M.; Slater, D.; Mills, R.; Teng, E.; Kamruzzaman, M.; Garcia-Beltran, W.F.; et al. Persistence and decay of human antibody responses to the receptor binding domain of SARS-CoV-2 spike protein in COVID-19 patients. *Sci. Immunol.* **2020**, *5*, eabe0367. [CrossRef]
11. Wajnberg, A.; Amanat, F.; Firpo, A.; Altman, D.R.; Bailey, M.J.; Mansour, M.; McMahon, M.; Meade, P.; Mendu, D.R.; Muellers, K.; et al. Robust neutralizing antibodies to SARS-CoV-2 infection persist for months. *Science* **2020**, *370*, 1227–1230. [CrossRef] [PubMed]
12. Dan, J.M.; Mateus, J.; Kato, Y.; Hastie, K.M.; Yu, E.D.; Faliti, C.E.; Grifoni, A.; Ramirez, S.I.; Haupt, S.; Frazier, A.; et al. Immunological memory to SARS-CoV-2 assessed for up to 8 months after infection. *Science* **2021**, *371*, eabf4063. [CrossRef] [PubMed]
13. Guo, L.; Ren, L.; Yang, S.; Xiao, M.; Chang, D.; Yang, F.; Dela Cruz, C.S.; Wang, Y.; Wu, C.; Xiao, Y.; et al. Profiling Early Humoral Response to Diagnose Novel Coronavirus Disease (COVID-19). *Clin. Infect. Dis.* **2020**, *71*, 778–785. [CrossRef]
14. Rongqing, Z.; Li, M.; Song, H.; Chen, J.; Ren, W.; Feng, Y.; Gao, G.F.; Song, J.; Peng, Y.; Su, B.; et al. Early Detection of Severe Acute Respiratory Syndrome Coronavirus 2 Antibodies as a Serologic Marker of Infection in Patients with Coronavirus Disease 2019. *Clin. Infect. Dis.* **2020**, *71*, 2066–2072. [CrossRef]
15. Zhang, G.; Nie, S.; Zhang, Z.; Zhang, Z. Longitudinal Change of Severe Acute Respiratory Syndrome Coronavirus 2 Antibodies in Patients with Coronavirus Disease 2019. *J. Infect. Dis.* **2020**, *222*, 183–188. [CrossRef] [PubMed]
16. Fang, F.C.; Naccache, S.N.; Greninger, A.L. The Laboratory Diagnosis of Coronavirus Disease 2019—Frequently Asked Questions. *Clin. Infect. Dis.* **2020**, *71*, 2996–3001. [CrossRef] [PubMed]
17. Centers for Disease Control and Prevention. Interim Guidelines for COVID-19 Antibody Testing in Clinical and Public Health Settings. Available online: https://www.cdc.gov/coronavirus/2019-ncov/lab/resources/antibody-tests-guidelines.html?deliveryName=USCDC_2067-DM29085 (accessed on 26 May 2020).
18. Babaha, F.; Rezaei, N. Primary Immunodeficiency Diseases in COVID-19 Pandemic: A Predisposing or Protective Factor? *Am. J. Med. Sci.* **2020**, *360*, 740–741. [CrossRef] [PubMed]
19. Gilardin, L.; Bayry, J.; Kaveri, S.V. Intravenous immunoglobulin as clinical immune-modulating therapy. *CMAJ* **2015**, *187*, 257–264. [CrossRef] [PubMed]
20. Marcus, N.; Frizinsky, S.; Hagin, D.; Ovadia, A.; Hanna, S.; Farkash, M.; Maoz-Segal, R.; Agmon-Levin, N.; Broides, A.; Nahum, A.; et al. Minor Clinical Impact of COVID-19 Pandemic on Patients with Primary Immunodeficiency in Israel. *Front. Immunol.* **2021**, *11*, 614086. [CrossRef]
21. Delavari, S.; Abolhassani, H.; Abolnezhadian, F.; Babaha, F.; Iranparast, S.; Ahanchian, H.; Moazzen, N.; Nabavi, M.; Arshi, S.; Fallahpour, M.; et al. Impact of SARS-CoV-2 Pandemic on Patients with Primary Immunodeficiency. *J. Clin. Immunol.* **2021**, *41*, 345–355. [CrossRef]

22. Meyts, I.; Bucciol, G.; Quinti, I.; Neven, B.; Fischer, A.; Seoane, E.; Lopez-Granados, E.; Gianelli, C.; Robles-Marhuenda, A.; Jeandel, P.Y.; et al. Coronavirus disease 2019 in patients with inborn errors of immunity: An international study. *J. Allergy Clin. Immunol.* **2021**, *147*, 520–531. [CrossRef] [PubMed]

23. Zhao, J.; Yuan, Q.; Wang, H.; Liu, W.; Liao, X.; Su, Y.; Wang, X.; Yuan, J.; Li, T.; Li, J.; et al. Antibody Responses to SARS-CoV-2 in Patients with Novel Coronavirus Disease 2019. *Clin. Infect. Dis.* **2020**, *71*, 2027–2034. [CrossRef] [PubMed]

24. Okba, N.M.A.; Müller, M.A.; Li, W.; Wang, C.; GeurtsvanKessel, C.H.; Corman, V.M.; Lamers, M.M.; Sikkema, R.S.; de Bruin, E.; Chandler, F.D.; et al. Severe Acute Respiratory Syndrome Coronavirus 2-Specific Antibody Responses in Coronavirus Disease Patients. *Emerg. Infect. Dis.* **2020**, *26*, 1478–1488. [CrossRef] [PubMed]

25. Petersen, L.R.; Sami, S.; Vuong, N.; Pathela, P.; Weiss, D.; Morgenthau, B.M.; Henseler, R.A.; Daskalakis, D.C.; Atas, J.; Patel, A.; et al. Lack of antibodies to SARS-CoV-2 in a large cohort of previously infected persons. *Clin. Infect. Dis.* **2020**, *1*, ciaa1685. [CrossRef] [PubMed]

26. Venkataraman, A.; Balasubramanian, S.; Putilibai, S.; Lakshan Raj, S.; Amperayani, S.; Senthilnathan, S.; Manoharan, A.; Sophi, A.; Amutha, R.; Sadasivam, K.; et al. Correlation of SARS-CoV-2 Serology and Clinical Phenotype Amongst Hospitalised Children in a Tertiary Children's Hospital in India. *J. Trop. Pediatr.* **2021**, *67*, fmab015. [CrossRef] [PubMed]

27. Bernatowska, E.; Pac, M.; Mikołuć, B.; Heropolitańska-Pliszk, E.; Pietrucha, B.; Wolska-Kuśnierz, B.; Kołtan, S.; Polska Grupa Robocza, ds. Pierwotnych Niedoborów Odporności. Standards of immunization in Primary Immunodeficiency-recommendations from the Polish Working Group on and the Committee for Immunology and Aetiology of Human Infections, Polish Academy of Science. *Stand. Med. Pediatr.* **2021**, *18*, 403–413.

 Journal of
Clinical Medicine

Article

Spectrum of Disease Manifestations in Patients with Selective Immunoglobulin E Deficiency

César Picado [1,2,3,*], Iñaki Ortiz de Landazuri [4], Alexandru Vlagea [4], Irina Bobolea [1,2,3], Ebymar Arismendi [1,2,3], Rosanel Amaro [1,2,3], Jacobo Sellarés [1,2,3], Joan Bartra [1,2], Raimon Sanmarti [2,5], José Hernandez-Rodriguez [2,6], José-Manuel Mascaró [2,7], Jordi Colmenero [2,8,9], Eva C. Vaquero [2,9,10] and Mariona Pascal [2,4]

1   Institut Clinic Respiratory, Hospital Clinic, Universitat de Barcelona, 08036 Barcelona, Spain; Bobolea@clinic.cat (I.B.); earismen@clinic.cat (E.A.); ramaro@clinic.cat (R.A.); sellares@clinic.cat (J.S.); JBARTRA@clinic.cat (J.B.)
2   Institut d'Investigacions Biomèdiques August Pi i Sunyer (IDIBAPS), 08036 Barcelona, Spain; sanmarti@clinic.cat (R.S.); jhernan@clinic.cat (J.H.-R.); mascaro@clinic.cat (J.-M.M.); jcolme@clinic.cat (J.C.); evaquero@clinic.cat (E.C.V.); MPASCAL@clinic.cat (M.P.)
3   Centro de Investigaciones Biomédicas en Red de Enfermedades Respiratorias (CIBERES), 28029 Madrid, Spain
4   Immunology Department, CDB. Hospital Clinic, Universitat de Barcelona, 08036 Barcelona, Spain; ortizdelan@clinic.cat (I.O.d.L.); vlagea@clinic.cat (A.V.)
5   Department of Rheumatology, Hospital Clinic, Universitat de Barcelona, 08036 Barcelona, Spain
6   Department of Autoimmune Diseases, Hospital Clinic, Universitat de Barcelona, 08036 Barcelona, Spain
7   Department of Dermatology, Hospital Clinic, Universitat de Barcelona, 08036 Barcelona, Spain
8   Liver Unit, Hospital Clinic, Universitat de Barcelona, 08036 Barcelona, Spain
9   Centro de Investigaciones en Red de Enfermedades Hepáticas y Digestivas (CIBEREHD), 28029 Madrid, Spain
10 Department of Gastroenterology, Hospital Clinic, Universitat de Barcelona, 08036 Barcelona, Spain
*   Correspondence: cpicado@ub.edu

Citation: Picado, C.; Ortiz de Landazuri, I.; Vlagea, A.; Bobolea, I.; Arismendi, E.; Amaro, R.; Sellarés, J.; Bartra, J.; Sanmarti, R.; Hernandez-Rodriguez, J.; et al. Spectrum of Disease Manifestations in Patients with Selective Immunoglobulin E Deficiency. *J. Clin. Med.* **2021**, *10*, 4160. https://doi.org/10.3390/jcm10184160

Academic Editors: Rita Consolini and Giorgio Costagliola

Received: 20 August 2021
Accepted: 12 September 2021
Published: 15 September 2021

Publisher's Note: MDPI stays neutral with regard to jurisdictional claims in published maps and institutional affiliations.

Abstract: Background: Selective IgE deficiency (SIgED) has been previously evaluated in selected patients from allergy units. This study investigates the effects of SIgED on the entire population in a hospital setting and sought to delineate in detail the clinical aspects of SIgED. Methods: A retrospective study of the data obtained from electronic medical records of 52 adult patients (56% female) with a mean age of 43 years and IgE levels of <2.0 kU/L with normal immunoglobulin (Ig) IgG, IgA, and IgM levels, seen at our hospital, without selection bias, from 2010 to 2019. Results: Recurrent upper respiratory infections were recorded in 18 (34.6%) patients, pneumonia was recorded in 16 (30.7%) patients, bronchiectasis was recorded in 16 (30.7%) patients, and asthma was recorded in 10 (19.2%) patients. Eighteen patients (34.6%) suffered autoimmune clinical manifestations either isolated (19%) or combining two or more diseases (15%), Hashimoto's thyroiditis being the most frequent (19%), which was followed by arthritis (10%) and thrombocytopenia and/or neutropenia (5.7%). Other less frequent associations were Graves' disease, primary sclerosing cholangitis, Sjögren's syndrome, and autoimmune hepatitis. Eczematous dermatitis (15.3%), chronic spontaneous urticaria (17.3%), and symptoms of enteropathy (21%) were also highly prevalent. Thirty percent of patients developed malignancies, with non-Hodgkin lymphomas (13.4%) being the most prevalent. Conclusions: The clinical manifestations of SIgED encompass a variety of infectious, non-infectious complications, and malignancy. Since it cannot be ruled out that some type of selection bias occurred in the routine assessment of IgE serum levels, prospective studies are required to better characterize SIgED and to determine whether it should be added to the list of antibody deficiencies.

Keywords: autoimmune diseases; Immunoglobulin E; Immunoglobulin deficiency; infections; malignancy

1. Introduction

A recently updated classification distributes innate errors of immunity into 10 groups, one of which is considered to be due to "antibody deficiencies" (Group 3) [1,2].

Immunoglobulins A (IgA), M (IgM), and G (IgG) are central in the humoral immune response and play a fundamental role in protecting against infections caused by all kinds of agents (viruses, bacteria, protozoa, parasites), and they represent the defense mediated by antibodies, which are part of the so-called acquired immunity [3].

Immunoglobulin E (IgE) has been conventionally related to the immune response against helminth infection, and its levels are particularly high in patients who suffer from a parasitic infestation [4]. IgE is also involved in type I hypersensitivity allergic reactions, which are diseases where it is also common to find high levels of specific IgE against allergens. Most IgE is found bound to its high-affinity receptor FcεRI located on the surface of mast cells and basophils. The binding of the allergen to the specific IgE/FcεRI complex triggers the degranulation of mast cells and basophils that release numerous substances (vasoactive, bronchoconstrictors, interleukins), which are ultimately responsible for the clinical manifestations of the allergic response (rhinitis, asthma, urticaria, angioedema, anaphylaxis) [5].

Various types of immunodeficiencies associated with low levels of one or a combination of IgA, IgG, and IgM immunoglobulins are recognized. [1,2]. The most studied combined form is known as "common variable immunodeficiency" (CVID), which is a disorder characterized by reduced serum levels of IgG, which can be combined with a reduction of IgA or IgM, or both, which is associated with recurrent sinopulmonary infections, autoimmune disorders, granulomatous diseases, and increased risk of malignancy and altered response of antibodies against infections [6,7].

Selective IgG deficiency (SIgGD) encompasses any subject with a serum IgG level below normal range with normal IgA and IgM levels. Studies comparing the SigGD and CVID patients found that the CVID group was more likely to have bronchiectasis, poorer responses to vaccines, and a higher incidence of autoimmune cytopenias, granulomas, splenomegaly, and lymphoid neoplasms than those with SigGD [8].

IgG subclass deficiency (IgGSD) is a heterogeneous subtype of primary immunodeficiency, which is defined as the triad of frequent or severe respiratory tract infections, subnormal levels of one or more of the four IgG subclasses, and decreased IgG response to pneumococcal polysaccharides. Many adults with IgGSD also have autoimmune conditions or atopy [9,10].

Selective IgA deficiency (SIgAD) [11,12], and selective IgM (SIgMD) [13,14] are diagnosed in a diverse group of patients, ranging from completely asymptomatic individuals to people with recurrent infections, allergic diseases, autoimmune processes, and malignant tumors.

The question is: are there any similar diseases associated with selective IgE deficiency (SIgED)? Conventionally, normal serum IgE values are considered to range between the technical detection limit ($\leq$2 kU/L) and up to 100 kU/L. An excess of IgE (>100 kU/L) can be established but, in contrast to the other immunoglobulins, there is no generally accepted minimum level to establish an IgE deficiency. In various studies in the literature, different cut-off points have been used to define IgE deficiency [15–19]. Most clinicians do not attribute any pathological significance to very low IgE values, even those that are unquantifiable ($\leq$2 kU/L), which are usually considered as "normal".

Low IgE is frequently associated with deficiencies in other immunoglobulins, particularly in patients with CVID [20–22]. Based on this observation, the use of routine IgE measurement has been proposed as the first step to detect the presence of CVID [21,22].

In the classification of primary immunodeficiencies attributed to antibody deficiency, the presence of low IgE values is mentioned, but it is always associated with deficiencies in some of the other immunoglobulins [1,2]. The possibility of an immunodeficiency associated only with an SIgED is not considered in the classification. However, a few studies have reported that an SIgED may be the biomarker of an immunodeficiency with a significant clinical impact that has been overlooked until now [23–26].

The studies analyzing the potential role of SIgED are retrospective and include a limited number of cases. Furthermore, most of the patients included in these studies were selected from allergy units [23,26] or were patients having any allergy-related symptoms and/or requesting antiallergy medications [25], which is a bias that could have limited the spectrum of diseases found associated with SIgED. Despite these limitations, it is worth noting that these studies show that individuals with a low level of IgE, with normal values for the other immunoglobulins, present recurrent respiratory infections, suffer from autoimmune diseases, and upper and lower airway diseases [23–26], similar to those described in patients with CVID [6–8], IgGSD [9] or with SIgGD [8], SIgAD [11,12], and SIgM D [13,14].

The predisposition to develop neoplasms in patients with antibody deficiency, either in combination or due to selective deficits of IgA or IgM, is widely documented [27]. In the same way, the scientific information supports that IgE deficiency is a predisposing factor for the development of malignancies [28].

The hypothesis of this study establishes that isolated IgE deficiency is associated with diseases similar to that described in other antibody deficiencies, but its clinical spectrum has been underestimated.

This study is the first to research the effects of SIgED in the entire population in a hospital setting with a 2-year follow up and sought to delineate in detail the clinical aspects of SIgED.

2. Patients and Methods

Any patient who was found to have an IgE concentration ≤ 2 kU/L with normal IgG, IgM, and IgA concentrations with at least 2-year follow up at our institution between January 2010 and December 2019 was included in the study. A total of 151 patients were analyzed, of whom 99 were excluded for different reasons shown in Figure 1. The remaining 52 patients with SIgED and regular follow-up in the hospital were included in the study. Medical records were reviewed and discussed together with the various specialists involved in their routine care. Of the 52 patients, 31 were female (56%), and the mean age was 43 years (range 18–87).

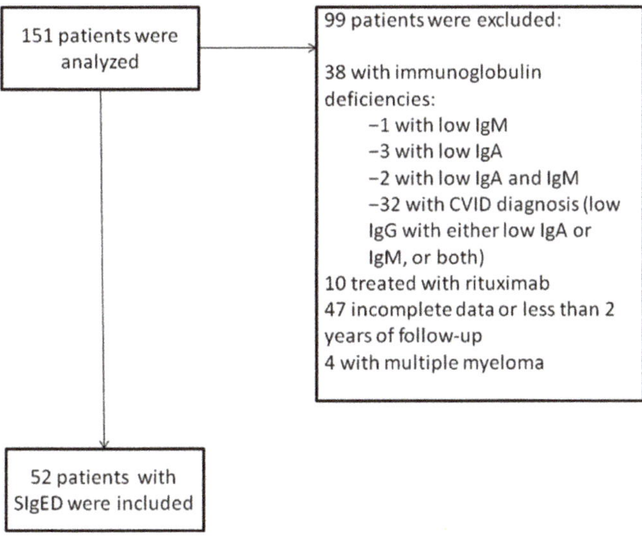

Figure 1. Flow chart of patients analyzed.

In patients with a suspicion of either a respiratory or food allergy, we routinely perform skin prick tests (SPTs) in our institution with a panel of commercial allergenic extracts of the most prevalent aeroallergens and food allergens in our area (Laboratorios LETI, Madrid, Spain). Foods suspected by the clinical history and not included in the standard panel are also usually tested with a commercial extract if available, or by prick-prick according to standard methods. Serum levels of IgG, IgA, IgM, and IgGs were measured by immunoturbidimetry (Atellica NEPH 630 Solution System. Siemens Healthineers, Germany). Serum total and specific IgE levels were measured by immunofluorescence enzyme immunoassay (ImmunoCAP, ThermoFisherScientific, Uppsala, Sweden). Serum IgA, IgM, and IgG values of the participants were: IgM 1.10 (range, 0.41–2.42) g/L (normal values (0.36–2.6 g/L); IgG 10.7 g/L (range, 7.10–13.30) g/L; and IgA 2.12 g/L (range 0.95–4.80) g/L (normal values 0.66–3.6 g/L). The study was approved by the Ethics Committee of the Hospital Clinic (Ethical Code: HCB/2021/0758).

3. Results

3.1. Infections

Three or more yearly upper respiratory infections (URI) (rhinorrhea, nasal congestion, and productive cough), requiring antibiotic therapy for at least two consecutive years, were recorded in 18 (34.6%) of the SIgED patients. Sixteen patients (30.7%) had suffered one or more episodes of pneumonia (range 1 to 3). Median IgG, IgM, and IgA levels were not significantly different in patients with or without URI or pneumonia (data not shown). In two patients, chronic respiratory infection with *M. avium* complex was identified. Recurrent episodes of otitis were recorded in four patients. Three patients had suffered from herpesvirus infections, and one had suffered from chronic pyelonephritis.

3.2. Lung Diseases

Twenty-nine patients underwent chest computed tomography (CT) scanning. The radiological study demonstrated the presence of bronchiectasis in 16 patients (30.7% of total sample) affecting between one and three lobes and mostly cylindrical, peribronchial thickening in three, air trapping in two, atelectasis in three, micronodules in three, cyst in three, pulmonary emphysema in three (all ex-smokers), ground glass opacities in six, and interstitial lung fibrosis in one. The chest CT scan was considered normal in six patients (all of them suffering from frequent respiratory infections). In two patients, the combination of chest CT findings (micronodules, cysts), bronchoalveolar lavage (BAL) fluid results (lymphocytic inflammation and multinucleated giant cell), and the lung biopsy of a nodule (lymphocytic infiltration) indicated the presence of lymphocytic interstitial lung disease. Ten (19.2%) patients were diagnosed with asthma with different levels of severity. Eight patients (15.3%), six of them associated with asthma, referred symptoms of allergic rhinitis, and four of them reported clinical symptoms apparently exacerbated seasonally (spring, autumn). In all patients, SPTs and specific IgE for common allergens) were negative (Table 1).

3.3. Autoimmune Diseases

Eighteen patients (34.6%) suffered autoimmune clinical manifestations, either isolated (19%) or combining two o more diseases (15%). Hypothyroidism was diagnosed in 10 patients (19.2%), eight secondary to Hashimoto's thyroiditis, and two resulted from previously treated hyperthyroidism (Graves' disease). Other less frequently found autoimmune diseases are shown in Table 1.

Table 1. Non-infectious complications.

	No. (%)
Airway/Lung diseases	
Bronchiectasis	16 (30.7)
Asthma	10 (19.2)
Rhinitis	8 (15.3)
Lymphocytic interstitial lung disease *	2 (3.8)
Interstitial lung fibrosis	1 (1.9)
Autoimmune diseases	
Hashimoto's disease	8 (15.4)
Arthritis (1 RA, 4 undifferentiated)	5 (9.6)
Thrombopenia (2 associated with neutropenia)	4 (7.7)
Neutropenia	3 (5.8)
Aphthous stomatitis	3 (5.8)
Graves' disease	2 (3.8)
Vitiligo	2 (3.8)
SLE	2 (3.8)
Alopecia	1 (1.9)
Acute hepatitis	1 (1.9)
Sjögren's syndrome	1 (1.9)
Gastrointestinal and liver diseases	
Symptoms of enteropathy **	11 (21)
PSC	2 (3.8)
Ulcerative colitis (associated to PSC)	2 (3.8)
Cirrhosis (secondary to PSC)	2 (3.8)
Other manifestations	
Chronic spontaneous urticaria	9 (17.3)
Eczematous dermatitis	8 (15.3)
Polyarthralgia	9 (17.3)
Fatigue	7 (13.4)
Arterial hypertension	13 (33.3)

PSC, primary sclerosing cholangitis; RA, rheumatoid arthritis; SLE, systemic lupus erythematous. * Diagnosis based on CT scan images; Bronchoalveolar lavage lung fluid findings and biopsy of a lung nodule. ** Chronic or intermittent diarrhea, abdominal pain and bloating.

3.4. Gastrointestinal and Liver Diseases

Symptoms of enteropathy such as intermittent or persistent chronic diarrhea, abdominal pain, and bloating were present in 11 patients (21%). Some patients associated their symptoms with the ingestion of certain foods. In all cases, both allergen SPTs and specific IgE studies with the putative culprit foods were negative. Fructose and lactose intolerance were assessed in four patients, and only one tested positive in the lactose test. Celiac disease was excluded in most (nine patients) but not all patients by anti-transglutaminase IgG serology. A gluten-free diet was tested in four patients with inconsistent or negative symptomatic response. Biopsies of colon mucosa and/or small intestine were obtained in six patients, and the histological findings were: intraepithelial lymphocytosis (four patients), lymphoid hyperplasia forming aggregates (one patient), and enteritis with chronic inflammation, eosinophilic infiltration, crypt distortion, and gland destruction (one patient). Acute severe autoimmune hepatitis was diagnosed in one patient who had required two liver transplantations. Two patients suffered from primary sclerosing cholangitis (PSC) progressing to cirrhosis requiring liver transplantation. The two patients also suffered from ulcerative colitis (Table 1).

3.5. Cutaneous Findings

Eczematous dermatitis (eight patients, 15.3%) associated with moderate or severe itching in most cases, chronic spontaneous urticaria (CSU) (nine patients, 17.3%), angiedema (four patients associated with CSU), and chronic leg ulcers (two patients) were present among the SIgED patients (Table 1).

3.6. Tumours

Sixteen (30.7%) patients developed malignancies including non-Hodgkin lymphomas (seven patients, 13.4%), chronic lymphocytic leukemia (two patients, one evolving from a lymphoma), and various types of malignant and non-malignant tumors, as shown in Table 2. Four patients developed more than one tumor.

Table 2. Lymphomas and other tumors.

	No. (%)
Lymphomas and Leukemias	
Diffuse Large B cell lymphoma	3
Follicular cell lymphoma	1
Burkitt lymphoma	1
Lymphocytic Lymphoma/CLL	1
Lymphoma B cell, not otherwise specified	1
CLL	1
Other	
Melanoma	3
Breast	2
Skin cancer (basal cell carcinoma)	2
Cholangiocarcinoma	1
Hepatocarcinoma	1
Gynaecological (endometrial carcinoma)	1
Clear cell renal carcinoma	1
Meningioma	1
Neurinoma	1

CCL = Chronic Lymphocytic Leukemia.

3.7. Other

Mastocytosis (two patients), monoclonal gammopathy of undetermined significance (MGUS) (two patients), fatigue (seven patients, 13.4%), and polyarthralgia (nine patients, 17.3%) were also reported by some patients as major complaints. Follicular hyperplasia affecting lymph nodes located in the mediastinum, armpits, groin, supraclavicular area, or abdomen were found in eight (15.3%) patients, which, when biopsied (three patients), showed a pattern of non-specific lymphoid reactivity, although one of them later evolved to a lymphoma. Arterial hypertension (13 patients, 33.3%) and ischemic heart disease (four patients, 7.7) were also documented.

3.8. SIgED and IgG Subclasses

Serum levels of IgG1, IgG2, IgG3, and IgG4 had been assessed in 14 patients and were normal in all but three patients: one with low IgG3 and two with low IgG4 (data not shown).

4. Discussion

It is generally accepted that low levels of IgA, IgM, and IgG predispose to respiratory bacterial and viral infections. The high incidence of URI and pneumonia in our patients with SIgED is in keeping with that reported for CVID [8,9,29], SIgGD [8], IgGSD [9,10], SIgAD [11,12], and SIgMD [13,14]. The mechanism by which an SIgED may also predispose to lung infections remains to be elucidated. IgE is usually related to protection against parasites [4], but its role in other infections is not usually considered, despite there being studies that have demonstrated the presence of specific IgE antibodies against viruses such as H1N1 influenza [30], respiratory syncytial [31], HIV1 [32], varicella [33], parvovirusB19 [34], and rhinovirus (RV) [35]. Anti-HIV1 IgE has been shown to inhibit HIV1 production in infected cell culture, the inhibitory effect being reversed when IgE was removed from the culture [32]. In a study involving children with HIV-1 infection, opportunistic infections were less frequent in children with high serum IgE levels than in those with low IgE levels [36].

It is generally assumed that IgE does not play any relevant role in the immune response against bacteria. However, there are studies reporting that IgE antibodies provide immunity against bacteria such as *Borrelia burgdorferi* [37]. It was recently discovered that the antibacterial activity of mast cells against *Staphylococcus aureus* (SA) in mice was markedly enhanced by the presence of IgE directed against bacterial components. Animal models deficient in IgE or FcεRI were unable to mount protective immune responses against SA infections [38]. Furthermore, other authors have found that SIgED deficiency predisposes to recurrent upper and lower airways with common respiratory bacteria such as *Haemophilus influenza, Moraxella catarrhalis*, and *Streptococcus pneumoniae* [23].

Taken together, these findings support the notion that a reduced synthesis of IgE may result in an immunodeficient response against virus and bacteria. As far as we know, the response to vaccines of patients with SIgED has never been studied.

Two of our patients (3.8%) had chronic *Mycobacterium avium* infection. The patients had clinical and radiological findings suggestive of 'Lady Windermere Syndrome' (LWS), which is characterized by chronic bronchiectasis in slender women, with scoliosis and/or *pectus excavatum*, and chronic productive cough. Multigenic variants with potential defects in proteins encoded by various genes might contribute to LWS by reducing both IFN-γ production and increasing transforming growth factor (TGF)-β levels in response to non-tuberculous mycobacterium (NTM) [39–42]. NTM infection has been reported in a very small percentage of patients with CVID (0–1%) [43]. Chronic respiratory infection with *M. avium* complex and bronchiectasis were identified in 5% of patients with SIgAD [44]. So far, in patients with SIgAD, SIgGD, and IgGSD, no NTM infections have been reported. The potential role of SIgED in NTM infection is unknown and should be evaluated in a larger series of patients with lungs infected with these pathogens.

It is generally assumed that in CVID patients, recurrent airway infections and persistent airway inflammation can lead to a vicious circle airway remodeling process resulting in bronchiectasis [43]. A recent analysis of existing data on the clinical presentation of CVID found that bronchiectasis was present in the CT scan in almost one-third of patients (28%, 95% CI 18–40) [29]. Bronchiectasis has been found in up to 14% of SIgAD patients and is more commonly reported when associated with IgG subclass deficiency [44–46]. We found that bronchiectasis was present in 30.7% of our SIgED patients, which is a percentage similar to that reported in CVID, which is an observation that suggests that the lack of IgE has a significant negative impact on the immune defense mechanisms of the lung. In contrast to our findings, bronchiectasis is not even mentioned in the few studies reporting the clinical manifestations present in SIgED patients, which is most probably due to the lack of CT scan evaluation in patients with frequent respiratory infections [23–26].

Viral infections are a strong risk factor for developing asthma in children, and they are major contributors to exacerbations of asthma in both children and adults [47]. The link between viruses and upper (rhinitis) and lower respiratory diseases (asthma) might explain the high percentage of patients with CVID that are diagnosed with asthma (25%, 95% CI 17–35) and rhinitis (18%, 95% CI 8–31) [29]. What is not yet clear is the mechanism underlying this association. Are CVID patients with asthma-like clinical symptoms a distinct hyperreactive airway phenotype? Or, are they subjects to genetic factors predisposing them to develop asthma, which is unmasked early by the presence of the immunodeficiency? Mutations in the TNFRSF13B gene have been found in CVID patients [48] and are also associated with an increased risk of asthma development [49].

CVID patients with asthma and rhinitis are often clinically characterized as allergic [29,43]. Interestingly, some of our patients reported nasal and bronchial allergic-like reactions—a few of the associated with seasonal exacerbations. This is not surprising, given the presence of ultralow serum levels of IgE, SPTs, and that the in vitro tests for serum-specific IgE against common allergens were negative in all patients. It is theoretically plausible that allergic-type symptoms could be due to the presence of IgE in the respiratory tract, which is something similar to so-called local allergic rhinitis (LAR) and local allergic asthma (LAA) [50,51]. These diseases are characterized by the negativity of the skin-prick test and serum-specific IgE for all relevant aeroallergens in a patient with upper and lower airway symptoms suggestive of allergy, and who tested positive in the nasal and bronchial allergen challenge [50,51]. Although the cells and the main sites of IgE production in humans remains to be fully characterized, it is assumed that IgE is produced in the peripheral blood and locally in various tissues, including the nose and lung [52]. One may speculate that airway mucosal IgE in patients with SIgED is still capable of developing respiratory allergic responses in a similar way to that described in LAR and LAA [51,52]. Interestingly, patients with CVID, IgE deficiency, and a history suggestive of allergic asthma with negative allergen SPTs did not show any bronchial reactions when subjected to an allergen challenge, but the exposure to allergens increased the airway response to histamine [53]. Whether the acquired airway hyperresponsiveness was due to a local IgE-dependent or another non-IgE related mechanism remains to be clarified. It is also unclear whether allergen-induced hyperresponsiveness can indirectly account for the symptoms of those IgE-deficient patients associated with allergen exposure. Allergy symptoms may have been confused with unspecific airway hyperreactivity-related clinical manifestations.

Similar to CVID, SIgAD has also been associated with allergic rhinoconjunctivitis and asthma [44,54,55]. However, the prevalence of these diseases shows large differences among studies, ranging from 13% [44] to 83% [55]. Furthermore, one age- and gender–matched survey found an increased prevalence of allergic rhinoconjunctivitis, but no differences were found in asthma prevalence between SIgAD and controls [55]. Thirty five percent of patients with SIgMD had atopic diseases, including allergic rhinitis and asthma [45], while allergic asthma and/or allergic rhinitis were the second commonest manifestations in patients with IgGSD without any subclass predominance [56].

In previous studies, SIgED has been found to be associated with a higher prevalence of non-allergic reactive airways disease (rhinorrhea, nasal congestion, dry cough, and/or wheezing) (73%) compared with controls (20%) [23], and with asthma or hyperreactive airway disease (26.5% vs. controls 6.8%) in children but not in adults [25]. The prevalence of asthma (19.2%) found in our study was higher than that reported in the adult Spanish population (range 10–16.7%) [57].

Interstitial lung disease (ILD) is a frequent (15–60%) non-infectious complication of CVID [58]. The histology of ILD in CVID shows heterogeneous and often mixed patterns, including lymphoid hyperplasia, lymphoid interstitial pneumonitis, follicular bronchiolitis, non-necrotizing granulomatous inflammation, organizing pneumonia, and interstitial fibrosis [59]. Granulomatous-lymphocytic interstitial lung disease (GLILD) is often used as a term to describe ILD with lymphocytic infiltrates and/or granulomata in CVID [59]. However, not all ILD in CVID have pulmonary granulomata, and therefore, the term does not fully cover the heterogeneous spectrum of the histopathology found in lung samples from CVID patients [59]. Approximately 20% of patients with ILD present polyclonal lymphocytic infiltration or non-malignant hyperplasia of the lymph nodes in addition to granuloma [60]. Monogenic disorders causing CVID-like diseases have also been reported in patients with ILD [61–63]. Patients with ILD have distinct clinical and immunological phenotypes in keeping with immune dysregulation, in contrast to those without ILD or those with bronchiectasis alone [64]. Recent studies have shown that ILD is also present in the lung of patients with selective immunoglobulin deficiencies, including SIgAD, SIgGD, and IgGSD, with a pattern of lymphoid proliferation and granulomata identical to that found in CVID [65–67]. Lung biopsies from CVID patients usually show some degree of fibrosis, which can be extensive, and is the predominant finding in up to 6.5% of cases [64]. The presence of extensive lung fibrosis is associated with a poor prognosis [68].

In our study, we found two patients with radiological, BAL fluid cytology, and histological lung findings suggestive of ILD, and one patient with clinical and radiological findings commonly associated with severe interstitial lung fibrosis, which caused her death.

Although not confirmed by biopsy, in four patients, the CT scan showed ground-glass opacities, pulmonary nodules, and mediastinal lymphadenopathy, which are images considered highly suggestive of ILD [64]. Taken together, our observations suggest adding SIgED to the immunodeficiencies potentially associated with ILD.

A substantial number of CVID patients (27%, 95% CI 22–32%) develop autoimmune manifestations [29]. Studies have shown that SIgGD [9,12] SIgMD [13,45,69], IgGSD [56], and SIgAD [11,12,44,55] are also associated with systemic and organ-specific autoimmune diseases. The clinical spectrum of autoimmunity in CVID and other selective immunodeficiencies is very wide and includes a plethora of hematologic (cytopenia, thrombocytopenic purpura, hemolytic anemia, Evans syndrome), and non-hematologic diseases (autoimmune thyroid diseases, rheumatoid arthritis, unspecific inflammatory arthritis, Sjögren´s syndrome, systemic lupus erythematous (SLE), autoimmune hepatitis) [65]. In our study, we found that SIgED was associated with hematologic and non-hematologic autoimmune diseases, with percentages similar to those described in other immunodeficiencies. Isolated and mixed autoimmune diseases were also significantly more frequent in adults and children with SIgED compared with control populations in previous studies [23,25]. As in our study, thyroid diseases (Hashimoto's thyroiditis and Grave's disease), cytopenias, SLE, and arthritis were autoimmune diseases reported in patients with SIgED [23,25]. Taken together, these findings support that autoimmunity is a relevant component of the clinical presentation of SIgED.

CSU, in some cases associated with angioedema, was frequently diagnosed in our patients. In contrast, CSU and angioedema are not usually listed among the more common clinical manifestations in CVID [29,65]. However, some reports point out that we should not overlook the association of CSU with CVID [70–73]. CSU has been found in 4.9% of patients with SIgAD compared with 0.9% in controls [74], and in up to 12% of patients with SIgMD [75]. A statistically significant prevalence of CSU was observed in patients with SIgED (19%) compared with controls (0.8%) in one study [25], while another study did not find any differences between patients (11%) and controls (11%) [23].

Eczematous dermatitis was also found in a high percentage of our patients. The "eczema group" is frequently (33.7%) diagnosed in patients with various primary immunodeficiencies [76], but it is not included among the most common manifestations of patients with CVID [29,64]. Interestingly, severe eczematous dermatitis is characteristic of diseases of the immune system associated with both autosomal dominant and autosomal recessive forms of hyper IgE syndrome [77]. In some of our patients, eczematous dermatitis was associated with severe itching requiring regular treatment with oral corticosteroids, in some cases complemented with immunosuppressive therapy. A previous study in patients with SIgED could not find any differences in skin rash complaints between patients and controls [23], while in another study, rashes diagnosed as psoriasis and seborrheic dermatitis were found to be significantly higher in SIgED patients than in controls [25].

Gastrointestinal symptoms that may mimic inflammatory bowel disease are very frequent in patients with CVID. Intermittent or persistent diarrhea (27%, 95% CI 21–34) [29,78], bloating (34%) [78], and abdominal pain (26%) [78], are the most common gastrointestinal symptoms. The enteropathy of CVID may affect any part of the gastrointestinal tract and is associated with various histological findings, including intraepithelial lymphocytosis (46%), a decreased number of plasma cells in the GI tract mucosa (62%), and lymphoid hyperplasia (38%) [78]. Many other histological findings, such as eosinophilic or lymphocytic enteritis, villous atrophy, collagenous enteritis, and granulomatous inflammation are less frequently found in biopsies [29,78]. Studies in patients with SIgMD show great variability in the prevalence of gastrointestinal manifestations, without clearly differentiating those that may be due to an enteropathy similar to that found in CVID [14,44,55,75]. Both chronic and recurrent diarrhea are more common among individuals with SigAD than in the control population. However, many of these cases are associated with either celiac disease or inflammatory bowel disease [56]. A comparison study shows more biopsy-confirmed enteropathy cases among CVID patients (7%) than among SigGD patients (3.2%) [8]. Gastrointestinal symptoms have not been reported associated with IgGSD [56]. Eleven (21%) of our patients reported gastrointestinal symptoms suggestive of enteropathy. However, only five had been assessed by endoscopy (9.6%), but all had histological findings (intraepithelial lymphocytosis, lymphoid hyperplasia, lymphocytic and eosinophilic enteritis) usually found in the enteropathy of CVID. Previous studies in SigED patients offer scant data on gastrointestinal symptoms, although they mention that some patients had been diagnosed with food allergy in some children [25], and inflammatory bowel disease and celiac disease in some adults [23], but without data from histological studies.

Liver diseases have been reported in up to 12.7% of CVID patients [67], ranging from elevated alkaline phosphatase to nodular regenerative hyperplasia (NRH), autoimmune hepatitis, liver cirrhosis, and primary sclerosing cholangitis (PSC) [79]. Some isolated cases of liver disease with NRH, and acute autoimmune hepatitis have been reported in patients with SIgAD [80] and SIgMD [81,82]. Acute severe autoimmune hepatitis (one patient) and PSC (two patients) progressing to cirrhosis were diagnosed in our SIgED patients. The three patients required liver transplantation. The two patients with PSC suffered from ulcerative colitis [83], and one of them developed a cholangiocarcinoma, which are both entities considered common complications in PSC [84].

Interestingly, high serum levels of IgE have been found associated with a lower incidence of biliary carcinoma in patients with PSC [85]. This is not an unexpected finding, since numerous epidemiological studies carried out in recent years have shown an inverse relationship between elevated IgE levels and malignant processes [28,86–89]. In keeping with these epidemiological observations, we found a strikingly high prevalence of malignancies (30%), with non-Hodgkin lymphomas (13.4%) topping the list in our patients. These findings are also very similar to those reported in some previous publications of SIgED patients, where a significant increase in malignant processes (lymphomas, lymphocytic leukemia, and epithelial cancers) was observed compared with the control group [25]. The high frequency of malignant processes in our study supports the relevant role played by IgE in antitumor surveillance detected in epidemiological studies.

The association between immunodeficiencies and cancer is well established. In CVID patients, the most commonly reported malignancies are non-Hodgkin lymphomas and various solid cancers (breast, colon, lung, gastric, ovarian, melanoma) [64,90,91]. A recent study has shown that the link between IgE and malignancies appears to be specific and independent of the presence of CVID in patients with IgE deficiency, which is a finding that lends further support to IgE's leading role in cancer development [92]. IgA deficiency is also associated with a moderately increased risk of cancer, with excess risks of gastrointestinal cancer not related to the presence of celiac disease [93]. Various types of cancer have been found in patients with SIgMD in some [44,69,94] but not all studies [94,95].

Fatigue as a major complaint was present in a high percentage (13.4%) of our patients. Chronic fatigue was also found significantly more frequently in patients with SIgED than in controls in previous studies, 3.8% vs. 0.3% [25] and 30% vs. 4% [23]. Fatigue is a very common complaint in patients with primary immunodeficiency disorders [96], particularly in patients with CVID (40%) [29], but it has only been reported in isolated patients with other selective immunoglobulin deficiencies. The prevalence of polyarthralgias in our patients was high (17.3%), but it was even less frequent than previously reported in patients with SIgED (32% vs. 7% in controls) [23]. SIgED associated with both arterial hypertension (37.7%) and ischemic heart disease (25.2%) was previously reported by E Magen et al. [97]. We found a similar prevalence of arterial hypertension (33.3%) but lower ischemic heart disease (7.7%). The mechanisms involved in these associations remain to be elucidated [97]. The same group has also reported that in comparison to a control group, a significantly larger proportion of patients with SIgED presented with duodenal ulcers (DU) (63.2% vs. 11.7%), who were positive for *Helicobacter pylori* (Hp) infection (47.4% vs. 11.7%) [98]. In our patients, the prevalence of DU was much lower (two patients, 3.8%, both positive for HP), which concurs with previous studies [23], including one from the group of Magen et al. [25], which did not find that SIgED can predispose to DU. The reasons that could explain the striking difference between their own two studies [25,98] are unclear and were not analyzed by the authors.

The number of patients in whom serum levels of IgG subclasses had been assessed was small and prevented us from evaluating the possible clinical impact when both deficiencies concur.

Currently, very little is known about the mechanisms responsible for the deficiency in IgE, either in isolation or associated with deficiencies in other immunoglobulins. Similarly, the link between low IgE and the high risk of developing malignancy has yet to be elucidated. No abnormalities in the mechanisms involved in IgE synthesis have been reported so far in patients with SIgED. One study looked at the gene encoding activation-induced cytidine deaminase, which is an enzyme involved in immunoglobulin class switching, but the researchers could not find any mutation in patients with SIgED [99].

Our study has several limitations, such as the small number of patients recruited and its retrospective nature. We cannot exclude that some bias may have influenced our results. For example, the high prevalence of CSU and eczema found in our patients may be due to the fact that they are diseases treated by dermatologists and allergists who often include IgE measurement in their routine work-up. Moreover, we chose to use a stringent diagnostic criterion (IgE <2.0 kU/L) to increase the specificity in the diagnosis of SIgED. However, it remains unclear whether patients with IgE close to this level could also carry a similarly increased risk of developing diseases. There should be further studies gathering clinical data with different stratified IgE levels before a definitive serum IgE level can be established as a diagnostic threshold immunodeficient risk. Studies are also necessary to elucidate the clinical impact of complementary immunodeficient profiles such as associated IgG subtypes and the response to vaccines.

In summary, the data reported to date suggest that SIgED is characterized by a high prevalence of recurrent respiratory infections, asthma, autoimmune diseases, and malignancies [23–26]. Our study expands the spectrum of diseases associated with SIgED by adding bronchiectasis, enteropathy, CSU, eczematous dermatitis, LID, and liver diseases (PSC and hepatitis) to the known list. Although it is not clear why these diseases were not detected in previous studies, it is likely that the discrepancies are due to differences in the method used to recruit patients. In previous studies, patients were selected from allergy services [23,26] or with allergy-related symptoms [25], while in ours, patients were recruited without bias from the general hospital base, which could explain the higher prevalence of diseases that are not usually treated in the allergy units. Prospective studies based on broader populations are needed to further examine the role of SIgED in the development of different pathologies usually associated with immunodeficiencies. The possible genetic basis of SIgED is currently unknown and remains to be investigated. Hopefully, these studies will reveal whether SIgED can be added to the current list of antibody deficiencies.

Author Contributions: Conceptualization: C.P., A.V., and M.P. Methodology: C.P. and M.P. Validation: I.O.d.L., A.V., I.B., E.A., R.A., J.S., J.B., R.S., J.H.-R., J.-M.M., J.C. and E.C.V. Formal analysis: C.P. Data curation: I.B., E.A., R.A., J.S., J.B., R.S., J.H.-R., J.-M.M., J.C. and E.C.V. Writing—original draft: C.P. Review and editing: I.O.d.L., A.V., I.B., E.A., R.A., J.S., J.B., R.S., J.H.-R., J.-M.M., J.C., E.C.V. and M.P. All authors have read and agreed to the published version of the manuscript.

Funding: This research received no external funding.

Institutional Review Board Statement: The study was approved by the Ethics Committee of the Hospital Clinic (Ethical Code: HCB/2021/0758).

Informed Consent Statement: Patient consent was waived as only anonymized data obtained from routine clinical care was used.

Conflicts of Interest: The authors declare no conflict of interest.

References

1. Bousfiha, A.; Jeddane, L.; Picard, C.; Al-Herz, W.; Ailal, F.; Chatila, T.; Cunningham-Rundles, C.; Etzioni, A.; Franco, J.; Holland, S.M.; et al. Human Inborn Errors of Immunity: 2019 Update of the IUIS Phenotypical Classification. *J. Clin. Immunol.* **2020**, *40*, 66–81. [CrossRef]
2. Tangye, S.G.; Al-Herz, W.; Bousfiha, A.; Chatila, T.; Cunningham-Rundles, C.; Etzioni, A.; Franco, J.L.; Holland, S.M.; Klein, C.; Morio, T.; et al. Human Inborn Errors of Immunity: 2019 Update on the Classification from the International Union of Immunological Societies Expert Committee. *J. Clin. Immunol.* **2020**, *40*, 24–64. [CrossRef]
3. Megha, K.B.; Mohanan, P.V. Role of immunoglobulin and antibodies in disease management. *Int. J. Biol. Mcromol.* **2021**, *169*, 28–38. [CrossRef]
4. Fitzsimmons, C.M.; Falcone, F.H.; Dunne, D.W. Helminth Allergens, Parasite-Specific IgE.; and Its Protective Role in Human Immunity. *Front. Immunol.* **2014**, *5*, 61. [CrossRef]
5. Plats-Mills, T.A.E.; Schuyler, A.J.; Erwin, E.A.; Commins, S.P.; Woodfolk, J.A. IgE in the diagnosis and treatment of allergic disease. *J. Allergy Clin. Immunol.* **2016**, *137*, 1662–1670. [CrossRef] [PubMed]
6. Ameratunga, R.; Allan, C.; Woon, S.T. Defining Common Variable Immunodeficiency Disorders in 2020. Immunol. *Allergy Clin. N. Am.* **2020**, *40*, 403–420. [CrossRef]
7. Chan, A.Y.; Torgerson, T.R. Primary immune regulatory disorders: A growing universe of immune dysregulation. *Curr. Opin. Allergy Clin. Immunol.* **2020**, *20*, 582–590. [CrossRef] [PubMed]
8. Filion, C.A.; Taylor-Black, S.; Maglione, P.J.; Radigan, L.; Cunningham-Rundles, C. Differentiation of Common Variable Immunodeficiency From IgG deficiency. *J. Allergy Clin. Immunol. Pract.* **2019**, *7*, 1277–1284. [CrossRef] [PubMed]
9. Barton, J.C.; Bertoli, L.F.; Barton, J.C. Comparisons of CVID and IgGSD: Referring physicians, autoimmune conditions, pneumovax reactivity, immunoglobulin levels, blood lymphocyte subsets, and HLA-A and -B typing in 432 adult index patients. *J. Immunol. Res.* **2014**, *2014*, 542706. [CrossRef]
10. Parker, A.R.; Skold, M.; Harding, S.; Barton, J.C.; Bertoli, L.F.; Barton, J.C. Pneumococcal vaccination responses in adults with subnormal IgG subclass concentrations. *BMC Immunol.* **2019**, *20*, 29. [CrossRef] [PubMed]
11. Swain, S.; Selmi, C.; Gershwin, M.E.; Teuber, S.S. The clinical implications of selective IgA deficiency. *J. Transl. Autoimmun* **2019**, *2*, 100025. [CrossRef]
12. Singh, K.; Chang, C.; Gershwin, M.E. IgA deficiency and autoimmunity. *Autoimmun Rev.* **2014**, *13*, 163–177. [CrossRef] [PubMed]
13. Campochiaro, C.; Atay, S.; Clark, K.E.N.; Ong, V.; Denton, C.P. Autoimmunity and immunodeficiency at the crossroad: Autoimmune disorders as the presenting feature of selective IgM deficiency. *BMJ Case Rep.* **2019**, *3*, e223180. [CrossRef] [PubMed]
14. Gupta, S.; Gupta, A. Selective IgM Deficiency-An Underestimated Primary Immunodeficiency. *Front. Immunol.* **2017**, *8*, 1056. [CrossRef] [PubMed]
15. Barton, J.C.; Barton, J.C.; Bertoli, L.F. Hypogammaglobulinemia E in 216 adults with IgG subclass deficiency and respiratory tract infections. *Ann. Allergy Asthma Immunol.* **2017**, *119*, 292–294. [CrossRef] [PubMed]
16. Gleich, G.J.; Averbeck, A.K.; Swedlund, H.A. Measurement of IgE in normal and allergic serum by radioimmunoassay. *J. Lab. Clin. Med.* **1971**, *77*, 690–698.
17. Polmar, S.H.; Waldmann, T.A.; Terry, W.D. IgE in immunodeficiency. *Am. J. Pathol.* **1972**, *69*, 499–512. [PubMed]
18. Stites, D.P.; Ishizaka, K.; Fudenberg, H.H. Serum IgE concentrations in hypogammaglobulinaemia and selective IgA deficiency. Studies on patients and family members. *Clin. Exp. Immunol.* **1972**, *10*, 391–397. [PubMed]
19. Elkuch, M.; Greiff, V.; Berger, C.T.; Bouchenaki, M.; Daikeler, T.; Bircher, A.; Navarini, A.; Heijnen, I.; Recher, M. Low immunoglobulin E flags two distinct types of immune dysregulation. *Clin. Exp. Immunol.* **2017**, *187*, 345–352. [CrossRef] [PubMed]
20. Unsworth, D.J.; Virgo, P.F.; Lock, R.J. Immunoglobulin E deficiency: A forgotten clue pointing to possible immunodeficiency? *Annals Clin. Biochem.* **2011**, *48*, 459–461. [CrossRef]
21. McVicker, S.; Karim, M.Y. IgE deficiency may indicate underlying hypogammaglobulinaemia? *J. Clin. Pathol.* **2014**, *67*, 832–833. [CrossRef]
22. Lawrence, M.G.; Palacios-Kibler, T.V.; Workman, L.J.; Schuyler, A.J.; Steinke, J.W.; Payne, S.; McGowan, E.C.; Patrie, J.; Fuleihan, R.L.; Sullivan, K.E.; et al. Low Serum IgE Is a Sensitive and Specific Marker for Common Variable immunodeficiency (CVID). *J. Clin. Immunol.* **2018**, *38*, 225–233. [CrossRef]
23. Smith, J.K.; Krishnaswamy, G.H.; Dykes, R.; Reynolds, S.; Berk, S.L. Clinical manifestations of IgE hypogammaglobulinemia. *Ann. Allergy Asthma Immunol.* **1997**, *78*, 313–318. [CrossRef]
24. Famuyiwa, F.; Rubinstein, I. Chronic sinopulmonary inflammatory diseases in adults with undetectable serum IgE in inner-city Chicago: A preliminary observation. *Lung* **2012**, *190*, 291–294. [CrossRef] [PubMed]
25. Magen, E.; Schlesinger, M.; David, M.; Ben-Zion, I.; Vardy, D. Selective IgE deficiency, immune dysregulation, and autoimmunity. *Allergy Asthma Proc.* **2014**, *35*, e27–e33. [CrossRef]
26. García Pavón Osorio, S.; Lopez Tiro, J.J.; Vera, J.G. IgE deficiency: A forgotten disease? *Rev. Allerg. Mex.* **2009**, *56*, 192–197.
27. Hauck, F.; Gennery, A.R.; Seidel, M.G. Editorial: The Relationship Between Cancer Predisposition and Primary Immunodeficiency. *Front. Immunol.* **2019**, *10*, 1781. [CrossRef] [PubMed]
28. Ferastraoaru, D.; Bax, H.J.; Bergmann, C.; Capron, M.; Castells, M.; Dombrowicz, D.; Fiebiger, E.; Gould, H.J.; Hartmann, K.; Jappe, U.; et al. AllergoOncology: Ultra-low IgE.; a potential novel biomarker in cancer-a Position Paper of the European Academy of Allergy and Clinical Immunology (EAACI). *Clin. Transl. Allergy* **2020**, *10*, 32. [CrossRef]

29. Janssen, L.M.A.; Van der Flier, M.; De Vries, E. Lessons Learned From the Clinical Presentation of Common Variable Immunodeficiency Disorders: A Systematic Review and Meta-Analysis. *Front. Immunol.* **2021**, *12*, 620709. [CrossRef]

30. Smith-Norowitz, T.A.; Kusonruksa, M.; Wong, D.; Norowitz, M.M.; Joks, R.; Durkin, H.G.; Bluth, M.H. Long-term persistence of IgE anti-influenza A HIN1 virus antibodies in serum of children and adults following influenza A vaccination with subsequent H1N1 infection: A case study. *J. Inflamm. Res.* **2012**, *5*, 111–116. [CrossRef]

31. Smith-Norowitz, T.A.; Mandal, M.; Joks, R.; Norowitz, L.T.; Weaver, D.; Durkin, H.G.; Bluth, M.H.; Kohlhoff, S. IgE anti-respiratory syncytial virus antibodies detected in serum of pediatric patients with Asthma. *Hum. Immunol.* **2015**, *76*, 519–524. [CrossRef] [PubMed]

32. Pellegrino, M.G.; Bluth, M.H.; Smith-Norowitz, T.; Fikrig, S.; Volsky, D.J.; Moallem, H.; Auci, D.L.; Nowakowski, M.; Durkin, H.G. HIV type 1-specific IgE in serum of long-term surviving children inhibits HIV type 1 production in vitro. *AIDS Res. Hum. Retrovir.* **2002**, *18*, 363–372. [CrossRef]

33. Smith-Norowitz, T.A.; Josekutty, J.; Silverberg, J.I.; Lev-Tov, H.; Norowitz, Y.M.; Kohlhoff, S.; Nowakowski, M.; Durkin, H.G.; Bluth, M.H. Long Term Persistence of IgE Anti-Varicella Zoster Virus in Pediatric and Adult Serum Post Chicken Pox Infection and after Vaccination with Varicella Virus Vaccine. *Int. J. Biomed. Sci.* **2009**, *5*, 353–358.

34. Smith-Norowitz, T.A.; Drew, H.; Norowitz, H.M.; Nowakowski, M.; Bluth, E.F.; Durkin, H.G.; Bluth, M.H. Detection of IgE anti-parvovirus antibodies in human breast milk. *Ann. Clin. Lab. Sci.* **2008**, *38*, 168–173. [PubMed]

35. Tam, J.S.; Jackson, W.T.; Hunter, D.; Proud, D.; Grayson, M.H. Rhinovirus specific IgE can be detected in human sera. *J. Allergy Clin. Immunol.* **2013**, *132*, 1241–1243. [CrossRef]

36. Secord, E.A.; Kleiner, G.I.; Auci, D.L.; Smith-Norowitz, T.; Chice, S.; Finkielstein, A.; Nowakowski, M.; Fikrig, S.; Durkin, H. IgE against HIV proteins in clinically healthy children with HIV disease. *J. Allergy Clin. Immunol.* **1996**, *98*, 979–984. [CrossRef]

37. Bluth, M.H.; Robin, J.; Ruditsky, M.; Norowitz, K.B.; Chice, S.; Pytlak, E.; Nowakowski, M.; Durkin, H.G.; Smith-Norowitz, T.A. IgE anti-Borrelia burgdorferi components (p18, p31, p34, p41, p45, p60) and increased blood CD8+CD60+ T cells in children with Lyme disease. *Scand. J. Immunol.* **2007**, *65*, 376–382. [CrossRef]

38. Starkl, P.; Watzenboeck, M.L.; Popov, L.M.; Zahalka, S.; Hladik, A.; Lakovits, K.; Radhouani, M.; Haschemi, A.; Marichal, T.; Reber, L.L.; et al. IgE Effector Mechanisms, in Concert with Mast Cells, Contribute to Acquired Host Defense against Staphylococcus aureus. *Immunity* **2020**, *53*, 1333. [CrossRef] [PubMed]

39. Szymanski, E.P.; Leung, J.M.; Fowler, C.J.; Haney, C.; Hsu, A.P.; Chen, F.; Duggal, P.; Oler, A.J.; McCormack, R.; Podack, E.; et al. Pulmonary nontuberculous Mycobacterial Infection. A Multisystem, Multigenic Disease. *Am. J. Respir. Crit. Care Med.* **2015**, *192*, 618–628. [CrossRef] [PubMed]

40. Becker, K.L.; Arts, P.; Jaeger, M.; Plantinga, T.S.; Gilissen, C.; Van Laarhoven, A.; Van Ingen, J.; Veltman, J.A.; Joosten, L.A.; Hoischen, A.; et al. MST1R mutation as a genetic cause of Lady Windermere syndrome. *Eur. Respir. J.* **2017**, *49*, 1601478. [CrossRef]

41. Ovrutsky, A.R.; Merkel, P.A.; Schonteich, E.; Bai, X.; Kinney, W.; Iseman, M.D.; Kart ija, M.; Knight, V.; Chan, E.D. Patients with non-tuberculous mycobacterial lung disease have elevated transforming growth factor-beta following ex vivo stimulation of blood with live Mycobacterium intracellulare. *Scand. J. Infect. Dis.* **2013**, *45*, 711–714. [CrossRef]

42. Greinert, U.; Schlaak, M.; Rüsch-Gerdes, S.; Flad, H.D.; Ernst, M. Low in vitro production of interferon-gamma and tumor necrosis factor-alpha in HIV-seronegative patients with pulmonary disease caused by nontuberculous mycobacteria. *J. Clin. Immunol.* **2000**, *20*, 445–452. [CrossRef] [PubMed]

43. Gupta, S.; Pattanaik, D.; Krishnaswamy, G. Common Variable Immune Deficiency and Associated Complications. *Chest* **2019**, *156*, 579–593. [CrossRef] [PubMed]

44. Aghamohammadi, A.; Cheraghi, T.; Gharagozlou, M.; Movahedi, M.; Rezaei, N.; Yeganeh, M.; Parvaneth, N.; Abolhassani, H.; Pourpak, Z.; Moin, M. IgA deficiency: Correlation between clinical and immunological phenotypes. *J. Clin. Immunol.* **2009**, *29*, 130–136. [CrossRef]

45. Lucuab-Fegurgur, D.L.; Gupta, S. Comprehensive clinical and immunological features of 62 adult patients with selective primary IgM deficiency. *Am J Clin. Exp. Immunol.* **2019**, *8*, 55–67.

46. Goldstein, M.F.; Hilditch, G.J.; Dvorin, D.J.; Belecanech, G.A. Immunoglobulin replacement for selective IgM immunodeficiency, bronchiectasis, and Asthma. *Ann. Allergy Asthma Immunol.* **2016**, *116*, 172–173. [CrossRef] [PubMed]

47. Mikhail, I.; Grayson, M.H. Asthma and viral infections: An intricate relationship. *Ann. Allergy Asthma Immunol.* **2019**, *123*, 352–358. [CrossRef]

48. De Valles-Ibáñez, G.; Esteve-Solé, A.; Piquer, M.; González-Navarro, E.A.; Hernandez-Rodriguez, J.; Laayouni, H.; Gonzalez-Roca, E.; Plaza-Martin, A.M.; Deyà-Martínez, A.; Martín-Nalda, A.; et al. Evaluating the Genetics of Common Variable Immunodeficiency: Monogenetic Model and Beyond. *Front. Immunol.* **2018**, *9*, 636. [CrossRef] [PubMed]

49. Janzi, M.; Melén, E.; Kull, I.; Wickman, M.; Hammarström, L. Rare mutations in TNFRSF13B increase the risk of asthma symptoms in Swedish children. *Genes Immun.* **2012**, *13*, 59–65. [CrossRef]

50. Eguiluz-Gracia, I.; Fernandez-Santamaria, R.; Testera-Montes, A.; Ariza, A.; Campo, P.; Prieto, A.; Perez-Sanchez, N.; Salas, M.; Mayorga, C.; Torres, M.J.; et al. Coexistence of nasal reactivity to allergens with and without IgE sensitization in patientswith allergic rhinitis. *Allergy* **2020**, *75*, 1689–1698. [CrossRef]

51. Campo, P.; Eguiluz-Gracia, I.; Plaza-Serón, M.C.; Salas, M.; José Rodríguez, M.; Pérez-Sánchez, N.; González, M.; Molina, A.; Mayorga, C.; Torres, M.J.; et al. Bronchial asthma triggered by house dust mites in patients with local allergic rhinitis. *Allergy* **2019**, *74*, 1502–1510. [CrossRef]

52. Eckl-Dorna, J.; Villazala-Merino, S.; Campion, N.J.; Byazrova, M.; Filatov, A.; Kudlay, D.; Karsonova, A.; Riabova, K.; Khaitov, M.; Karaulov, A.; et al. Tracing IgE-Producing Cells in Allergic Patients. *Cells* **2019**, *8*, 994. [CrossRef]

53. Agondi, R.C.; Barros, M.T.; Kokron, C.M.; Cohon, A.; Oliveira, A.K.; Kalil, J.; Giavina-Bianchi, P. Can patients with common variable immunodeficiency have allergic rhinitis? *Am. J. Rhinol. Allergy* **2013**, *27*, 79–83. [CrossRef]

54. Edwards, E.; Razvi, S.; Cunningham-Rundles, C. IgA deficiency: Clinical correlates and responses to pneumococcal vaccine. *Clin. Immunol.* **2004**, *111*, 93–97. [CrossRef]

55. Jorgensen, G.H.; Gardulf, A.; Sigurdsson, M.I.; Sigurdardottir, S.T.; Thorsteinsdottir, I.; Gudmundsson, S.; Hammarström, L.; Ludviksson, B.R. Clinical symptoms in adults with selective IgA deficiency: A case-control study. *J. Clin. Immunol.* **2013**, *33*, 742–747. [CrossRef]

56. Khokar, A.; Gupta, S. Clinical and Immunological Features of 78 Adult Patients with Primary Selective IgG Subclass Deficiencies. *Arch. Immunol. Ther. Exp.* **2019**, *67*, 325–334. [CrossRef] [PubMed]

57. GEMA 4.4—Guía Española Para el Manejo del Asma. 2019. Available online: https://www.semg.es/index.php/consensos-guias-y-protocolos/316-gema-4-4-guia-espa~nola-para-el-manejo-del-asma (accessed on 1 May 2020).

58. Baumann, U.; Routes, J.M.; Soler-Palacín, P.; Jolles, S. The Lung in Primary Immunodeficiencies: New Concepts in Infection and Inflammation. *Front. Immunol.* **2018**, *9*, 1837. [CrossRef] [PubMed]

59. Rao, N.; Mackinnon, A.C.; Routes, J.M. Granulomatous and lymphocytic interstitial lung disease: A spectrum of pulmonary histopathologic lesions in common variable immunodeficiency—histologic and immunohistochemical analyses of 16 cases. *Hum. Pathol.* **2015**, *46*, 1306–1314. [CrossRef]

60. Tam, J.S.; Routes, J.M. Common variable immunodeficiency. *Am. J. Rhinol. Allergy* **2013**, *27*, 260–265. [CrossRef] [PubMed]

61. Kuehn, H.S.; Ouyang, W.; Lo, B.; Deenick, E.K.; Niemela, J.E.; Avery, D.T.; Schickel, J.-N.; Tran, D.Q.; Stoddard, J.; Zhang, Y.; et al. Immune dysregulation in human subjects with heterozygous germline mutations in CTLA4. *Science* **2014**, *345*, 1623–1627. [CrossRef] [PubMed]

62. Lopez-Herrera, G.; Tampella, G.; Pan-Hammarstrom, Q.; Herholz, P.; Trujillo- Vargas, C.M.; Phadwal, K.; Simon, A.K.; Moutschen, M.; Etzioni, A.; Mory, A.; et al. Deleterious mutations in LRBA are associated with a syndrome of immune deficiency and autoimmunity. *Am. J. Hum. Genet.* **2012**, *90*, 986–1001. [CrossRef]

63. Buchbinder, D.; Baker, R.; Lee, Y.N.; Ravell, J.; Zhang, Y.; McElwee, J.; Nugent, D.; Coonrod, E.M.; Durtschi, J.D.; Augustine, N.H.; et al. Identification of patients with RAG mutations previously diagnosed with common variable immunodeficiency disorders. *J. Clin. Immunol.* **2015**, *35*, 119–124. [CrossRef] [PubMed]

64. Ho, H.E.; Cunningham-Rundles, C. Non-infectious Complications of Common Variable Immunodeficiency: Updated Clinical Spectrum, Sequelae, and Insights to Pathogenesis. *Front. Immunol.* **2020**, *11*, 149. [CrossRef] [PubMed]

65. Schussler, E.; Beasley, M.B.; Maglione, P.J. Lung Disease in Primary Antibody Deficiencies. *J. Allergy Clin. Immunol. Pract.* **2016**, *4*, 1039–1052. [CrossRef]

66. Larsen, B.T.; Smith, M.L.; Tazelaar, H.D.; Yi, E.S.; Ryu, J.H.; Churg, A. GLILD Revisited: Pulmonary Pathology of Common Variable and Selective IgA Immunodeficiency. *Am. J. Surg. Pathol.* **2020**, *44*, 1073–1081. [CrossRef]

67. Shin, J.J.; Liauw, D.; Siddiqui, S.; Lee, J.; Chung, E.J.; Steele, R.; Hsu, F.I.; Price, C.; Kang, I. Immunological and Clinical Phenotyping in Primary Antibody Deficiencies: A Growing Disease Spectrum. *J. Clin. Immunol.* **2020**, *40*, 592–601. [CrossRef] [PubMed]

68. Popa, V.; Colby, T.V.; Reich, S.B. Pulmonary interstitial disease in Ig deficiency. *Chest* **2002**, *122*, 1594–1603. [CrossRef]

69. Goldstein, M.F.; Goldstein, A.L.; Dunsky, E.H.; Dvorin, D.J.; Belecanech, G.A.; Shamir, K. Selective IgM immunodeficiency: Retrospective analysis of 36 adult patients with review of the literature. *Ann. Allergy Asthma Immunol.* **2006**, *97*, 717–730. [CrossRef]

70. Altschul, A.; Cunningham-Rundles, C. Chronic urticaria and angioedema as the first presentations of common variable immunodeficiency. *J. Allergy Clin. Immunol.* **2002**, *110*, 664–665. [CrossRef]

71. Smith, A.A.; Humphrey, J.; McAuley, J.B.; Tharp, M.D. Common variable immunodeficiency presenting as chronic urticaria. *J. Am. Acad. Dermatol.* **2008**, *59*, S40–S41. [CrossRef] [PubMed]

72. Comberiati, P.; Costagliola, G.; Carli, N.; Legitimo, A.; D'Elios, S.; Consolini, R.; Petroni, D.G. Refractory Chronic Spontaneous Urticaria Treated With Omalizumab in an Adolescent With Common Variable Immunodeficiency. *Front. Immunol.* **2019**, *10*, 1700. [CrossRef]

73. Villatoro Santos, C.; Yacoub, M. Chronic urticaria and common variable immunodeficiency (CVID): An association to remember. *BMJ Case Rep.* **2021**, *14*, e239733. [CrossRef]

74. Magen, E.; Masalha, A.; Waitman, D.A.; Kahan, N.; Viner, I.; Klassov, L.; Vardy, D. Prevalence of dermatologic diseases among patients with selective immunoglobulin A deficiency. *Allergy Asthma Proc.* **2017**, *38*, 70–77. [CrossRef] [PubMed]

75. Chovancova, Z.; Kralickova, P.; Pejchalova, A.; Bloomfield, M.; Pechvatalova, J.; Vlkova, M.; PNechvatalova, J.; Vlkova, M.; Lizman, J. Selective IgM Deficiency: Clinical and Laboratory Features of 17 Patients and a Review of the Literature. *J. Clin. Immunol.* **2017**, *37*, 559–574. [CrossRef]

76. Mauracher, A.A.; Gujer, E.; Bachmann, L.M.; Güsewell, S.; Pachlopnik Schmid, J. Patterns of Immune dysregulation in Primary Immunodeficiencies: A Systematic Review. *J. Allergy Clin. Immunol. Pract.* **2021**, *9*, 792–802.e10. [CrossRef]

77. Bergerson, J.R.E.; Freeman, A.F. An Update on Syndromes with a Hyper-IgE Phenotype. *Immunol. Allergy Clin. North Am.* **2019**, *39*, 49–61. [CrossRef]

78. Jørgensen, S.F.; Reims, H.M.; Frydenlund, D.; Holm, K.; Paulsen, V.; Michelsen, A.E.; Jørgensen, K.K.; Osnes, L.T.; Bratlie, J.; Eide, T.J.; et al. A Cross-Sectional Study of the Prevalence of Gastrointestinal Symptoms and Pathology in Patients with Common Variable Immunodeficiency. *Am. J. Gastroenterol.* **2016**, *111*, 1467–1475. [CrossRef] [PubMed]

79. Song, J.; Lleo, A.; Yang, G.X.; Zhang, W.; Bowlus, C.L.; Gershwin, M.E.; Lung, P.S.C. Common Variable Immunodeficiency and Liver Involvement. *Clin. Rev. Allergy Immunol.* **2018**, *55*, 340–351. [CrossRef] [PubMed]

80. Li, Y.; Xu, C.; Lin, N.; Zhao, Y.; Song, J. Nodular regenerative hyperplasia and portal hypertension are the characteristics of liver abnormalities in patients with selective immunoglobulin A deficiency. *Clin. Res. Hepatol. Gastroenterol.* **2020**, *44*, e123–e125. [CrossRef] [PubMed]

81. Sano, A.; Inoue, J.; Kakazu, E.; Ninomiya, M.; Iwata, T.; Morosawa, T.; Takai, S.; Nakamura, T.; Masamun, A. Acute-onset Autoimmune Hepatitis in a Patient with Selective Immunoglobulin M Deficiency. *Intern. Med.* **2019**, *1*, 2185–2190. [CrossRef]

82. Arahata, M.; Tajiri, K.; Nomoto, K.; Tsuneyama, K.; Minami, S.; Shimizu, Y. A novel type of selective immunoglobulin m deficiency in a patient with autoimmune liver cirrhosis with recurrent hepatocellular carcinoma: A case report and review of the literature. *Int. Arch. Allergy Immunol.* **2013**, *161*, 91–96. [CrossRef]

83. Dyson, J.K.; Beuers, U.; Jones, D.E.J.; Lohse, A.W.; Hudson, M. Primary sclerosing cholangitis. *Lancet* **2018**, *391*, 2547–2559. [CrossRef]

84. Song, J.; Li, Y.; Bowlus, C.L.; Yang, G.; Leung, P.S.C.; Gershwin, M.E. Cholangiocarcinoma in Patients with Primary Sclerosing Cholangitis (PSC): A Comprehensive Review. *Clin. Rev. Allergy Immunol.* **2020**, *58*, 134–149. [CrossRef] [PubMed]

85. Hirano, K.; Tada, M.; Mizuno, S.; Isayama, H.; Takahara, N.; Nagano, R.; Hamada, T.; Miyabayashi, K.; Ito, Y.; Mohri, D. Lower incidence of biliary carcinoma in patients with primary sclerosing cholangitis and high serum levels of immunoglobulin E. *Clin. Gastroenterol. Hepatol.* **2012**, *10*, 79–83. [CrossRef]

86. Ferastraoaru, D.; Rosenstreich, D. IgE deficiency and prior diagnosis of malignancy: Results of the 2005–2006 National Health and Nutrition Examination Survey. *Ann. Allergy Asthma Immunol.* **2018**, *121*, 613–618. [CrossRef]

87. Ferastraoaru, D.; Goodman, B.; Rosenstreich, D. Higher rates of malignancy in patients with immunoglobulin E deficiency and negative immediate hypersensitivity skin tests. *Ann. Allergy Asthma Immunol.* **2021**, *126*, 194–195. [CrossRef]

88. Ferastraoaru, D.; Schwartz, D.; Rosenstreich, D. Increased Malignancy Rate in Children With IgE Deficiency: A Single-center Experience. *J. Pediatr. Hematol. Oncol.* **2021**, *43*, e472–e477. [CrossRef] [PubMed]

89. Nieters, A.; Łuczyńska, A.; Becker, S.; Becker, N.; Vermeulen, R.; Overvad, K.; Aleksandrova, K.; Boeing, H.; Lagiou, P.; Trichopoulos, D.; et al. Prediagnostic immunoglobulin E levels and risk of chronic lymphocytic leukemia, other lymphomas and multiple myeloma-results of the European Prospective Investigation into Cancer and Nutrition. *Carcinogenesis* **2014**, *35*, 2716–2722. [CrossRef] [PubMed]

90. Resnick, E.S.; Moshier, E.L.; Godbold, J.H.; Cunningham-Rundles, C. Morbidity and mortality in common variable immune deficiency over 4 decades. *Blood* **2012**, *119*, 1650–1657. [CrossRef]

91. Cunningham-Rundles, C.; Lieberman, P.; Hellman, G.; Chaganti, R.S. Non-Hodgkin lymphoma in common variable immunodeficiency. *Am. J. Hematol.* **1991**, *37*, 69–74. [CrossRef]

92. Ferastraoaru, D.; Gross, R.; Rosenstreich, D. Increased malignancy incidence in IgE deficient patients not due to concomitant Common Variable Immunodeficiency. *Ann. Allergy Asthma Immunol.* **2017**, *119*, 267–273. [CrossRef] [PubMed]

93. Ludvigsson, J.F.; Neovius, M.; Ye, W.; Hammarström, L. IgA deficiency and risk of cancer: A population-based matched cohort study. *J. Clin. Immunol.* **2015**, *35*, 182–188. [CrossRef]

94. Caka, C.; Cimen, O.; Kahyaoğlu, P.; Tezcan, İ.; Cagdas, D. Selective IgM deficiency: Follow-up and outcome. *Pediatr. Allergy Immunol.* **2021**, *32*, 1327–1334. [CrossRef] [PubMed]

95. Ni, J.; Zhang, J.; Chen, Q.; Chen, Y.; Liu, J. The epidemiology and clinical features of selective immunoglobulin M deficiency: A single-center study in China. *J. Clin. Lab. Anal.* **2020**, *34*, e23289. [CrossRef] [PubMed]

96. Hajjar, J.; Guffey, D.; Minard, C.G.; Orange, J.S. Increased incidence of fatigue in patients with primary immunodeficiency disorders: Prevalence and associations within the US immunodeficiency network registry. *J. Clin. Immunol.* **2017**, *37*, 153–165. [CrossRef]

97. Magen, E.; Mishal, J.; Vardy, D. Selective IgE deficiency and cardiovascular diseases. *Allergy Asthma Proc.* **2015**, *36*, 225–229. [CrossRef]

98. Magen, E.; Waitman, D.A.; Goldstein, N.; Schlesinger, M.; Dickstein, Y.; Kahan, N.R. Helicobacter pylori infection in patients with selective immunoglobulin a deficiency. *Clin. Exp. Immunol.* **2016**, *184*, 332–337. [CrossRef]

99. Roa, S.; Isidoro-Garcia, M.; Davila, I.; Laffond, E.; Lorente, F.; Gonzalez-Sarmiento, R. Molecular analysis of activation-induced cytidine deaminase gene in immunoglobulin-E deficient patients. *Clin. Dev. Immunol.* **2008**, *2008*, 146715. [CrossRef]

Journal of
Clinical Medicine

Review

The Epidemiology and Clinical Presentations of Atopic Diseases in Selective IgA Deficiency

Izabela Morawska [1], Sara Kurkowska [2], Dominika Bębnowska [3], Rafał Hrynkiewicz [3], Rafał Becht [4], Adam Michalski [1], Hanna Piwowarska-Bilska [2], Bożena Birkenfeld [2], Katarzyna Załuska-Ogryzek [5], Ewelina Grywalska [1], Jacek Roliński [1] and Paulina Niedźwiedzka-Rystwej [3,*]

1. Department of Clinical Immunology and Immunotherapy, Medical University of Lublin, Chodźki 4a St., 20-093 Lublin, Poland; izabelamorawska19@gmail.com (I.M.); adam.96.michalski@gmail.com (A.M.); ewelina.grywalska@gmail.com (E.G.); jacek.rolinski@gmail.com (J.R.)
2. Department of Nuclear Medicine, Pomeranian Medical University, Unii Lubelskiej 1 St., 71-252 Szczecin, Poland; sarakurkowska95@gmail.com (S.K.); hannap@pum.edu.pl (H.P.-B.); birka@pum.edu.pl (B.B.)
3. Institute of Biology, University of Szczecin, Felczaka 3c St., 71-412 Szczecin, Poland; dominika.bebnowska@usz.edu.pl (D.B.); rafal.hrynkiewicz@usz.edu.pl (R.H.)
4. Clinical Department of Oncology, Chemotherapy and Cancer Immunotherapy, Pomeranian Medical University of Szczecin, Unii Lubelskiej 1, 71-252 Szczecin, Poland; rafal.becht@pum.edu.pl
5. Department of Pathophysiology, Medical University of Lublin, Jaczewskiego 8b St., 20-090 Lublin, Poland; zaluskakatarzynaa@gmail.com
*   Correspondence: paulina.niedzwiedzka-rystwej@usz.edu.pl

Citation: Morawska, I.; Kurkowska, S.; Bębnowska, D.; Hrynkiewicz, R.; Becht, R.; Michalski, A.; Piwowarska-Bilska, H.; Birkenfeld, B.; Załuska-Ogryzek, K.; Grywalska, E.; et al. The Epidemiology and Clinical Presentations of Atopic Diseases in Selective IgA Deficiency. *J. Clin. Med.* **2021**, *10*, 3809. https://doi.org/10.3390/jcm10173809

Academic Editors: Rita Consolini and Giorgio Costagliola

Received: 3 August 2021
Accepted: 23 August 2021
Published: 25 August 2021

Abstract: Selective IgA deficiency (sIgAD) is the most common primary immunodeficiency disease (PID), with an estimated occurrence from about 1:3000 to even 1:150, depending on population. sIgAD is diagnosed in adults and children after the 4th year of age, with immunoglobulin A level below 0.07 g/L and normal levels of IgM and IgG. Usually, the disease remains undiagnosed throughout the patient's life, due to its frequent asymptomatic course. If symptomatic, sIgAD is connected to more frequent viral and bacterial infections of upper respiratory, urinary, and gastrointestinal tracts, as well as autoimmune and allergic diseases. Interestingly, it may also be associated with other PIDs, such as IgG subclasses deficiency or specific antibodies deficiency. Rarely sIgAD can evolve to common variable immunodeficiency disease (CVID). It should also be remembered that IgA deficiency may occur in the course of other conditions or result from their treatment. It is hypothesized that allergic diseases (e.g., eczema, rhinitis, asthma) are more common in patients diagnosed with this particular PID. Selective IgA deficiency, although usually mildly symptomatic, can be difficult for clinicians. The aim of the study is to summarize the connection between selective IgA deficiency and atopic diseases.

Keywords: atopic diseases; atopy; allergy; selective IgA deficiency; primary immunodeficiency

1. Introduction

Primary immunodeficiency diseases (PIDs) are a heterogeneous group of congenital diseases with various clinical manifestations and different models of inheritance (X-linked, AR, polygenetic), caused by the impairment or loss of at least one function of the immune system. They weaken the body's defenses, increasing the frequency of infections as well as the risk of autoimmune and proliferative diseases, including cancers [1].

PIDs can affect various elements of the immune system. As a result of next-generation sequencing and a better understanding of the molecular and immunological mechanisms, which affect the immune system, researchers can identify new genes and disorders. According to the latest data, ten basic types of PID can be distinguished: humoral and cellular response deficiency, PID's with associated or syndromic features, predominantly antibody deficiencies, immune dysregulation, congenital defects of phagocyte number and/or function [2].

Early diagnosis is of major importance and might be life-saving in patients with some PID. Recurrent or severe infections should raise a suspicion for immunodeficiency. The National Primary Immunodeficiency Resource Center developed a list of ten warning signs of PID [3]. Besides, Cunningham-Rundles et al. developed an immunodeficiency-related (IDR) score to assess the likelihood of finding immunodeficiency [4]. According to the recent work of Bahrami et al. the mean diagnostic delay among primary immunodeficient patients was 2.05 ± 1.7 years [5]. This delay is especially prominent in antibody deficiency defects and therefore requires special attention.

An unusual and challenging disease in the group of antibody deficiencies is selective IgA deficiency (sIgAD). Selective IgA deficiency is the most common primary immunodeficiency disease with an estimated occurrence from about 1:3000 to even 1:150, depending on the population, diagnosed more often in males [6,7]. The course of the disease is very varied, as most cases are asymptomatic, but recurrent infections, allergies, autoimmune diseases, and an increased risk of cancer may occur [7,8]. Besides the decreased level of serum IgA, patients with sIgAD suffer also from a deficiency of secretory IgA [9]. This facilitates the passage through the mucosal barrier for aeroallergens and food antigens, which makes these patients prone to develop allergies. Sometimes allergies can be even the first presentation of sIgAD. Aghamohammadi and colleagues reported that 40.5% of patients had allergic symptoms as the first manifestation of the disease [10]. Therefore, the suspicion of sIgAD should raise not only patients with recurrent infections but also with other clinical manifestations.

2. IgA—Structure

IgA is a class of immunoglobulins characterized by the presence of an alpha heavy chain. The daily synthesis of immunoglobulin A exceeds the total production of all other immunoglobulins [11]. In the human body, there are two subclasses of this immunoglobulin: IgA1 and IgA2. The most important difference between them lays in the structure of their hinge region and the number of the glycosylation sites [12].

In serum, IgA1 is predominant, accounting for as much as 90%, while in mucosal tissues, both subclasses are more evenly distributed, comprising 40% IgA1 and 60% IgA2. [13]. In human blood, IgA occurs mostly in monomeric form, while secretory IgA (SIgA) present on the surface of mucous membranes usually occurs in the form of dimers, much less often as trimers and tetramers [14,15]. Dimeric SIgA antibodies, covalently linked by a J-linking chain, are secreted onto the mucosal surface with their characteristic secretory complement (SC) [15].

3. IgA—Function

The majority of total IgA in the human body occurs in the mucosal tissues with a proven great role in the immune response. Serum level of IgA is 2–3 mg/mL, and it is the second most prevalent circulating immunoglobulin after IgG. However, until recently, the role of plasma IgA was still unclear. Now, we have some evidence that serum IgA has some immunological functions, which are independent of the role of secretory IgA.

Serum monomeric immunoglobulin A acting through Fc alpha receptor I (FcαRI) has important immunomodulatory functions [16,17]. FcαRI is expressed on cells of the myeloid lineage, including monocytes, neutrophils, eosinophils, some macrophages, intestinal dendritic cells, and Kupffer cells [18]. Its role is associated with activation of different signaling pathways, immunoreceptor tyrosine-based activation motif (ITAM), and ITAM inhibitory (ITAMi) [19]. Once a multimeric ligand binds FcαRI, activation of an inflammatory response through ITAM signaling takes place. On contrary, monovalent ligand, like monomeric IgA, acts through ITAMi signaling, which results in an anti-inflammatory response (Figure 1).

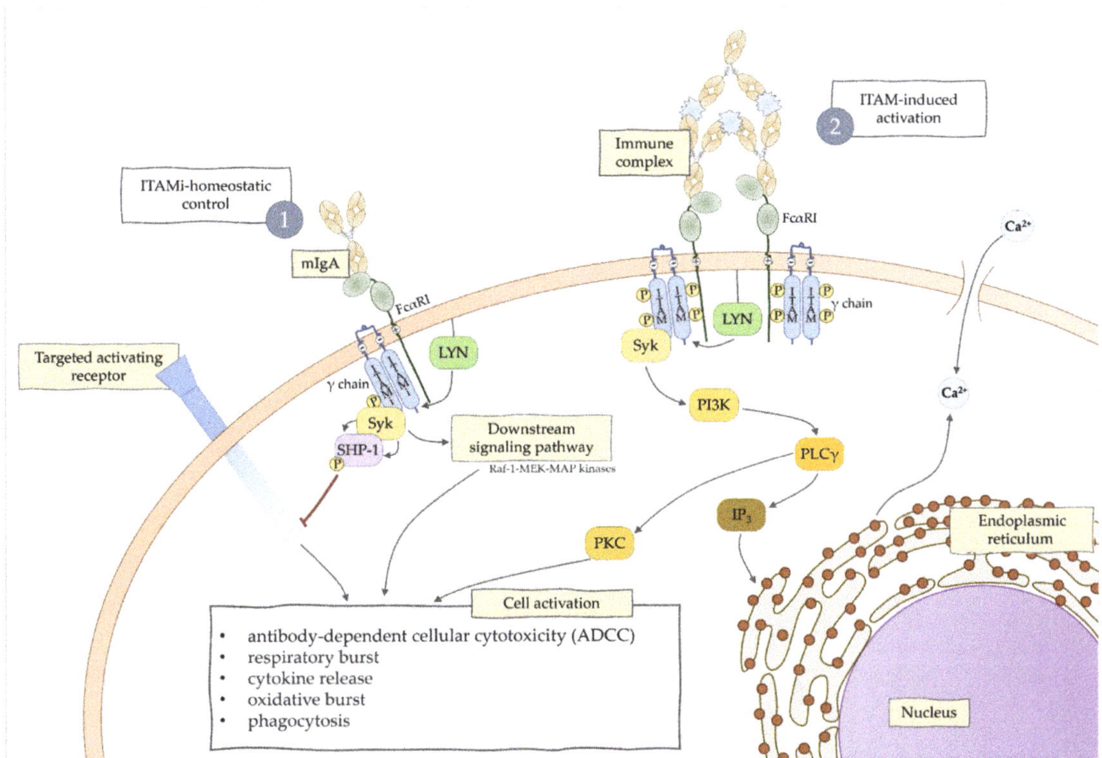

Figure 1. Regulation of immune responses by FcαRI, including ITAM-induced activation and ITAMi-control.

Instead, secretory IgA due to such a numerous representation within the mucous membranes is considered the body's first line of defense against harmful external factors [20]. It has been proven that they can eliminate pathogens, for example, by adsorption of food antigens, agglutination of bacteria, inhibition of epithelial adhesion to mucous membranes [16,18,21]. Reports indicate the ability to neutralize and inhibit the release of viral and bacterial particles, neutralize toxins and enzymes produced by numerous pathogens [22,23].

IgA has been shown to exert an anti-inflammatory function by inducing the expression of anti-inflammatory cytokines such as IL-10 and inhibiting releasing pro-inflammatory cytokines such as IL-6 [24–26]. Moreover, IgA activates complement only in a limited amount, but this class of immunoglobulin can block the activation of the complement mediated by IgG [21,27]. It is known that IgA silences some responses after bacterial cell activation, like for example, oxidative burst activity, phagocytosis, as well as chemotaxis [28]. Effector functions of IgA are complex. As mentioned before, IgA may interfere directly with immune cells of the myeloid lineage using FcαRI. After the interaction of IgA with monocyte-derived dendritic cells, antigen presentation, maturation, and production of IL-10 may occur [29]. Monocytes also have the ability to produce IL-10 after IgA ligation and to inhibit the production of IL-6 and TNF-alfa [30]. SIgA is important in eosinophil activation and degranulation and is more potent at stimulation of the release of reactive oxygen species than IgE, as well as it regulates oxidative burst and cytokine release by human alveolar macrophages [31,32]. Moreover, binding IgA with mannan-binding lectin (MBL) results in complement activation, which is a part of antimicrobial defense [33]. SIgA may act as a competitive blocker of IgG-mediated complement activation [30]. A significant aspect influencing the proper development of humoral immunity is an adequate stimulation of the immune system and, as a result, the ability to maintaining an appropriate balance between the cellular and humoral response.

4. Pathogenesis of IgA Deficiency

The pathogenesis of the disease is not yet fully understood. It is possible that sIgAD can be caused by the overlap of some of these mechanisms [34].

One of them, which has been widely described, is the presence of errors in the differentiation of IgA$^+$ plasmablasts causing a low number of IgA-secretory cells, difficulties with switching IgA to SIgA, and a low number of mRNA in B-cells producing IgA [7,35–38]. Another mechanism concerns cytokines that are involved in IgA production (IL-10, IL-4, IL-6, IL-12, IL-21) and is caused by dysregulation of their pathways, especially in secondary lymphoid organs [8,9,39–44]. In patients with sIgAD, IL-10 is proven to be crucial in the differentiation of the B cells to IgA-secreting cells. Furthermore, it has a synergic effect with IL-4 [43,45]. Another cytokine that causes IgA production is TNF-B, it also possesses the ability to act as an isotype "switch" factor for IgA production [46,47]. Lowered level of TGF-B may lead to a low IgA level in patients with sIgAD [9]. Il-21 stimulation is even more effective in inducing IgA production than IL-4 and IL-10, as well as it prevents CD19 + B cells spontaneous apoptosis [39]. This increased apoptosis could cause a reduction in survival of B cells and, therefore, decreased production of normal levels of IgA immunoglobulin [48]. Additionally, it is possible that T-cells impairment is connected with sIgAD. Soheili et al. suggest a direct association between decreased level of T regulatory (T_{reg}) and the severity of clinical presentation of sIgAD [49]. In this study, patients were divided into two groups—G1 with a lower-than-cut-point T_{reg} value, where there was a higher risk of developing autoimmunity and class switching recombination defects, and G2 with a higher T_{reg} value, where only one person had autoimmunity and no one had the described antibodies defect. The link between T_{reg} cells and IgA production is complex and multifaceted. T_{reg} cells colonize the intestinal mucosa where they produce TGF-beta and IL-10, which are essential in the production of IgA. Reduced amount of T_{reg} negatively affects the amount of IgA$^+$ B lymphocytes, and restoration of the correct amount of T_{reg}, consequently, restores normal IgA production in the intestines [50–52] Interestingly, according to the meta-analysis by Bronson et al. there is a multiple gene linkage between the "Intestinal Immune Network for IgA production" and "T_{reg}" [53]. Besides, the highest levels of APRIL (a proliferation-inducing ligand), which is connected with IgA-synthesis as a compensatory mechanism, are observed in patients with sIgAD [20,54]. It has been proved that there is a genetic background to sIgAD [55–57].

Moreover, level of these immunoglobulins may be influenced by drugs that are often used in everyday practice—non-steroid anti-inflammatory drugs (NSAIDs), angiotensin convertase enzyme inhibitors (ACEI), several anti-epileptic drugs, or drugs used in rheumatology. They can even trigger iatrogenic isolated sIgAD [58,59]. Moreover, some of the viral infections, e.g., EBV, hepatitis type C may induce post-infection IgA deficiency [60,61].

5. Clinical Presentation of sIgAD

Based on clinical presentation, sIgAD patients can be classified into different phenotypes. Yazdani et al. [8] in their work from 2015 divided these phenotypes into five main categories: asymptomatic, minor infections, autoimmunity, allergy, severe. It was reported that there is no correlation between serum IgA levels and clinical phenotype and disease severity [34].

Diagnosis of sIgAD is a diagnosis of exclusion. Immunologists should take into consideration infection-induced or drug-induced IgAD, as well as drug-induced IgAD/IgG2 subclass deficiency [62–64]. Important factors in establishing the diagnosis of IgA deficiency are family background and other laboratory parameters, which are relevant in order to differentiate sIgAD from CVID (lowered IgA/IgG and sometimes IgM levels), secondary hypogammaglobulinemia (moderately low levels of IgA), single-gene primary immunodeficiencies, hypoglobulinemia due to the protein loss as the result of enteropathy or nephrotic syndrome and malignancies such as thymoma, myeloma, and chronic lymphocytic leukemia [64,65].

5.1. Asymptomatic

Most of the patients are asymptomatic. The estimate number is 60% [8], but it varies in different studies. These patients might develop some clinical manifestations; therefore, they should undergo regular evaluations [34]. A 22-year study based on a cohort of 184 pediatric SIgAD patients, performed by Lougaris et al. [66] shows how the clinical presentation of the disease can vary with time. They assessed laboratory parameters and long-term health status of patients, 62% of whom had symptoms at the time of diagnosis. Allergic complications during follow-up were additionally developed in 16%, autoimmune diseases excluding celiac disease in 9%, and celiac disease developed in 11% of previously disease-free patients. During the follow-up period, 4% of patients achieved age-appropriate IgA levels, 9% of patients achieved partial IgA deficiency diagnosis and 2% of patients developed CVID.

5.2. Minor Infections

Children with recurrent and severe infections present a diagnostic challenge [67]. Four or more ear infections and two or more serious sinus infections or episodes of pneumonia within one year are warning signs for primary immunodeficiencies in children [68].

Secretory IgA plays an important role in maintaining the equilibrium of the body, as it takes part in the mucosal immune system and serves as the interface between the body and the microbiome. In the human body, the largest mucosal systems are the gastrointestinal tract and respiratory system, and therefore, decreased level of IgA will affect mostly them.

Symptomatic patients with minor infection usually present recurrent upper respiratory tract infections (40–90%), mainly viral, less frequently bacterial (with encapsulated bacteria etiology, such as Streptococcus pneumoniae, Hemophilus Influenzae) [7,8,69,70]. Bacterial bronchitis and pneumonia are much less common, but these infections may be complicated by bronchiectasis [71]. Infections of the ear, sinuses, conjunctiva, nose, and throat mucosa may occur. Most often, these infections are mild, not requiring hospitalization and their treatment does not differ significantly from the treatment of a patient without sIgAD. It was found that patients with sIgAD have a compensatory increase in secretory IgM level [9,72], however, these IgM cannot replace all functions of IgA [73]. In the intestines of sIgAD patients, there is 65–75% of Ig-containing plasma cells with the ability to produce IgM in comparison to about 6% in healthy volunteers, possibly due to the homology in structure and function between those two isotypes [74–77]. Besides, patients with sIgAD suffer from urinary tract infections (UTI) and gastrointestinal tract infections with viruses and bacteria. Moreover, intermittent or chronic diarrhea due to Giardia Lamblia is common, because the attachment and proliferation of this parasite on the gastrointestinal mucous are facilitated due to lack of IgA [78,79]. It is important to mention that in patients with recurrent UTI, bronchitis and pneumonia, defects in the urinary and respiratory systems should be excluded.

The diagnosis of selective IgA deficiency mostly does not significantly influence the therapeutic management of patients. Treatment of infections should be adequate to their etiology, patient's age, and clinical condition. Treatment of comorbidities and prevention of complications remain the basis. There are no clear guidelines that would suggest the need for longer and more aggressive antibiotic therapy in this disease than in patients without sIgAD. There is also no consensus on the use of antibacterial prophylaxis in this immunodeficiency, but its usage was suggested in more severe cases, at least periodically [34,48,80]. Vaccinations play a significant role in minimizing the risk of infections [48]. It is advisable to extend the standard calendar with vaccination against *S. pneumoniae*, *N. meningitidis*, *H. influenzae*, and annual vaccination against influenza [34]. It is not typically recommended to initiate IgG (i.v, s.c) replacement therapy in patients without the coexistence of other immune-related diseases, acute, severe infections, or coexistence of specific antibodies deficiency [81]. According to one of the latest meta-analyses covering the effects of oral probiotics, parabiotics, and synbiotics on immunoglobulin levels, it has been shown that their supplementation increases significantly salivary IgA secretion, without a significant effect on the level of other immunoglobulins and with no effect on the serum IgA [82]. In addition, there are reports of an increase in the amount of IgA$^+$ cells in the intestines of lamina propria in mice after oral ingestion of Lactobacillus-based preparations [83,84]. One prospective, randomized study demonstrated the validity of the use of oral immunomodulator bacterial extract (OM-85 BV) in patients with sIgAD and/or IgG subclass deficiency, resulting in a lower one-year infection rate [85]. A suggestion has been made to use oral IgA in patients with sIgAD, since this deficiency is associated with dysbiosis and chronic inflammation, and the present inflammation is inversely correlated with systemic anticommensal IgG response, which acts as "second line of defense" [86,87].

The importance of the IgA was raised again because of the ongoing global pandemic of coronavirus disease 2019 (COVID-19). When looking for the reasons for the varied course of the disease, questions arose on whether deficiency of IgA could be the reason for disease severity, vaccine failure, and prolonged viral shedding [88]. As mentioned above, the prevalence of sIgAD differs in various countries and the same was found for COVID-19. Naito et al. compared the number of cases of COVID-19 with the prevalence of selective IgA deficiency in different countries [89]. They found "a strong positive correlation between the frequency of sIgAD and the COVID-19 infection rate per population". It was then concluded that one of the factors contributing to the low death rate from COVID-19 infection in Japan could be the low incidence of sIgAD in the country. As primary immunodeficiencies are a group of rare diseases, there is little data on the coexistence of sIgAD and COVID-19 infection. Nevertheless, literature data showing an extremely significant effect of class A immunoglobulins on early protection against SARS-CoV-2 virus also suggest a potentially more severe/complicated course of the disease [88,90]. This thesis is supported by the aforementioned literature data: a positive correlation between a high number of COVID-19 infections and a high incidence of sIgAD has been demonstrated, and an inverse relationship was observed in the extreme example of Japan [88]. In Israel, during two "so-called" waves, 20 patients with PID were affected by COVID-19 and none of them was diagnosed with sIgAD [91]; importantly, the relationship between the development of autoimmune diseases in the course of COVID-19 in patients with sIgAD—AIHA and Guillain-Baree syndrome [92]. Researchers also point to the risk of a poor response against SARS-CoV-2 after immunization in this group of patients [88].

5.3. Autoimmunity

There is an association between IgA deficiency and a higher prevalence of autoimmune disease [93,94]. Based on extended research in that field, the prevalence of autoimmune disease in this group rises to 31.7% [95]. According to Azizi et al. the median age of the onset of the first episode of autoimmunity was 7 [95]. Among diseases with higher prevalence in sIgAD subjects, we differentiate systemic lupus erythematosus, hypo- and hyperthyroidism, type 1 diabetes mellitus, Crohn's disease, ulcerative colitis, rheumatoid arthritis, juvenile idiopathic arthritis, ankylosing spondylitis, and vitiligo. Whereas other diseases like scleroderma, celiac disease, autoimmune hepatitis, immune thrombocytopenic purpura, and autoimmune hemolytic anemia, occur less often but still with higher prevalence than in the general population [96].

The mechanism of autoimmunity in sIgAD is still not fully understood. There are six hypotheses that try to explain these phenomena, each based on a different mode of autoimmunity, such as human leukocyte antigen, cytogenic, monogenic, molecular mimicry, lingering inflammation and immune complexes, dysregulation of molecular pathways [96]. Some studies suggest that various mechanisms are likely to play concurrently. It has been found that there is also a higher incidence of autoimmunity in first-degree relatives of sIgAD patients [97].

In a recent case reported by Pfeuffer et al., authors stated that the presence of other acute diseases could induce autoimmunity in SIgAD patients [98]. In their case, it was Guillain-Barré syndrome induced by COVID-19.

5.4. Allergy

Allergy has long been a component of immune deficiency; however, allergic burden differs in different types of immunological disorders. Both PID and allergy are associated with impaired reactions of the immune system. In the case of PID, malfunctioning of some of its components will lead to infectious susceptibility. Atopy, on the other hand, is a hypersensitivity reaction of the immune system and a form of misdirected immunity.

The true prevalence of allergy among patients with sIgAD is still under debate since studies from different countries present inconsistent results. Therefore, it suggests that the prevalence varies depending on the ethnic background. There is even controversy in the scientific world whether the coexistence of IgA deficiency and allergic diseases is, in fact, true. Most publications support the relationship between sIgAD or low-IgA levels and allergic diseases [10,99–103], but some researchers deny it [104,105]. This connection has been the subject of medical research for over 50 years. In 1975, Buckley suggested that about half of the patients with sIgAD presented atopic diseases and related findings have been published later by Kemola [106,107]. A similar observation has been done in Ankara more recently, in 2017, where 45.7% of the patients diagnosed with sIgAD presented one of the following: asthma, rhinitis, eczema, atopic dermatitis, and interestingly the prevalence of allergy in a close family of this patients rose up to 43.2% [101]. Aghamohammadi, in his study on Iranian patients, revealed that allergy was observed in 84% of patients with sIgAD [10]. In a study from China, 17.6% of patients had allergic symptoms, however, most of them were allergic reactions to drugs (mostly penicillin) [108]. These results were inconsistent with typical allergies reported in other countries, such as asthma, rhinitis, food allergy, and atopic dermatitis [108]. On the other hand, there is a study with a prevalence of allergy in children with sIgAD on the percentage of 13% [70].

Many clinicians point out the frequent coexistence of IgA deficiency with bronchial asthma, allergic rhinitis, and atopic dermatitis in everyday practice but it is unclear whether it is the immunoglobulin A deficiency that promotes an allergic reaction, or the allergic reaction weakening the mucous membranes and consequently leading to a secondary IgA deficiency. There are plenty of possible explanations of this phenomenon. The connection between IgA deficiency and allergies may be caused by increased levels of circulating antigens, due to increased permeability at mucosal surfaces. It could also be a result of the inability to induce ITAMi signaling, due to decreased level of monomeric serum IgA, which, consequently, causes overactivation of the immunological system [79,109]. Another hypothesized mechanism is the deficiency of TGF-beta response. TGF-beta has properties to induce IgA synthesis, as well as inhibiting proliferation of Th2-cells. Th2-response is involved in the pathogenesis of atopic diseases [110–112]. Interestingly, there are allergen-specific A immunoglobulins, but their role in the pathogenesis of allergic diseases is unclear [113]. We do not know whether they are responsible for exacerbation or silencing the symptoms, but what we know is that they are observed in healthy people without allergic symptoms and low or undetectable IgE-levels [114,115]. Moreover, children with a tendency to allergic diseases have a more pronounced physiological IgA deficiency in the neonatal period and the lower these concentrations are, the greater is the severity of symptoms (although they usually remain within the reference values for age).

The American Academy of Allergy, Asthma & Immunology (AAAAI) and the American College of Allergy, Asthma & Immunology (ACAAI) developed practice parameters to guide the management of primary immunodeficiencies [116]. It is stated there that atopic diseases should be treated aggressively in patients with sIgAD. Since allergic inflammation facilitates the development of respiratory tract infections, it is crucial to treat allergy using all standard modalities, like avoidance of allergens, medication, and immunotherapy [116]. Our clinical experience shows that treating atopic diseases in patients with immunodeficiency is difficult and requires special attention and scrupulousness. In addition to the commonly used anti-histamine drugs, beta-mimetics, and glucocorticosteroids, in the case of treatment-resistant atopic diseases, biological drugs such as omalizumab and dupilumab might be helpful. Omalizumab is an anti-IgE antibody that is FDA approved for the treatment of moderate to severe allergic asthma, while dupilumab is an IL-4 receptor blocking antibody and is FDA approved for the treatment of moderate to severe atopic dermatitis in patients with the refractory disease [117–119]. The use of omalizumab in a young adult patient diagnosed with CVID, who suffered from chronic spontaneous urticaria, and did not respond either to an immunoglobulin substitution in immunomodulatory doses, anti-H1 and anti-H2 antihistamines, as well as leukotriene receptor antagonists, has been described. Only the inclusion of omalizumab resulted in a significant improvement in the condition of the skin and quality of life [120]. However, data suggest that such treatment could carry a risk of possible side effects—Banh et al. described a case of a 24-year-old patient diagnosed with asthma and CVID, where treatment with omalizumab might have increased the level of white blood cells and elevated myeloid cell count. Serious disorders, e.g., malignancy or severe infections were excluded. Importantly blood test results returned to normal levels shortly after drug discontinuation [121]. The use of dupilumab has been described in the context of a patient with CVID suffering from severe skin lesions such as erythematous-squamous and generalized infiltrated rash with exacerbation in sun-exposed zones and severe recurrent infections, in whom no improvement in skin condition was observed after treatment with glucocorticoids or cyclosporine. The introduction of dupilumab resulted in a reduction in the severity of skin lesions and the addition of IgG replacement therapy lowered the frequency of infections [122]. In our opinion, based on experience from other immunodeficiencies with predominantly antibody deficiency, it is possible to use the above-mentioned monoclonal antibodies in the treatment of severe allergic complications in patients with SIgAD.

5.4.1. Food Allergy

The prevalence of food allergy in patients with PIDs was examined using the US Immunodeficiency Network (USIDNET). Surprisingly, it was lower than that in the general population. However, for some specific types of PID, like sIgAD, the prevalence was increased and it was found to be 25% [123], but there were only four patients with sIgAD in the registry. It is consistent with the study performed by Aghamohammadi et al. where the prevalence of food allergy among patients with sIgAD was 22% [10]. Across all studies in this review, the prevalence of food allergy among patients with SIgAD is presented in Table 1. Another study reports an increased risk of parentally reported food hypersensitivity at 4 years of age among children with sIgAD [124]. Moreover, the authors did not find any association between IgAD and increased levels of specific IgE, which could suggest that hypersensitivity in IgAD children is not IgE-mediated [124].

The majority of patients with deficiency of secretory IgA have substitution with secretory IgM. However, it might not guarantee proper mucosal protection and might allow food antigens to pass through the gastrointestinal mucosa and predispose to develop a food allergy. Another possible explanation connected with eczema and food allergy is the hypothesis that, due to the IgA-deficiency to gastrointestinal antigens in the gut, there is no antigen immunological-exclusion, which consists of antigen binding to SIgA at the level of the mucosal surface, and, consequently, blocking the absorption of the antigen [9,125].

Recent years showed that there is a strong connection between microbiota and allergy development. For example, in 2009, researchers found that children with allergy not only had lower salivary SIgA levels but also less differentiated bacterial microenvironment [126,127]. A study from 2018 focused on the effects of IgA deficiency on human gut microbiota composition [128]. They found out that patients with sIgAd have an altered gut microbiota composition compared to healthy patients. Moreover, the secretion of IgM cannot fully compensate for the lack of SIgA. It is therefore suggested that IgA plays a critical role in controlling stable gut microbial community. A different study from the same year showed only mild loss in microbial diversity in sIgAD subjects [129].

It was also found that serum IgA plays a role in suppressing IgE-mediated food allergy. IgE-mediated food allergy is a common cause of enteric disease, and, in the study conducted by Strait et al. concerning IgE-mediated systemic anaphylaxis induced by ingested allergens, it has been found that both serum antigen-specific IgG and IgA antibodies can protect against severe IgE-mediated allergic reaction [130]. This suggests that decreased serum IgA antibody levels might predispose to increased intestinal mucosal permeability and absorption of ingested antigens, therefore, increasing the risk of severe food allergy [131].

Table 1. Food allergy and sIgAD.

Year	Country	Sample Size	Disease Prevalence among SIgAD (%)	Reference	Diagnostic Tools
2009	Iran	37	22	[10]	The allergy status was evaluated by skin prick test, using 14 common standard allergen extracts
2012	Spain	330	4.2	[132]	Retrospective study of patients records
2017	Turkey	81	1.2	[101]	Skin prick tests + serum IgE measurements. Food allergy diagnosis was confirmed with an oral food challenge test.
2020	Iran	166	3.6	[133]	Data about clinical presentations were collected based on a detailed questionnaire

5.4.2. Asthma

Asthma is a chronic inflammatory disease of the respiratory system characterized by bronchial hyperresponsiveness and reversible airflow obstruction. It is one of the most common chronic illnesses in childhood and its etiology in this group is vastly associated with atopy. Some studies report that asthmatic patients are more likely to have a diagnosis of sIgAD/CVID than non-asthmatic individuals [134]. In a study on an Iranian group, the prevalence of asthma among sIgAD patients was 51% [10], while in the general Iranian population it is 22–23% [135,136]. In the study on a Spanish group, asthma was observed in only 12.4% patients [97]. On the other hand, no difference in prevalence was found comparing sIgAD patients and control group in the case-control study of Jorgensen et al. [137]. The prevalence of asthma among patients with SIgAD is presented in Table 2.

Papadopoulou et al. state that the insufficient protection provided by the respiratory mucosa deprived of IgA in children with sIgAD makes them prone to develop bronchial hyperresponsiveness and consequently asthma [138]. In a different study, a high number of IgA-specific salivatory antibodies has been connected to a lower risk of late-onset wheezing in sensitized infants [139]. Furthermore, sIgAD may be connected with TNFRSF13B gene variants as one of the genetic susceptibilities. This gene encodes the transmembrane activator and calcium modulator and cyclophilin ligand interactor (TACI), which is the tumor necrosis factor receptor (TNFR) expressed on activated B cells and macrophages and is involved in isotype class switching to IgA [54,140,141]. Moschese et al. investigated the prevalence of TNFRSF13B mutations in 56 patients with absolute and partial sIgAD reporting 20% prevalence in this group [142]. Furthermore, researchers suggest that the mutation in these genes increases the risk of asthma development up to 2.5 fold, despite the IgE levels [69]. Moreover, the studies on the mice model proved that treating with antigen-specific IgA may protect animals from hyperresponsiveness as well as eosinophilic inflammation in airways [143]. Additionally, since mice do not express FcαRI [144], studies on human FcαRI transgenic mice were used in studies on the asthma model. It was found that by targeting FcαRI, IgA has been established as a strong inhibitor of asthma development [145].

Some studies reported a higher prevalence of respiratory tract infections among patients with sIgAD and allergy compared to those with sIgAD without any manifestation of allergy disease [99,133,146]. It suggests that allergic patients are more susceptible to respiratory tract infections.

Table 2. Asthma and sIgAD.

Year	Country	Sample Size	Disease Prevalence among SIgAD (%)	Reference	Diagnostic Tools
2008	Brazil	126	48.4	[79]	Diagnostic criteria of allergic diseases were not defined in the paper
2009	Iran	37	51	[10]	Lung function was evaluated according to the American Thoracic Society guidelines, using a computerized pneumotachograph
2010	Israel	63	23.8	[99]	Retrospective study of patients records
2012	Spain	330	12.4	[132]	Retrospective study of patients records
2013	USA	39	23	[134]	Asthma status was determined based on predetermined criteria for asthma
2013	Iceland	32	18.8	[137]	Self-administered questionnaire + interview performed by physician + lung function tests using spirometry
2017	Turkey	81	34.6	[101]	Asthma status was determined based on the Global Initiative for Asthma guidelines
2019	Italy	103	10.7	[142]	Patients' clinical data were collected at enrolment and every 6–12 months for 5 years. Diagnostic criteria for allergic diseases were not defined in the paper
2020	Iran	166	6.6	[133]	Data about clinical presentations were collected based on a detailed questionnaire

5.4.3. Atopic Dermatitis

A variety of primary immunodeficiencies have cutaneous manifestations. In the case of sIgAD, nonspecific cutaneous finding is eczematous dermatitis. Here, similarly to other allergic manifestations, there is a huge variety in the prevalence reported in different studies, which could be caused by ethnic diversity and, also, by different algorithms for atopic dermatitis diagnosis. Therefore, in a study performed by Aghamohammadi, the prevalence was 52% [10], but in the study of Magen, it was only 4.6%, however, it was still higher than in the control group [147]. The prevalence of atopic dermatitis among patients with SIgAD is presented in Table 3.

Moreover, Orivari et al., showed that the levels of secretory IgA in breast milk were inversely associated with the development of atopic dermatitis up to 2 and 4 years [148] among breastfeeding children. In a different study, though, such connection was not found [149].

Moreover, people with higher IgA levels and Staphylococcus aureus colonization in the gastrointestinal tract are less susceptible to the development of eczema [150].

Table 3. Atopic dermatitis and sIgAD.

Year	Country	Sample Size	Disease Prevalence among SIgAD (%)	Reference	Diagnostic Tools
2008	Brazil	126	2.4	[79]	Diagnostic criteria of allergic diseases were not defined in the paper
2009	Iran	37	49	[10]	The allergy status was evaluated by skin prick test, using 14 common standard allergen extracts
2010	Israel	63	3.2	[99]	Retrospective study of patients records
2012	Spain	330	3.6	[132]	Retrospective study of patients records
2015	Italy	102	57.84	[151]	Diagnosis was based on Hanifin-Rajka criteria and on skin biopsies where applicable
2017	Turkey	81	11.1	[101]	Diagnosis was based on Hanifin-Rajka criteria
2017	Israel	347	4.6	[147]	Retrospective study of patients records. Criteria for diagnosis of chronic spontaneous urticaria according to EAACI, GA2LEN, EDF and WAO guidelines
2019	Italy	103	12.6	[142]	Patients' clinical data were collected at enrolment and every 6–12 months for 5 years. Diagnostic criteria for allergic diseases were not defined in the paper

5.4.4. Allergic Rhinitis and Conjunctivitis

Serum IgA level in children under the age of 4 with positive skin-prick test was significantly lower than in healthy population, also allergic rhinitis and eczema were connected with a low level of salivary IgA [152]. The frequency of allergic rhinitis among patients with sIgAD in a study performed in Turkey was 27.2% [101], while the prevalence of allergic rhinitis in Turkish school-age children was 16.9% [153]. The presence of allergic rhinitis was only accepted if it was diagnosed by a physician. Furthermore, in a different study, the prevalence of allergic-rhinoconjunctivitis tended to be increased in the sIgAD group and was reported to be 37.5% [137]. Across all studies in this review, the prevalence of allergic rhinitis and conjunctivitis among patients with SIgAD is presented in Table 4.

Table 4. Allergic rhinitis/conjunctivitis and sIgAD.

Year	Country	Sample Size	Disease Prevalence Among SIgAD (%)	Reference	Diagnostic Tools
2008	Brazil	126	53.2 (AR)	[79]	Diagnostic criteria of allergic diseases were not defined in the paper
2009	Iran	37	40 (AR/C)	[10]	The allergy status was evaluated by skin prick test, using 14 common standard allergen extracts
2010	Israel	63	12.7 (AR)	[99]	Retrospective study of patients records
2012	Spain	330	9 (AR)	[132]	Retrospective study of patients records
2013	Iceland	32	37.5 (AR/C)	[137]	Self-administered questionnaire + interview performed by physician + skin prick tests
2017	Turkey	81	27.2 (AR)	[101]	Presence of allergic rhinitis was only accepted if it was diagnosed by a physician
2019	Italy	103	18.4 (AR) 9.7 (C)	[142]	Patients' clinical data were collected at enrolment and every 6–12 months for 5 years. Diagnostic criteria for allergic diseases were not defined in the paper

5.5. Severe

As opposed to other primary immunodeficiencies, sIgAD rarely presents with severe manifestations. Therefore, differential diagnosis with other possible immunological disorders should be performed.

Patients with this phenotype suffer from recurrent and severe infections even in lower respiratory tracts [8]. One of the severe complications of severe respiratory infections is bronchiectasis. In such cases, it is crucial to eliminate other immunodeficiencies such as IgG2subclass, specific antibody deficiencies, and mannan-binding lectin deficiency [79].

Patients with this phenotype should be provided with extra care. In case of recurrent infections, prophylactic antibiotics should be considered, especially during autumn and winter [8]. The usage of IVIG replacement therapy in these patients is extremely controversial [116,154]. Usually, this treatment is recommended for individuals with both IgA deficiency and concomitant IgG2 subclass deficiency [155,156]. To determine if this treatment would be beneficial, the IgG antibody responses to protein and polysaccharide vaccines should be evaluated first [116].

6. Complications

In a prospective cohort study that examined mortality among patients with sIgAD turn out that they have an increased risk of death in the first 10 years after diagnosis [157]. Afterward, the mortality is similar to that of the general population. The most common causes of death include malignancy and cardiovascular diseases. There are a few life-threatening complications of sIgAD. Even if their prevalence is not high, they should be known for physicians to provide proper help for their patients. Among those included in the literature, we differentiate progression to CVID, transfusion-related anaphylaxis, and malignancy.

6.1. IgA Deficiency and CVID

Common variable immunodeficiency (CVID) is an immune disorder characterized by decreased serum levels of both IgG and IgA, with or without a decreased level of IgM, and poor antibody vaccine response or low switched memory B cells less than 70% of age-appropriate normal [158]. CVID most often presents with recurrent infections of the respiratory and gastrointestinal tract [159]. Symptomatic sIgAD and CVID have many similar features, moreover, some patients with sIgAD progress to CVID, especially if autoimmunity or IgG subclass deficiency is observed [149–164].

In patients with sIgAD, there is a significantly lower number of class-switched memory B cells and transitional B cells [165]. Preprint of another study showed an increased percentage of naive B cells and decreased percentage of switched memory B cells. Only one parameter correlated with the severity of the disease—CD21low cells. They were increased in patients with severe SIgAD as compared to those with mild severity [166]. Increased level of CD21low was previously described but without correlation to clinical status [167].

In CVID, there is a classification based on B-cell phenotype, which divides CVID patients into B − group and B + group, depending on the CD19 expression (lower or higher > 1%). B + patients may be further divided into groups smB + or smB-, based on a proportion of switched memory B-cell percentage (lower or higher >2%). Recently, an increase in transitional B cells and CD21low B-cells is used as a base to subdivide groups [168]. Some of the B-cell phenotype findings are similar between CVID and SIgAD. There is an interesting observation of an increased CD21low cells number in patients with severe sIgAD; a higher level of those cells in CVID patients is connected with autoimmune phenomena [169,170].

The major histocompatibility complex (MHC) represents the most common genetic susceptibility locus for CVID. However, non-MHC-associated single-gene mutations have been identified. These include the genes for ICOS, BAFF-R, TACI, CD19, CD21, CD81, CD20, LRBA, PKC-Delta, NF-kB1, NF-kB2, IL-21 [171]. Defects of these genes represent only approximately 2–10% of patients with CVID [172]. Some authors state that a common genetic basis for IgAD and CVID can suggest that at least in some cases, IgAD and CVID may be part of a spectrum of diseases caused by a common genetic factor—for example, a mutation in the TACI—transmembrane activator CAML (calcium modulator and cyclophilin ligand) [173]. Another, slightly different thesis is the presence of autoantibodies against BAFF, APRIL, or IL21 as a common ground for CVID and sIgAD [174]. Both in CVID and sIgAD, there are reports of an increased level of BAFF and APRIL [175]. Increased apoptosis is also one of the mechanisms reported for both of these disorders [48]. An interesting observation indicating a similar genetic background of both diseases is their coexistence in families [176].

Besides, the relation between human leukocyte antigen (HLA) A1, B8, DR3, DQ2, or any part of this haplotype and IgA deficiency could indicate progression to CVID [159,161]. In patients with severe clinical manifestations, HLA typing could be helpful for the prediction of progression to CVID [160]. Moreover, in sIgAD subjects with simultaneous Ig subclass deficiency and bronchiectasis, the presence of hematologic autoimmunity could be another predictor of progression to CVID [94].

6.2. Transfusion Selective IgA Deficiency

One of the most dangerous complications of sIgAD is an anaphylactic transfusion reaction. It has been found that some patients with sIgAD are sensitized, which means that there are anti-IgA antibodies (IgG or IgE) present in their blood [34,177–179]. These autoreactive antibodies were found in 20–40% of patients with sIgAD [180]. After transfusion of blood containing IgA in such individuals, there is a risk of anaphylactoid reaction mediated by these immunoglobulins.

Rachid and Bonilla reviewed the articles reporting reactions to immunoglobulin products in patients with sIgAD [181]. The severity of adverse reactions with anti-IgA antibodies depends among others on the isotype (IgG or IgE), its specificity and serum concentration, the method of measurement. IgG anti-IgA antibodies are found in approximately one-third of sIgAD patients [181]. But only a few studies have reported anaphylactic transfusion reaction associated with IgE class. It has been also reported that IgE anti-IgA is less frequently studied than IgG anti-IgA. However, when both have been studied together, anti-IgA of the IgE class occurs much less frequently than IgG [182]. Burks et al. reported two patients (one with CVID and another with sIgAD) with IgE anti-IgA and IgG anti-IgA [177]. One of them had anaphylaxis with IVIG and another with IgA-deficient plasma. Ferreira et al. found IgE anti-IgA1 in a patient with CVID, which also had IgG anti-IgA [183].

The diagnosis of IgA-related anaphylaxis is made after transfusion-related anaphylaxis by measuring the levels of IgA and anti-IgA. The mechanism of anti-IgA production remains unexplained [184] and the clinically significant threshold of anti-IgA is still unknown [185]. However, the incidence of anaphylactic blood transfusion reactions is not very high and it occurs in one in 20,000–50,000 transfusions [186]. Moreover, some studies suggest that transfusion reactions occur less commonly than previously thought [187]. Only 17.5% of all blood samples coming from patients after transfusion reactions contained an IgA antibody, which indicates the presence of some other triggers [188].

There are no evidence-based guidelines regarding the proper approach while performing transfusion to patients with sIgAD. It is evident that patients with a history of anaphylactic transfusion reactions should not receive IgA-containing blood products [96]. These patients can receive blood products coming from donors with IgA deficiency, washed red blood cells, or platelet components. It is performed to remove residual plasma before transfusion and to decrease the risk of anaphylactic transfusion reactions in such patients [189]. There are also some cases that illustrate successful desensitization to IgA using IgA-enriched immunoglobulin preparations as a source of antigen [190].

6.3. Malignancy

There is a relationship between the occurrence of immune disorders and the overall risk of malignancy. This relationship is evident in some immunity disorders such as

CVID [191,192]. It has been shown that the risk of malignant lymphoma among these patients is increased by 30 times, while the risk of gastric cancer is 47 times higher [193]. In the case of IgA deficiency, the association with the incidence of cancer is not that clear. There are studies that report the different incidences of malignancies among patients with sIgAD, especially adenocarcinoma of the gastrointestinal tract, and lymphomas. Such studies require a long-term follow-up to diagnose this kind of evolution, therefore there are not much data regarding this topic. In a review of 330 patients, the authors report a 1.5% prevalence of malignancy, and five patients who presented neoplasms had Hodgkin lymphoma, acute lymphoid leukemia, Wilms tumor, Burkitt lymphoma, and ganglioneuroma [132]. Another study that included 63 children from Israeli reports a much higher frequency of malignancy (4.8%) [99] and the following malignancies were present: astrocytoma, adenocarcinoma of the colon, Hodgkin's lymphoma, neuroblastoma. On the other hand, a combined Danish and Swedish study including 386 patients with sIgAD did show an elevated incidence of cancer compared to a healthy cohort, however, this increase was non-significant [194]. In 2015, Ludvigsson et al. performed a prospective nationwide population-based cohort study with 2320 individuals with IgA deficiency [195]. They concluded that there is a moderately increased risk of cancer, especially gastrointestinal one, and that the risk is highest after diagnosis of sIgAD.

7. Atopic Diseases in Other PIDs

Tuano et al. described the prevalence of asthma, allergic rhinitis, atopic dermatitis, and food allergy in a cohort of 2923 patients with PID in US population [123]. Atopic dermatitis and food allergy were most common in patients with CVID, combined immunodeficiency (CID), and hyper IgE syndrome. Patients with CID and sIgAD presented a higher percentage of food allergy symptoms than the healthy population; 33.3% in CID and 25% in SIgAD [123]. In CVID the prevalence rates of asthma, rhinitis, and documented food allergy have been established as 37.5%, 55.5%, and 11.25% respectively [196]. In the case of patients with hypogammaglobulinemia, prevalence rates of asthma, rhinitis, and atopic dermatitis were established at 20%, 22%, and 9% respectively. Interestingly, Szczawinska-Poplonyk assessed the incidence of food allergy as 74% in the pediatric population [131,197,198].

8. Conclusions

sIgAD is an antibody deficiency and it usually remains undiagnosed throughout the patient's life, due to its frequent asymptomatic course. If symptomatic, sIgAD is connected to more frequent viral and bacterial infections of upper respiratory, urinary, and gastrointestinal tracts, as well as allergic and autoimmune diseases. It was suggested that allergic diseases (e.g., eczema, rhinitis, asthma) are more common in patients diagnosed with this particular PID, however, the prevalence and severity of allergic manifestations can be associated with ethnic background.

Since there is a controversy in the scientific world whether the coexistence of IgA deficiency and allergic diseases is in fact true, further studies on a large group should be carried out. Atopy in sIgAD subjects is common, but is also possible that it is overlooked. Therefore, it is necessary to follow diagnostic criteria to make a diagnosis of any atopic disease. The possible reasons for different prevalence which is observed in different studies could be caused by different diagnostic criteria or inclusion of patients based on parentally reported symptoms in the children population.

Moreover, these patients can present with simultaneous atopic and infectious manifestations which can intensify the symptoms; therefore, atopic diseases should be treated aggressively in patients with sIgAD. It is necessary to provide these patients with a proper multi-disciplinary team of physicians.

Moreover, it is important to emphasize that the course of the disease may change and there are some serious complications of this disorder, among which there are progression to CVID, transfusion-related anaphylaxis, and malignancy. Although they do not happen very often, sIgAD remains the most common PID disease, therefore physicians should be aware of all possible complications to provide the best care for their patients.

Author Contributions: Conceptualization, I.M., A.M., E.G., K.Z.-O., and J.R.; methodology, S.K., D.B., R.H., and P.N.-R.; validation, R.B., H.P.-B., and B.B.; formal analysis, I.M., S.K., E.G., J.R., and P.N.-R.; investigation, I.M., S.K., D.B., R.H., R.B., and A.M.; data curation, I.M., S.K., D.B., E.G., and J.R.; writing—original draft preparation, I.M., S.K., D.B., R.H., R.B., A.M., H.P.-B., B.B., and K.Z.-O.; writing—review and editing, E.G., J.R., and P.N.-R.; visualization, R.H.; funding acquisition, J.R. and E.G.; supervision, P.N.-R., E.G., and J.R. All authors have read and agreed to the published version of the manuscript.

Funding: This work was supported by Research Grant No. 2016/21/B/NZ6/02279 of the Polish National Science Centre (NCN) and Research Grant No. DS460 of the Medical University of Lublin.

Conflicts of Interest: The authors declare no conflict of interest.

References

1. Mayor, P.C.; Eng, K.H.; Singel, K.L.; Abrams, S.I.; Odunsi, K.; Moysich, K.B.; Fuleihan, R.; Garabedian, E.; Lugar, P.; Ochs, H.D.; et al. Cancer in primary immunodeficiency diseases: Cancer incidence in the United States Immune Deficiency Network Registry. *J. Allergy Clin. Immunol.* **2018**, *141*, 1028–1035. [CrossRef]
2. Bousfiha, A.; Jeddane, L.; Picard, C.; Al-Herz, W.; Ailal, F.; Chatila, T.; Cunningham-Rundles, C.; Etzioni, A.; Franco, J.L.; Holland, S.M.; et al. Human inborn errors of immunity: 2019 Update of the IUIS phenotypical classification. *J. Clin. Immunol.* **2020**, *40*, 66–81. [CrossRef]
3. National Primary Immunodeciency Resource Center. Available online: http://www.info4pi.org./library/educational-materials/10-warning-signs (accessed on 21 July 2021).
4. Cunningham-Rundles, C.; Sidi, P.; Estrella, L.; Doucette, J. Identifying undiagnosed primary immunodeficiency diseases in minority subjects by using computer sorting of diagnosis codes. *J. Allergy Clin. Immunol.* **2004**, *113*, 747–755. [CrossRef] [PubMed]
5. Bahrami, A.; Sayyahfar, S.; Soltani, Z.; Khodadost, M.; Moazzami, B.; Rezaei, N. Evaluation of the frequency and diagnostic delay of primary immunodeficiency disorders among suspected patients based on the 10 warning sign criteria: A cross-sectional study in Iran. *Allergol. Immunopathol.* **2020**, *48*, 711–719. [CrossRef]
6. Weber-Mzell, D.; Kotanko, P.; Hauer, A.C.; Goriup, U.; Haas, J.; Lanner, N.; Erwa, W.; Ahmaida, I.A.; Haitchi-Petnehazy, S.; Stenzel, M.; et al. Gender, age and seasonal effects on iga deficiency: A study of 7293 caucasians. *Eur. J. Clin. Investig.* **2004**, *34*, 224–228. [CrossRef] [PubMed]
7. Yel, L. Selective IgA deficiency. *J. Clin. Immunol.* **2010**, *30*, 10–16. [CrossRef]
8. Yazdani, R.; Latif, A.; Tabassomi, F.; Abolhassani, H.; Azizi, G.; Rezaei, N.; Aghamohammadi, A. Clinical phenotype classification for selective immunoglobulin A deficiency. *Expert Rev. Clin. Immunol.* **2015**, *11*, 1245–1254. [CrossRef] [PubMed]
9. Cunningham-Rundles, C. Physiology of IgA and IgA deficiency. *J. Clin. Immunol.* **2001**, *21*, 303–309. [CrossRef]
10. Aghamohammadi, A.; Cheraghi, T.; Gharagozlou, M.; Movahedi, M.; Rezaei, N.; Yeganeh, M.; Parvaneh, N.; Abolhassani, H.; Pourpak, Z.; Moin, M. IgA Deficiency: Correlation between clinical and immunological phenotypes. *J. Clin. Immunol.* **2008**, *29*, 130–136. [CrossRef]
11. Brandtzaeg, P.; Prydz, H. Direct evidence for an integrated function of J chain and secretory component in epithelial transport of immunoglobulins. *Nature* **1984**, *311*, 71–73. [CrossRef]
12. Steffen, U.; Koeleman, C.A.; Sokolova, M.V.; Bang, H.; Kleyer, A.; Rech, J.; Unterweger, H.; Schicht, M.; Garreis, F.; Hahn, J.; et al. IgA subclasses have different effector functions associated with distinct glycosylation profiles. *Nat. Commun.* **2020**, *11*, 1–12. [CrossRef]
13. De Sousa-Pereira, P.; Woof, J.M. IgA: Structure, function, and developability. *Antibodies* **2019**, *8*, 57. [CrossRef] [PubMed]
14. Wines, B.D.; Hogarth, P.M. IgA receptors in health and disease. *Tissue Antigens* **2006**, *68*, 103–114. [CrossRef] [PubMed]
15. Woof, J.M.; Kerr, M.A. The function of immunoglobulin A in immunity. *J. Pathol.* **2006**, *208*, 270–282. [CrossRef]
16. Monteiro, R.C. The role of IgA and IgA Fc receptors as anti-inflammatory agents. *J. Clin. Immunol.* **2010**, *30*, S61–S64. [CrossRef] [PubMed]
17. Davis, S.K.; Selva, K.J.; Kent, S.J.; Chung, A.W. Serum IgA Fc effector functions in infectious disease and cancer. *Immunol. Cell Biol.* **2020**, *98*, 276–286. [CrossRef]
18. Bakema, J.; van Egmond, M. The human immunoglobulin A Fc receptor FcαRI: A multifaceted regulator of mucosal immunity. *Mucosal Immunol.* **2011**, *4*, 612–624. [CrossRef]
19. Ben Mkaddem, S.; Rossato, E.; Heming, N.; Monteiro, R.C. Anti-inflammatory role of the IgA Fc receptor (CD89): From autoimmunity to therapeutic perspectives. *Autoimmun. Rev.* **2013**, *12*, 666–669. [CrossRef]
20. Cerutti, A.; Rescigno, M. The biology of intestinal immunoglobulin A responses. *Immunity* **2008**, *28*, 740–750. [CrossRef]
21. Gloudemans, A.K.; Lambrecht, B.N.; Smits, H.H. Potential of immunoglobulin A to prevent allergic asthma. *Clin. Dev. Immunol.* **2013**, *2013*, 542091. [CrossRef]

22. Williams, R.C.; Gibbons, R.J. Inhibition of bacterial adherence by secretory immunoglobulin A: A mechanism of antigen disposal. *Science* **1972**, *177*, 697–699. [CrossRef]

23. Mazanec, M.B.; Nedrud, J.G.; Kaetzel, C.S.; Lamm, M.E. A three-tiered view of the role of IgA in mucosal defense. *Immunol. Today* **1993**, *14*, 430–435. [CrossRef]

24. Wolf, H.M.; Fischer, M.B.; Pühringer, H.; Samstag, A.; Vogel, E.; Eibl, M.M. Human serum IgA downregulates the release of inflammatory cytokines (tumor necrosis factor-alpha, interleukin-6) in human monocytes. *Blood* **1994**, *83*, 1278–1288. [CrossRef] [PubMed]

25. Olas, K.; Butterweck, H.; Teschner, W.; Schwarz, H.P.; Reipert, B. Immunomodulatory properties of human serum immunoglobulin A: Anti-inflammatory and pro-inflammatory activities in human monocytes and peripheral blood mononuclear cells. *Clin. Exp. Immunol.* **2005**, *140*, 478–490. [CrossRef] [PubMed]

26. Pilette, C.; Detry, B.; Guisset, A.; Gabriels, J.; Sibille, Y. Induction of interleukin-10 expression through Fcalpha receptor in human monocytes and monocyte-derived dendritic cells: Role of p38 MAPKinase. *Immunol. Cell Biol.* **2010**, *88*, 486–493. [CrossRef] [PubMed]

27. Woof, J.M.; Russell, M.W. Structure and function relationships in IgA. *Mucosal Immunol.* **2011**, *4*, 590–597. [CrossRef]

28. Russell, M.W.; Sibley, D.A.; Nikolova, E.B.; Tomana, M.; Mestecky, J. IgA antibody as a non-inflammatory regulator of immunity. *Biochem. Soc. Trans.* **1997**, *25*, 466–470. [CrossRef] [PubMed]

29. Geissmann, F.; Launay, P.; Pasquier, B.; Lepelletier, Y.; Leborgne, M.; Lehuen, A.; Brousse, N.; Monteiro, R.C. A subset of human dendritic cells expresses IgA Fc receptor (CD89), which mediates internalization and activation upon cross-linking by IgA complexes. *J. Immunol.* **2001**, *166*, 346–352. [CrossRef]

30. Pfaffenbach, G.; Lamm, M.E.; Gigli, I. Activation of the guinea pig alternative complement pathway by mouse IgA immune complexes. *J. Exp. Med.* **1982**, *155*, 231–247. [CrossRef]

31. Pleass, R.J.; Lang, M.L.; Kerr, M.A.; Woof, J.M. IgA is a more potent inducer of NADPH oxidase activation and degranulation in blood eosinophils than IgE. *Mol. Immunol.* **2007**, *44*, 1401–1408. [CrossRef] [PubMed]

32. Ouadrhiri, Y.; Pilette, C.; Monteiro, R.C.; Vaerman, J.P.; Sibille, Y. Effect of IgA on respiratory burst and cytokine release by human alveolar macrophages: Role of ERK1/2 mitogen-activated protein kinases and NF-kappaB. *Am. J. Respir. Cell Mol. Biol.* **2002**, *26*, 315–332. [CrossRef]

33. Roos, A.; Bouwman, L.H.; van Gijlswijk-Janssen, D.J.; Faber-Krol, M.C.; Stahl, G.L.; Daha, M.R. Human IgA activates the complement system via the mannan-binding lectin pathway. *J. Immunol.* **2001**, *167*, 2861–2868. [CrossRef] [PubMed]

34. Swain, S.; Selmi, C.; Gershwin, M.E.; Teuber, S.S. The clinical implications of selective IgA deficiency. *J. Transl. Autoimmun.* **2019**, *2*, 100025. [CrossRef] [PubMed]

35. Cipe, F.E.; Doğu, F.; Güloğlu, D.; Aytekin, C.; Polat, M.; Biyikli, Z.; Ikincioğullari, A. B-cell subsets in patients with transient hypogammaglobulinemia of infancy, partial IgA deficiency, and selective IgM deficiency. *J. Investig. Allergol. Clin. Immunol.* **2013**, *23*, 94–100.

36. Aghamohammadi, A.; Abolhassani, H.; Biglari, M.; Abolmaali, S.; Moazzami, K.; Tabatabaeiyan, M.; Asgarian-Omran, H.; Parvaneh, N.; Mirahmadian, M.; Rezaei, N. Analysis of switched memory B cells in patients with IgA deficiency. *Int. Arch. Allergy Immunol.* **2011**, *156*, 462–468. [CrossRef] [PubMed]

37. Lawton, A.R.; Royal, S.A.; Self, K.S.; Cooper, M.D. IgA determinants on B-lymphocytes in patients with deficiency of circulating IgA. *J. Lab. Clin. Med.* **1972**, *80*, 26–33. [PubMed]

38. Wang, Z.; Yunis, D.; Irigoyen, M.; Kitchens, B.; Bottaro, A.; Alt, F.W.; Alper, C.A. Discordance between IgA switching at the DNA level and IgA expression at the mRNA level in IgA-deficient patients. *Clin. Immunol.* **1999**, *91*, 263–270. [CrossRef]

39. Borte, S.; Pan-Hammarström, Q.; Liu, C.; Sack, U.; Borte, M.; Wagner, U.; Graf, D.; Hammarström, L. Interleukin-21 restores immunoglobulin production ex vivo in patients with common variable immunodeficiency and selective IgA deficiency. *Blood* **2009**, *114*, 4089–4098. [CrossRef]

40. Ramsay, A.J.; Husband, A.J.; Ramshaw, I.A.; Bao, S.; Matthaei, K.I.; Koehler, G.; Kopf, M. The role of interleukin-6 in mucosal IgA antibody responses in vivo. *Science* **1994**, *264*, 561–563. [CrossRef]

41. Kowalczyk, D.; Mytar, B.; Zembala, M. Cytokine production in transient hypogammaglobulinemia and isolated IgA deficiency. *J. Allergy Clin. Immunol.* **1997**, *100*, 556–562. [CrossRef]

42. Kowalczyk, D.; Baran, J.; Webster, A.D.; Zembala, M. Intracellular cytokine production by Th1/Th2 lymphocytes and monocytes of children with symptomatic transient hypogammaglobulinaemia of infancy (THI) and selective IgA deficiency (SIgAD). *Clin. Exp. Immunol.* **2002**, *127*, 507–512. [CrossRef]

43. Marconi, M.; Plebani, A.; Avanzini, M.A.; Maccario, R.; Pistorio, A.; Duse, M.; Stringa, M.; Monafo, V. IL-10 and IL-4 co-operate to normalize in vitro IgA production in IgA-deficient (IgAD) patients. *Clin. Exp. Immunol.* **1998**, *112*, 528–532. [CrossRef]

44. Müller, F.; Aukrust, P.; Nilssen, D.E.; Frøland, S.S. Reduced serum level of transforming growth factor-beta in patients with IgA deficiency. *Clin. Immunol. Immunopathol.* **1995**, *76*, 203–208. [CrossRef]

45. Brière, F.; Bridon, J.M.; Chevet, D.; Souillet, G.; Bienvenu, F.; Guret, C.; Martinez-Valdez, H.; Banchereau, J. Interleukin 10 induces B lymphocytes from IgA-deficient patients to secrete IgA. *J. Clin. Investig.* **1994**, *94*, 97–104. [CrossRef]

46. Van Vlasselaer, P.; Punnonen, J.; de Vries, J.E. Transforming growth factor-beta directs IgA switching in human B cells. *J. Immunol.* **1992**, *148*, 2062–2067.

47. Defrance, T.; Vanbervliet, B.; Brière, F.; Durand, I.; Rousset, F.; Banchereau, J. Interleukin 10 and transforming growth factor beta cooperate to induce anti-CD40-activated naive human B cells to secrete immunoglobulin A. *J. Exp. Med.* **1992**, *175*, 671–682. [CrossRef]

48. Yazdani, R.; Fatholahi, M.; Ganjalikhani-Hakemi, M.; Abolhassani, H.; Azizi, G.; Hamid, K.M.; Rezaei, N.; Aghamohammadi, A. Role of apoptosis in common variable immunodeficiency and selective immunoglobulin A deficiency. *Mol. Immunol.* **2016**, *71*, 1–9. [CrossRef]

49. Soheili, H.; Abolhassani, H.; Arandi, N.; Khazaei, H.A.; Shahinpour, S.; Hirbod-Mobarakeh, A.; Rezaei, N.; Aghamohammadi, A. Evaluation of natural regulatory T cells in subjects with selective IgA deficiency: From senior idea to novel opportunities. *Int. Arch. Allergy Immunol.* **2013**, *160*, 208–214. [CrossRef]

50. Mestecky, J.; McGhee, J.R. Immunoglobulin A (IgA): Molecular and cellular interactions involved in IgA biosynthesis and immune response. *Adv. Immunol.* **1987**, *40*, 153–245. [CrossRef]

51. Cong, Y.; Feng, T.; Fujihashi, K.; Schoeb, T.R.; Elson, C.O. A dominant, coordinated T regulatory cell-IgA response to the intestinal microbiota. *Proc. Natl. Acad. Sci. USA* **2009**, *106*, 19256–19261. [CrossRef]

52. Zhang, J.; van Oostrom, D.; Li, J.; Savelkoul, H. Innate mechanisms in selective IgA deficiency. *Front. Immunol.* **2021**, *12*, 649112. [CrossRef]

53. Bronson, P.G.; Chang, D.; Bhangale, T.; Seldin, M.F.; Ortmann, W.; Ferreira, R.C.; Urcelay, E.; Pereira, L.F.; Martin, J.; Plebani, A. Common variants at PVT1, ATG13-AMBRA1, AHI1 and CLEC16A are associated with selective IgA deficiency. *Nat. Genet.* **2016**, *48*, 1425–1429. [CrossRef] [PubMed]

54. Castigli, E.; Wilson, S.A.; Garibyan, L.; Rachid, R.; Bonilla, F.; Schneider, L.; Geha, R.S. TACI is mutant in common variable immunodeficiency and IgA deficiency. *Nat. Genet.* **2005**, *37*, 829–834. [CrossRef] [PubMed]

55. Ortiz, J.; Fernández-Arquero, M.; Urcelay, E.; López-Mejías, R.; Ferreira, A.; Fontán, G.; de la Concha, E.G.; Martínez, A. Interleukin-10 polymorphisms in Spanish IgA deficiency patients: A case-control and family study. *BMC Med. Genet.* **2006**, *7*, 56. [CrossRef] [PubMed]

56. López-Mejías, R.; Martínez, A.; Del Pozo, N.; Fernández-Arquero, M.; Ferreira, A.; Urcelay, E.; Fontán, G.; de la Concha, E.G.; Núñez, C. Interleukin-6 gene variation in Spanish patients with immunoglobulin-A deficiency. *Hum. Immunol.* **2008**, *69*, 301–305. [CrossRef] [PubMed]

57. Abolhassani, H.; Aghamohammadi, A.; Hammarström, L. Monogenic mutations associated with IgA deficiency. *Expert Rev. Clin. Immunol.* **2016**, *12*, 1321–1335. [CrossRef]

58. Hammarström, L.; Smith, C.I.; Berg, C.I. Captopril-induced IgA deficiency. *Lancet* **1991**, *337*, 436. [CrossRef]

59. Murphy, E.A.; Morris, A.J.; Walker, E.; Lee, F.D.; Sturrock, R.D. Cyclosporine A induced colitis and acquired selective IgA deficiency in a patient with juvenile chronic arthritis. *J. Rheumatol.* **1993**, *20*, 1397–1398.

60. Ilan, Y.; Shouval, D.; Ashur, Y.; Manns, M.; Naparstek, Y. IgA deficiency associated with chronic hepatitis C virus infection. A cause or an effect? *Arch. Intern. Med.* **1993**, *153*, 1588–1592. [CrossRef]

61. Saulsbury, F.T. Selective IgA deficiency temporally associated with Epstein-Barr virus infection. *J. Pediatr.* **1989**, *115*, 268–270. [CrossRef]

62. Truedsson, L.; Baskin, B.; Pan, Q.; Rabbani, H.; Vořechovský, I.; Smith, C.I.; Hammarström, L. Genetics of IgA deficiency. *APMIS* **1995**, *103*, 833–842. [CrossRef]

63. Maeoka, Y.; Hara, T.; Dejima, S.; Takeshita, K. IgA and IgG2 deficiency associated with zonisamide therapy: A case report. *Epilepsia* **1997**, *38*, 611–613. [CrossRef]

64. Rawla, P.; Joseph, N. IgA deficiency. In *StatPearls*; StatPearls Publishing: Treasure Island, FL, USA, 2021.

65. Hammarström, L.; Vorechovsky, I.; Webster, D. Selective IgA deficiency (SIgAD) and common variable immunodeficiency (CVID). *Clin. Exp. Immunol.* **2000**, *120*, 225–231. [CrossRef]

66. Lougaris, V.; Sorlini, A.; Monfredini, C.; Ingrasciotta, G.; Caravaggio, A.; Lorenzini, T.; Baronio, M.; Cattalini, M.; Meini, A.; Ruggeri, L. Clinical and laboratory features of 184 italian pediatric patients affected with selective IgA deficiency (SIgAD): A longitudinal single-center study. *J. Clin. Immunol.* **2019**, *39*, 470–475. [CrossRef]

67. Alkhater, S.A. Approach to the child with recurrent infections. *J. Fam. Community Med.* **2009**, *16*, 77–82.

68. Arkwright, P.D.; Gennery, A.R. Ten warning signs of primary immunodeficiency: A new paradigm is needed for the 21st century. *Ann. N. Y. Acad. Sci.* **2011**, *1238*, 7–14. [CrossRef]

69. Janzi, M.; Melén, E.; Kull, I.; Wickman, M.; Hammarström, L. Rare mutations in TNFRSF13B increase the risk of asthma symptoms in Swedish children. *Genes Immun.* **2012**, *13*, 59–65. [CrossRef]

70. Edwards, E.; Razvi, S.; Cunningham-Rundles, C. IgA deficiency: Clinical correlates and responses to pneumococcal vaccine. *Clin. Immunol.* **2004**, *111*, 93–97. [CrossRef]

71. Ozkan, H.; Atlihan, F.; Genel, F.; Targan, S.; Gunvar, T. IgA and/or IgG subclass deficiency in children with recurrent respiratory infections and its relationship with chronic pulmonary damage. *J. Investig. Allergol. Clin. Immunol.* **2005**, *15*, 69–74.

72. Brandtzaeg, P.; Farstad, I.N.; Johansen, F.E.; Morton, H.C.; Norderhaug, I.N.; Yamanaka, T. The B-cell system of human mucosae and exocrine glands. *Immunol. Rev.* **1999**, *171*, 45–87. [CrossRef]

73. Brandtzaeg, P.; Karlsson, G.; Hansson, G.; Petruson, B.; Björkander, J.; Hanson, L.A. The clinical condition of IgA-deficient patients is related to the proportion of IgD- and IgM-producing cells in their nasal mucosa. *Clin. Exp. Immunol.* **1987**, *67*, 626–636.

74. Pabst, R.; Russell, M.W.; Brandtzaeg, P. Tissue distribution of lymphocytes and plasma cells and the role of the gut. *Trends Immunol.* **2008**, *29*, 206–210. [CrossRef]

75. Savilahti, E. IgA deficiency in children. Immunoglobulin-containing cells in the intestinal mucosa, immunoglobulins in secretions and serum IgA levels. *Clin. Exp. Immunol.* **1973**, *13*, 395–406.

76. Mestecky, J.; Russell, M.W. Specific antibody activity, glycan heterogeneity and polyreactivity contribute to the protective activity of S-IgA at mucosal surfaces. *Immunol. Lett.* **2009**, *124*, 57–62. [CrossRef] [PubMed]

77. Brandtzaeg, P.; Fjellanger, I.; Gjeruldsen, S.T. Immunoglobulin M: Local synthesis and selective secretion in patients with immunoglobulin A deficiency. *Science* **1968**, *160*, 789–791. [CrossRef]

78. Eren, M.; Saltik-Temizel, I.N.; Yüce, A.; Cağlar, M.; Koçak, N. Duodenal appearance of giardiasis in a child with selective immunoglobulin A deficiency. *Pediatr. Int.* **2007**, *49*, 409–411. [CrossRef]

79. Jacob, C.M.; Pastorino, A.C.; Fahl, K.; Carneiro-Sampaio, M.; Monteiro, R.C. Autoimmunity in IgA deficiency: Revisiting the role of IgA as a silent housekeeper. *J. Clin. Immunol.* **2008**, *1*, S56–S61. [CrossRef]

80. Gustafson, R.; Gardulf, A.; Granert, C.; Hansen, S.; Hammarström, L. Prophylactic therapy for selective IgA deficiency. *Lancet* **1997**, *350*, 865. [CrossRef]

81. Vignesh, P.; Rawat, A.; Singh, S. An Update on the Use of immunomodulators in primary immunodeficiencies. *Clin. Rev. Allergy Immunol.* **2017**, *52*, 287–303. [CrossRef]

82. Kazemi, A.; Soltani, S.; Nasri, F.; Clark, C.; Kolahdouz-Mohammadi, R. The effect of probiotics, parabiotics, synbiotics, fermented foods and other microbial forms on immunoglobulin production: A systematic review and meta-analysis of clinical trials. *Int. J. Food Sci. Nutr.* **2021**, *72*, 632–649. [CrossRef]

83. Tsai, Y.T.; Cheng, P.C.; Liao, J.W.; Pan, T.M. Effect of the administration of Lactobacillus paracasei subsp. paracasei NTU 101 on Peyer's patch-mediated mucosal immunity. *Int. Immunopharmacol.* **2010**, *10*, 791–798. [CrossRef]

84. Vinderola, G.; Perdigón, G.; Duarte, J.; Farnworth, E.; Matar, C. Effects of the oral administration of the exopolysaccharide produced by Lactobacillus kefiranofaciens on the gut mucosal immunity. *Cytokine* **2006**, *36*, 254–260. [CrossRef]

85. Genel, F.; Kutukculer, N. Prospective, randomized comparison of OM-85 BV and a prophylactic antibiotic in children with recurrent infections and immunoglobulin A and/or G subclass deficiency. *Curr. Ther. Res. Clin. Exp.* **2003**, *64*, 600–615. [CrossRef] [PubMed]

86. Fadlallah, J.; Sterlin, D.; Fieschi, C.; Parizot, C.; Dorgham, K.; El Kafsi, H.; Autaa, G.; Ghillani-Dalbin, P.; Juste, C.; Lepage, P.; et al. Synergistic convergence of microbiota-specific systemic IgG and secretory IgA. *Allergy Clin. Immunol.* **2019**, *143*, 1575.e4–1585.e4. [CrossRef] [PubMed]

87. Berbers, R.M.; Franken, I.A.; Leavis, H.L. Immunoglobulin A and microbiota in primary immunodeficiency diseases. *Curr. Opin. Allergy Clin. Immunol.* **2019**, *19*, 563–570. [CrossRef]

88. Quinti, I.; Mortari, E.P.; Fernandez Salinas, A.; Milito, C.; Carsetti, R. IgA antibodies and IgA deficiency in SARS-CoV-2 infection. *Front. Cell Infect. Microbiol.* **2021**, *11*, 655896. [CrossRef]

89. Naito, Y.; Takagi, T.; Yamamoto, T.; Watanabe, S. Association between selective IgA deficiency and COVID-19. *J. Clin. Biochem. Nutr.* **2020**, *67*, 122–125. [CrossRef]

90. Sterlin, D.; Mathian, A.; Miyara, M.; Mohr, A.; Anna, F.; Claër, L.; Quentric, P.; Fadlallah, J.; Devilliers, H.; Ghillani, P.; et al. IgA dominates the early neutralizing antibody response to SARS-CoV-2. *Sci. Transl. Med.* **2021**, *13*, eabd2223. [CrossRef]

91. Marcus, N.; Frizinsky, S.; Hagin, D.; Ovadia, A.; Hanna, S.; Farkash, M.; Maoz-Segal, R.; Agmon-Levin, N.; Broides, A.; Nahum, A.; et al. Minor clinical impact of COVID-19 pandemic on patients with primary immunodeficiency in Israel. *Front. Immunol.* **2021**, *11*, 614086. [CrossRef]

92. Meyts, I.; Bucciol, G.; Quinti, I.; Neven, B.; Fischer, A.; Seoane, E.; Lopez-Granados, E.; Gianelli, C.; Robles-Marhuenda, A.; Jeandel, P.Y.; et al. Coronavirus disease 2019 in patients with inborn errors of immunity: An international study. *J. Allergy Clin. Immunol.* **2021**, *147*, 520–531. [CrossRef]

93. Sarmiento, E.; Mora, R.; Rodríguez-Mahou, M.; Rodríguez-Molina, J.; Fernández-Cruz, E.; Carbone, J. Enfermedad autoinmune en inmunodeficiencias primarias de anticuerpos. *Allergol. Immunopathol.* **2005**, *33*, 69–73. [CrossRef]

94. Abolhassani, H.; Gharib, B.; Shahinpour, S.; Masoom, S.N.; Havaei, A.; Mirminachi, B.; Arandi, N.; Torabi-Sagvand, B.; Khazaei, H.A.; Mohammadi, J.; et al. Autoimmunity in patients with selective IgA deficiency. *J. Investig. Allergol. Clin. Immunol.* **2015**, *25*, 112–119. [PubMed]

95. Azizi, G.; Tavakol, M.; Rafiemanesh, H.; Kiaee, F.; Yazdani, R.; Heydari, A.; Abouhamzeh, K.; Anvari, P.; Mohammadikhajehdehi, S.; Sharifia, L.; et al. Autoimmunity in a cohort of 471 patients with primary antibody deficiencies. *Expert Rev. Clin. Immunol.* **2017**, *13*, 1099–1106. [CrossRef]

96. Odineal, D.D.; Gershwin, M.E. The Epidemiology and clinical manifestations of autoimmunity in selective IgA deficiency. *Clin. Rev. Allergy Immunol.* **2020**, *58*, 107–133. [CrossRef]

97. Mohammadi, J.; Ramanujam, R.; Jarefors, S.; Rezaei, N.; Aghamohammadi, A.; Gregersen, P.K.; Hammarström, L. IgA deficiency and the MHC: Assessment of relative risk and microheterogeneity within the HLA A1 B8, DR3 (8.1) haplotype. *J. Clin. Immunol.* **2010**, *30*, 138–143. [CrossRef]

98. Pfeuffer, S.; Pawlowski, M.; Joos, G.S.; Minnerup, J.; Meuth, S.G.; Dziewas, R.; Wiendl, H. Autoimmunity complicating SARS-CoV-2 infection in selective IgA-deficiency. *Neurol. Neuroimmunol. Neuroinflamm.* **2020**, *7*, e881. [CrossRef] [PubMed]

99. Shkalim, V.; Monselize, Y.; Segal, N.; Zan-Bar, I.; Hoffer, V.; Garty, B.Z. Selective IgA deficiency in children in Israel. *J. Clin. Immunol.* **2010**, *30*, 761–765. [CrossRef] [PubMed]
100. Kaufman, H.S.; Hobbs, J.R. Immunoglobulin deficiencies in an atopic population. *Lancet* **1970**, *2*, 1061–1063. [CrossRef]
101. Erkoçoğlu, M.; Metin, A.; Kaya, A.; Özcan, C.; Akan, A.; Civelek, E.; Çapanoğlu, M.; Giniş, T.; Kocabaş, C.N. Allergic and autoimmune disorders in families with selective IgA deficiency. *Turk. J. Med. Sci.* **2017**, *47*, 592–598. [CrossRef]
102. Gleeson, M.; Cripps, A.W.; Clancy, R.L.; Hensley, M.J.; Henry, R.J.; Wlodarczyk, J.H. The significance of transient mucosal IgA deficiency on the development of asthma and atopy in children. *Adv. Exp. Med. Biol.* **1995**, *371*, 861–864.
103. Van Asperen, P.P.; Gleeson, M.; Kemp, A.S.; Cripps, A.W.; Geraghty, S.B.; Mellis, C.M.; Clancy, R.L. The relationship between atopy and salivary IgA deficiency in infancy. *Clin. Exp. Immunol.* **1985**, *62*, 753–757. [PubMed]
104. Ammann, A.J.; Hong, R. Selective IgA deficiency and autoimmunity. *Clin. Exp. Immunol.* **1970**, *7*, 833–838.
105. Burgio, G.R.; Duse, M.; Monafo, V.; Ascione, A.; Nespoli, L. Selective IgA deficiency: Clinical and immunological evaluation of 50 pediatric patients. *Eur. J. Pediatr.* **1980**, *133*, 101–106. [CrossRef]
106. Buckley, R.H.; Fiscus, S.A. Serum IgD and IgE concentrations in immunodeficiency diseases. *J. Clin. Investig.* **1975**, *55*, 157–165. [CrossRef] [PubMed]
107. Klemola, T. Deficiency of immunoglobulin A. *Ann. Clin. Res.* **1987**, *19*, 248–257. [PubMed]
108. Wang, W.; Yao, T.; Zhang, T.; Quan, M.; Wang, C.; Wang, C.; Zhang, L.; Tang, X.; Jian, S.; Song, H. Selective immunoglobulin A deficiency (SIgAD) primarily leads to recurrent infections and autoimmune diseases: A retrospective study of Chinese patients in the past 40 years. *Genes Dis.* **2019**, *7*, 115–121. [CrossRef] [PubMed]
109. Breedveld, A.; van Egmond, M. IgA and FcαRI: Pathological roles and therapeutic opportunities. *Front. Immunol.* **2019**, *10*, 553. [CrossRef] [PubMed]
110. Shirakawa, T.; Enomoto, T.; Shimazu, S.; Hopkin, J.M. The inverse association between tuberculin responses and atopic disorder. *Science* **1997**, *275*, 77–79. [CrossRef]
111. Robinson, D.S.; Larché, M.; Durham, S.R. Tregs and allergic disease. *J. Clin. Investig.* **2004**, *114*, 1389–1397. [CrossRef]
112. Holgate, S.T. Innate and adaptive immune responses in asthma. *Nat. Med.* **2012**, *18*, 673–683. [CrossRef]
113. Pilette, C.; Durham, S.R.; Vaerman, J.P.; Sibille, Y. Mucosal immunity in asthma and chronic obstructive pulmonary disease: A role for immunoglobulin A? *Proc. Am. Thorac. Soc.* **2004**, *1*, 125–135. [CrossRef] [PubMed]
114. Jutel, M.; Akdis, C.A. Immunological mechanisms of allergen-specific immunotherapy. *Allergy* **2011**, *66*, 725–732. [CrossRef] [PubMed]
115. Platts-Mills, T.A.; Woodfolk, J.A. Allergens and their role in the allergic immune response. *Immunol. Rev.* **2011**, *242*, 51–68. [CrossRef]
116. Bonilla, F.A.; Khan, D.A.; Ballas, Z.K.; Chinen, J.; Frank, M.M.; Hsu, J.T.; Keller, M.; Kobrynski, L.J.; Komarow, H.D.; Mazer, B.; et al. Practice parameter for the diagnosis and management of primary immunodeficiency. *J. Allergy Clin. Immunol.* **2015**, *136*, 1186–1205.e78. [CrossRef] [PubMed]
117. FDA Label for Omalizumab. Available online: https://www.accessdata.fda.gov/drugsatfda_docs/label/2016/103976s5225lbl.pdf (accessed on 23 August 2021).
118. Oosterhaven, J.; Romeijn, G.; Schuttelaar, M. Dupilumab treatment of very severe refractory atopic hand eczema. *JAMA Dermatol.* **2018**, *154*, 969–970. [CrossRef]
119. Simpson, E.L.; Bieber, T.; Guttman-Yassky, E.; Beck, L.A.; Blauvelt, A.; Cork, M.J.; Silverberg, J.I.; Deleuran, M.; Kataoka, Y.; Lacour, J.P.; et al. Two Phase 3 Trials of Dupilumab versus Placebo in Atopic Dermatitis. *N. Engl. J. Med.* **2016**, *375*, 2335–2348. [CrossRef]
120. Comberiati, P.; Costagliola, G.; Carli, N.; Legitimo, A.; D'Elios, S.; Consolini, R.; Peroni, D.G. Refractory chronic spontaneous urticaria treated with omalizumab in an adolescent with common variable immunodeficiency. *Front. Immunol.* **2019**, *10*, 1700. [CrossRef]
121. Banh, H.L.; Trevoy, J.; Pabst, H.; Beach, J.; Vethanayagam, D. Persistent elevation of peripheral blood myeloid cell counts associated with omalizumab therapy. *Am. J. Health Syst. Pharm.* **2012**, *69*, 302–306. [CrossRef]
122. Votquenne, N.; Dupire, G.; Michel, O.; Said, B.B. Dupilumab for severe generalized eczematous eruption complicating common variable immunodeficiency. *Eur. J. Dermatol.* **2021**, *31*, 93–94. [CrossRef]
123. Tuano, K.S.; Orange, J.S.; Sullivan, K.; Cunningham-Rundles, C.; Bonilla, F.A.; Davis, C.M. Food allergy in patients with primary immunodeficiency diseases: Prevalence within the US Immunodeficiency Network (USIDNET). *J. Allergy Clin. Immunol.* **2015**, *135*, 273–275. [CrossRef]
124. Janzi, M.; Kull, I.; Sjöberg, R.; Wan, J.; Melén, E.; Bayat, N.; Ostblom, E.; Pan-Hammarström, Q.; Nilsson, P.; Hammarström, L. Selective IgA deficiency in early life: Association to infections and allergic diseases during childhood. *Clin. Immunol.* **2009**, *133*, 78–85. [CrossRef] [PubMed]
125. Sloper, K.S.; Brook, C.G.; Kingston, D.; Pearson, J.R.; Shiner, M. Eczema and atopy in early childhood: Low IgA plasma cell counts in the jejunal mucosa. *Arch. Dis. Child* **1981**, *56*, 939–942. [CrossRef] [PubMed]
126. Sjögren, Y.M.; Jenmalm, M.C.; Böttcher, M.F.; Björkstén, B.; Sverremark-Ekström, E. Altered early infant gut microbiota in children developing allergy up to 5 years of age. *Clin. Exp. Allergy* **2009**, *39*, 518–526. [CrossRef] [PubMed]
127. Fagerås, M.; Tomičić, S.; Voor, T.; Björkstén, B.; Jenmalm, M.C. Slow salivary secretory IgA maturation may relate to low microbial pressure and allergic symptoms in sensitized children. *Pediatr. Res.* **2011**, *70*, 572–577. [CrossRef] [PubMed]

128. Catanzaro, J.R.; Strauss, J.D.; Bielecka, A.; Porto, A.F.; Lobo, F.M.; Urban, A.; Schofield, W.B.; Palm, N.W. IgA-deficient humans exhibit gut microbiota dysbiosis despite secretion of compensatory IgM. *Sci. Rep.* **2019**, *9*, 13574. [CrossRef]

129. Fadlallah, J.; El Kafsi, H.; Sterlin, D.; Juste, C.; Parizot, C.; Dorgham, K.; Autaa, G.; Gouas, D.; Almeida, M.; Lepage, P.; et al. Microbial ecology perturbation in human IgA deficiency. *Sci. Transl. Med.* **2018**, *10*, eaan1217. [CrossRef]

130. Strait, R.T.; Mahler, A.; Hogan, S.; Khodoun, M.; Shibuya, A.; Finkelman, F.D. Ingested allergens must be absorbed systemically to induce systemic anaphylaxis. *J. Allergy Clin. Immunol.* **2011**, *127*, 982.e1–989.e1. [CrossRef]

131. Szczawinska-Poplonyk, A.; Breborowicz, A.; Ossowska, L. Food allergy in children with hypogammaglobulinemia. *Pediatria. Pol.* **2012**, *87*, 444–448. [CrossRef]

132. Domínguez, O.; Giner, M.T.; Alsina, L.; Martín, M.A.; Lozano, J.; Plaza, A.M. Fenotipos clínicos asociados a la deficiencia selectiva de IgA: Revisión de 330 casos y propuesta de un protocolo de seguimiento. *An. Pediatr.* **2012**, *76*, 261–267. [CrossRef]

133. Delavari, S.; Shad, M.T.; Shariati, S.; Salami, F.; Rasouli, S. Allergy in patients with selective IgA Deficiency. *Immunol. Genet. J.* **2020**, *3*, 54–63. [CrossRef]

134. Urm, S.H.; Yun, H.D.; Fenta, Y.A.; Yoo, K.H.; Abraham, R.S.; Hagan, J.; Juhn, Y.J. Asthma and risk of selective IgA deficiency or common variable immunodeficiency: A population-based case-control study. *Mayo Clin. Proc.* **2013**, *88*, 813–821. [CrossRef]

135. Golshan, M.; Mohamad-Zadeh, Z.; Zahedi-Nejad, N.; Rostam-Poor, B. Prevalence of asthma and related symptoms in primary school children of Isfahan, Iran, in 1998. *Asian Pac. J. Allergy Immunol.* **2001**, *19*, 163–170.

136. Golshan, M.; Mohammad-Zadeh, Z.; Khanlar-Pour, A.; Iran-Pour, R. Prevalence of asthma and related symptoms in junior high school children in Isfahan, Iran. *Monaldi Arch. Chest Dis.* **2002**, *57*, 19–24.

137. Jorgensen, G.H.; Gardulf, A.; Sigurdsson, M.I.; Sigurdardottir, S.T.; Thorsteinsdottir, I.; Gudmundsson, S.; Hammarström, L.; Ludviksson, B.R. Clinical symptoms in adults with selective IgA deficiency: A case-control study. *J. Clin. Immunol.* **2013**, *33*, 742–747. [CrossRef] [PubMed]

138. Papadopoulou, A.; Mermiri, D.; Taousani, S.; Triga, M.; Nicolaidou, P.; Priftis, K.N. Bronchial hyper-responsiveness in selective IgA deficiency. *Pediatr. Allergy Immunol.* **2005**, *16*, 495–500. [CrossRef] [PubMed]

139. Sandin, A.; Björkstén, B.; Böttcher, M.F.; Englund, E.; Jenmalm, M.C.; Bråbäck, L. High salivary secretory IgA antibody levels are associated with less late-onset wheezing in IgE-sensitized infants. *Pediatr. Allergy Immunol.* **2011**, *22*, 477–481. [CrossRef]

140. Chang, S.K.; Arendt, B.K.; Darce, J.R.; Wu, X.; Jelinek, D.F. A role for BLyS in the activation of innate immune cells. *Blood* **2006**, *108*, 2687–2694. [CrossRef]

141. Rachid, R.; Castigli, E.; Geha, R.S.; Bonilla, F.A. TACI mutation in common variable immunodeficiency and IgA deficiency. *Curr. Allergy Asthma Rep.* **2006**, *6*, 357–362. [CrossRef] [PubMed]

142. Moschese, V.; Chini, L.; Graziani, S.; Sgrulletti, M.; Gallo, V.; Di Matteo, G.; Ferrari, S.; Di Cesare, S.; Cirillo, E.; Pession, A.; et al. Follow-up and outcome of symptomatic partial or absolute IgA deficiency in children. *Eur. J. Pediatr.* **2019**, *178*, 51–60. [CrossRef]

143. Schwarze, J.; Cieslewicz, G.; Joetham, A.; Sun, L.K.; Sun, W.N.; Chang, T.W.; Hamelmann, E.; Gelfand, E.W. Antigen-specific immunoglobulin-A prevents increased airway responsiveness and lung eosinophilia after airway challenge in sensitized mice. *Am. J. Respir. Crit. Care Med.* **1998**, *158*, 519–525. [CrossRef] [PubMed]

144. Monteiro, R.C.; Van De Winkel, J.G. IgA Fc receptors. *Annu. Rev. Immunol.* **2003**, *21*, 177–204. [CrossRef]

145. Pasquier, B.; Launay, P.; Kanamaru, Y.; Moura, I.C.; Pfirsch, S.; Ruffié, C.; Hénin, D.; Benhamou, M.; Pretolani, M.; Blank, U.; et al. Identification of FcalphaRI as an inhibitory receptor that controls inflammation: Dual role of FcRgamma ITAM. *Immunity* **2005**, *22*, 31–42. [CrossRef]

146. Koopman, L.P.; Smit, H.A.; Heijnen, M.L.; Wijga, A.; van Strien, R.T.; Kerkhof, M.; Gerritsen, J.; Brunekreef, B.; de Jongste, J.C.; Neijens, H.J. Respiratory infections in infants: Interaction of parental allergy, child care, and siblings—The PIAMA study. *Pediatrics* **2001**, *108*, 943–948. [CrossRef] [PubMed]

147. Magen, E.; Masalha, A.; Waitman, D.A.; Kahan, N.; Viner, I.; Klassov, L.; Vardy, D. Prevalence of dermatologic diseases among patients with selective immunoglobulin A deficiency. *Allergy Asthma Proc.* **2017**, *38*, 70–77. [CrossRef]

148. Orivuori, L.; Loss, G.; Roduit, C.; Dalphin, J.C.; Depner, M.; Genuneit, J.; Lauener, R.; Pekkanen, J.; Pfefferle, P.; Riedler, J.; et al. Soluble immunoglobulin A in breast milk is inversely associated with atopic dermatitis at early age: The PASTURE cohort study. *Clin. Exp. Allergy* **2014**, *44*, 102–112. [CrossRef]

149. Pesonen, M.; Kallio, M.J.; Siimes, M.A.; Savilahti, E.; Ranki, A. Serum immunoglobulin A concentration in infancy, but not human milk immunoglobulin A, is associated with subsequent atopic manifestations in children and adolescents: A 20-year prospective follow-up study. *Clin. Exp. Allergy* **2011**, *41*, 688–696. [CrossRef]

150. Lundell, A.C.; Hesselmar, B.; Nordström, I.; Saalman, R.; Karlsson, H.; Lindberg, E.; Aberg, N.; Adlerberth, I.; Wold, A.E.; Rudin, A. High circulating immunoglobulin A levels in infants are associated with intestinal toxigenic Staphylococcus aureus and a lower frequency of eczema. *Clin. Exp. Allergy* **2009**, *39*, 662–670. [CrossRef] [PubMed]

151. Gualdi, G.; Lougaris, V.; Baronio, M.; Vitali, M.; Tampella, G.; Moratto, D.; Tanghetti, P.; Monari, P.; Calzavara-Pinton, P.; Plebani, A. Burden of skin disease in selective IgA deficiency and common variable immunodeficiency. *J. Investig. Allergol. Clin. Immunol.* **2015**, *25*, 369–371.

152. Lúdvíksson, B.R.; Arason, G.J.; Thorarensen, O.; Ardal, B.; Valdimarsson, H. Allergic diseases and asthma in relation to serum immunoglobulins and salivary immunoglobulin A in pre-school children: A follow-up community-based study. *Clin. Exp. Allergy* **2005**, *35*, 64–69. [CrossRef] [PubMed]

153. Civelek, E.; Cakir, B.; Boz, A.B.; Yuksel, H.; Orhan, F.; Under, A.; Sekerel, B.E. Extent and burden of allergic diseases in elementary schoolchildren: A national multicenter study. *J. Investig. Allergol. Clin. Immunol.* **2010**, *20*, 280–288. [PubMed]
154. Lilic, D.; Sewell, W.A. IgA deficiency: What we should-or should not-be doing. *J. Clin. Pathol.* **2001**, *54*, 337–338. [CrossRef] [PubMed]
155. Orange, J.S.; Hossny, E.M.; Weiler, C.R.; Ballow, M.; Berger, M.; Bonilla, F.A.; Buckley, R.; Chinen, J.; El-Gamal, Y.; Mazer, B.D.; et al. Use of intravenous immunoglobulin in human disease: A review of evidence by members of the Primary Immunodeficiency Committee of the American Academy of Allergy, Asthma and Immunology. *J. Allergy Clin. Immunol.* **2006**, *117*, S525–S553. [CrossRef] [PubMed]
156. Langereis, J.D.; van der Flier, M.; de Jonge, M.I. Limited innovations after more than 65 years of immunoglobulin replacement therapy: Potential of IgA- and IgM-enriched formulations to prevent bacterial respiratory tract infections. *Front. Immunol.* **2018**, *9*, 1925. [CrossRef] [PubMed]
157. Ludvigsson, J.F.; Neovius, M.; Hammarström, L. IgA deficiency and mortality: A population-based cohort study. *J. Clin. Immunol.* **2013**, *33*, 1317–1324. [CrossRef]
158. Seidel, M.G.; Kindle, G.; Gathmann, B.; Quinti, I.; Buckland, M.; van Montfrans, J.; Scheible, R.; Rusch, S.; Gasteiger, L.M.; Grimbacher, B.; et al. The European Society for Immunodeficiencies (ESID) registry working definitions for the clinical diagnosis of inborn errors of immunity. *J. Allergy Clin. Immunol. Pract.* **2019**, *7*, 1763–1770. [CrossRef]
159. Aghamohammadi, A.; Mohammadi, J.; Parvaneh, N.; Rezaei, N.; Moin, M.; Espanol, T.; Hammarstrom, L. Progression of selective IgA deficiency to common variable immunodeficiency. *Int. Arch. Allergy Immunol.* **2008**, *147*, 87–92. [CrossRef]
160. Cheraghi, T.; Aghamohammadi, A.; Mirminachi, B.; Keihanian, T.; Hedayat, E.; Abolhassani, H.; Sagvand, B.T.; Rezaei, N. Prediction of the evolution of common variable immunodeficiency: HLA typing for patients with selective IgA deficiency. *J. Investig. Allergol. Clin. Immunol.* **2014**, *24*, 198–200.
161. Litzman, J.; Burianova, M.; Thon, V.; Lokaj, J. Progression of selective IgA deficiency to common variable immunodeficiency in a 16 year old boy. *Allergol. Immunopathol.* **1996**, *24*, 174–176.
162. Español, T.; Catala, M.; Hernandez, M.; Caragol, I.; Bertran, J.M. Development of a common variable immunodeficiency in IgA-deficient patients. *Clin. Immunol. Immunopathol.* **1996**, *80*, 333–335. [CrossRef]
163. Gutierrez, M.G.; Kirkpatrick, C.H. Progressive immunodeficiency in a patient with IgA deficiency. *Ann. Allergy Asthma Immunol.* **1997**, *79*, 297–301. [CrossRef]
164. Carvalho Neves Forte, W.; De Carvalho, F.F., Jr.; Damaceno, N.; Vidal Perez, F.; Gonzales Lopes, C.; Mastroti, R.A. Evolution of IgA deficiency to IgG subclass deficiency and common variable immunodeficiency. *Allergol. Immunopathol.* **2000**, *28*, 18–20.
165. Lemarquis, A.L.; Einarsdottir, H.K.; Kristjansdottir, R.N.; Jonsdottir, I.; Ludviksson, B.R. Transitional B cells and TLR9 responses are defective in selective IgA deficiency. *Front Immunol.* **2018**, *9*, 909. [CrossRef] [PubMed]
166. Bagheri, Y.; Shad, T.M.; Namazi, S.; Azizi, G.; Hosseini, A.; Saeidi, M.; Tofighi, F.; Salami, F.; Pashangzadeh, S.; Delavari, S.; et al. B cells and T cells abnormalities in patients with selective IgA deficiency. *Authorea* **2021**. [CrossRef]
167. Grosserichter-Wagener, C.; Franco-Gallego, A.; Ahmadi, F.; Moncada-Vélez, M.; Dalm, V.A.; Rojas, J.L.; Orrego, J.C.; Correa Vargas, N.; Hammarström, L.; Schreurs, M.W.; et al. Defective formation of IgA memory B cells, Th1 and Th17 cells in symptomatic patients with selective IgA deficiency. *Clin. Transl. Immunol.* **2020**, *9*, e1130. [CrossRef] [PubMed]
168. Wehr, C.; Kivioja, T.; Schmitt, C.; Ferry, B.; Witte, T.; Eren, E.; Vlkova, M.; Hernandez, M.; Detkova, D.; Bos, P.R.; et al. The EUROclass trial: Defining subgroups in common variable immunodeficiency. *Blood* **2008**, *111*, 77–85. [CrossRef] [PubMed]
169. Warnatz, K.; Denz, A.; Dräger, R.; Braun, M.; Groth, C.; Wolff-Vorbeck, G.; Eibel, H.; Schlesier, M.; Peter, H.H. Severe deficiency of switched memory B cells (CD27(+)IgM(-)IgD(-)) in subgroups of patients with common variable immunodeficiency: A new approach to classify a heterogeneous disease. *Blood* **2002**, *99*, 1544–1551. [CrossRef]
170. Mouillot, G.; Carmagnat, M.; Gérard, L.; Garnier, J.L.; Fieschi, C.; Vince, N.; Karlin, L.; Viallard, J.F.; Jaussaud, R.; Boileau, J.; et al. B-cell and T-cell phenotypes in CVID patients correlate with the clinical phenotype of the disease. *J. Clin. Immunol.* **2010**, *30*, 746–755. [CrossRef]
171. Robert, R.; Thomas, F.; William, S.; Harry, S.; Anthony, F.; Cornelia, W. *Clinical Immunology: Principles and Practice*; Elsevier: Amsterdam, The Netherlands, 2019; ISBN 978-0-70207-040-2.
172. Bogaert, D.J.; Dullaers, M.; Lambrecht, B.N.; Vermaelen, K.Y.; De Baere, E.; Haerynck, F. Genes associated with common variable immunodeficiency: One diagnosis to rule them all? *J. Med. Genet.* **2016**, *53*, 575–590. [CrossRef]
173. Castigli, E.; Geha, R.S. TACI, isotype switching, CVID and IgAD. *Immunol. Res.* **2007**, *38*, 102–111. [CrossRef]
174. Pott, M.C.; Frede, N.; Wanders, J.; Hammarström, L.; Glocker, E.-O.; Glocker, C.; Tahami, F.; Grimbacher, B. Autoantibodies against BAFF, APRIL or IL21—An alternative pathogenesis for antibody-deficiencies? *BMC Immunol.* **2017**, *18*, 34. [CrossRef]
175. Jin, R.; Kaneko, H.; Suzuki, H.; Arai, T.; Teramoto, T.; Fukao, T.; Kondo, N. Age-related changes in BAFF and APRIL profiles and upregulation of BAFF and APRIL expression in patients with primary antibody deficiency. *Int. J. Mol. Med.* **2008**, *21*, 233–238. [CrossRef] [PubMed]
176. Vorechovský, I.; Zetterquist, H.; Paganelli, R.; Koskinen, S.; Webster, A.D.; Björkander, J.; Smith, C.I.; Hammarström, L. Family and linkage study of selective IgA deficiency and common variable immunodeficiency. *Clin. Immunol. Immunopathol.* **1995**, *77*, 185–192. [CrossRef] [PubMed]
177. Burks, A.W.; Sampson, H.A.; Buckley, R.H. Anaphylactic reactions after gamma globulin administration in patients with hypogammaglobulinemia. Detection of IgE antibodies to IgA. *N. Engl. J. Med.* **1986**, *314*, 560–564. [CrossRef]

178. Wei, C.H.; Lin, M.; Hsieh, R.K.; Tzeng, C.H.; Liu, J.H.; Fan, S.; Chiou, T.J.; Chen, P.M. Selective IgA deficiency and anaphylactoid transfusion reaction: A case report. *Zhonghua Yi Xue Za Zhi* **1996**, *57*, 165–168.
179. Kumar, N.D.; Sharma, S.; Sethi, S.; Singh, R.P. Anaphylactoid transfusion reaction with anti-IgA antibodies in an IgA deficient patient: A case report. *Indian J. Pathol. Microbiol.* **1993**, *36*, 282–284. [PubMed]
180. Klartag, A.; Chen, C.C.; Dougherty, J.P.; Ron, Y. Cross-reactive antibodies induced by xenogeneic IgA can cause selective IgA deficiency. *Autoimmunity* **2010**, *43*, 164–171. [CrossRef]
181. Rachid, R.; Bonilla, F.A. The role of anti-IgA antibodies in causing adverse reactions to gamma globulin infusion in immunodeficient patients: A comprehensive review of the literature. *J. Allergy Clin. Immunol.* **2012**, *129*, 628–634. [CrossRef]
182. Rachid, R.; Castells, M.; Cunningham-Rundles, C.; Bonilla, F.A. Association of anti-IgA antibodies with adverse reactions to γ-globulin infusion. *J. Allergy Clin. Immunol.* **2011**, *128*, 228.e1–230.e1. [CrossRef]
183. Ferreira, A.; Garcia Rodriguez, M.C.; Lopez-Trascasa, M.; Salcedo, D.P.; Fontan, G. Anti-IgA antibodies in selective IgA deficiency and in primary immunodeficient patients treated with gamma-globulin. *Clin. Immunol. Immunopathol.* **1988**, *47*, 199–207. [CrossRef]
184. Feng, M.L.; Zhao, Y.L.; Shen, T.; Huang, H.; Yin, B.; Liu, R.Z.; Qian, K.C.; Liu, D.Z. Prevalence of immunoglobulin A deficiency in Chinese blood donors and evaluation of anaphylactic transfusion reaction risk. *Transfus. Med.* **2011**, *21*, 338–343. [CrossRef]
185. Yuan, S.; Goldfinger, D. A readily available assay for anti-immunoglobulin A: Is this what we have been waiting for? *Transfusion* **2008**, *48*, 2048–2050. [CrossRef]
186. Schrijvers, D. Management of anemia in cancer patients: Transfusions. *Oncologist* **2011**, *3*, 12–18. [CrossRef] [PubMed]
187. Sandler, S.G.; Eder, A.F.; Goldman, M.; Winters, J.L. The entity of immunoglobulin A-related anaphylactic transfusion reactions is not evidence based. *Transfusion* **2015**, *55*, 199–204. [CrossRef] [PubMed]
188. Sandler, S.G.; Mallory, D.; Malamut, D.; Eckrich, R. IgA anaphylactic transfusion reactions. *Transfus. Med. Rev.* **1995**, *9*, 1–8. [CrossRef]
189. Sandler, S.G. How I manage patients suspected of having had an IgA anaphylactic transfusion reaction. *Transfusion* **2006**, *46*, 10–13. [CrossRef] [PubMed]
190. Kiani-Alikhan, S.; Yong, P.F.; Grosse-Kreul, D.; Height, S.E.; Mijovic, A.; Suddle, A.R.; Ibrahim, M.A. Successful desensitization to immunoglobulin A in a case of transfusion-related anaphylaxis. *Transfusion* **2010**, *50*, 1897–1901. [CrossRef] [PubMed]
191. Cunningham-Rundles, C.; Siegal, F.P.; Cunningham-Rundles, S.; Lieberman, P. Incidence of cancer in 98 patients with common varied immunodeficiency. *J. Clin. Immunol.* **1987**, *7*, 294–299. [CrossRef]
192. Kiaee, F.; Azizi, G.; Rafiemanesh, H.; Zainaldain, H.; Sadaat Rizvi, F.; Alizadeh, M.; Jamee, M.; Mohammadi, S.; Habibi, S.; Sharifi, L.; et al. Malignancy in common variable immunodeficiency: A systematic review and meta-analysis. *Expert Rev. Clin. Immunol.* **2019**, *15*, 1105–1113. [CrossRef] [PubMed]
193. Kinlen, L.J.; Webster, A.D.; Bird, A.G.; Haile, R.; Peto, J.; Soothill, J.F.; Thompson, R.A. Prospective study of cancer in patients with hypogammaglobulinaemia. *Lancet* **1985**, *1*, 263–266. [CrossRef]
194. Mellemkjaer, L.; Hammarstrom, L.; Andersen, V.; Yuen, J.; Heilmann, C.; Barington, T.; Bjorkander, J.; Olsen, J.H. Cancer risk among patients with IgA deficiency or common variable immunodeficiency and their relatives: A combined Danish and Swedish study. *Clin. Exp. Immunol.* **2002**, *130*, 495–500. [CrossRef]
195. Ludvigsson, J.F.; Neovius, M.; Ye, W.; Hammarström, L. IgA deficiency and risk of cancer: A population-based matched cohort study. *J. Clin. Immunol.* **2015**, *35*, 182–188. [CrossRef] [PubMed]
196. Bjelac, J.A.; Blanch, M.B.; Fernandez, J. Allergic disease in patients with common variable immunodeficiency at a tertiary care referral center. *Ann. Allergy Asthma Immunol.* **2018**, *120*, 90–92. [CrossRef]
197. Dadkhah, M.; Aghamohammadi, A.; Movahedi, M.; Gharagozlou, M. Atopic manifestations: Dermatitis, allergic rhinitis and asthma in patients with hypogammaglobulinemia. *Iran. J. Pediatr.* **2015**, *25*, e2786. [CrossRef] [PubMed]
198. Szczawinska-Poplonyk, A. An overlapping syndrome of allergy and immune deficiency in children. *J. Allergy* **2012**, *2012*, 658279. [CrossRef] [PubMed]

Journal of
Clinical Medicine

Article

Patients with Common Variable Immunodeficiency Complicated by Autoimmune Phenomena Have Lymphopenia and Reduced Treg, Th17, and NK Cells

Ewa Więsik-Szewczyk [1,*], Elżbieta Rutkowska [2], Iwona Kwiecień [2], Marcelina Korzeniowska [1], Dariusz Sołdacki [1,3] and Karina Jahnz-Różyk [1]

[1] Department of Internal Medicine, Pulmonology, Allergy and Clinical Immunology, Military Institute of Medicine, Szaserów 128, 04-141 Warsaw, Poland; mkorzeniowska1@wim.mil.pl (M.K.); dariusz.soldacki@gmail.com (D.S.); kjrozyk@wim.mil.pl (K.J.-R.)

[2] Laboratory of Hematology and Flow Cytometry, Department of Internal Medicine and Hematology, Military Institute of Medicine, Szaserów 128, 04-141 Warsaw, Poland; erutkowska@wim.mil.pl (E.R.); ikwiecien@wim.mil.pl (I.K.)

[3] Department of Clinical Immunology, Medical University of Warsaw, 02-691 Warsaw, Poland

* Correspondence: ewa.w.szewczyk@gmail.com

Abstract: Most patients with primary immune deficiency suffer from recurrent infections; however, paradoxical autoimmune phenomena can also manifest. The aim of this study was to identify immunological markers of autoimmune phenomena associated with common variable immunodeficiency (CVID). The study included 33 adults with CVID divided into two groups: (1) those with noninfectious autoimmune complications (CVID-C (n = 24)) and (2) those with only infectious symptoms (CVID-OI (n = 9)). Flow cytometry of peripheral blood was performed and compared with systemic lupus erythematosus (SLE) patients (n = 17) and healthy controls (n = 20). We found that all lymphocytes were lower in CVID-C and SLE. NK cells were lowest in CVID-C. Th17 cells were significantly reduced in CVID-C and SLE. Tregs were significantly lower in CVID-C and SLE. Bregs did not significantly differ between any groups. Class-switched memory B cells were significantly lower in CVID-C and CVID-OI. Lastly, plasmablasts were significantly higher in SLE. Among the T cell subsets, CVID-C patients had lower naive and recent thymic emigrant CD4+ T cells. In conclusion, reduced Treg, Th17, and NK cells are features of CVID with autoimmune complications, and class-switched memory B cells can help distinguish patients with different causes of autoimmunity. Future studies are needed to confirm whether reductions of Treg, Th17, and NK cells might be a biomarker of more complicated CVID cases.

Keywords: B cell maturation; CD4+ cells; T cell maturation; primary immune deficiency; autoimmunity; connective tissue diseases

Citation: Więsik-Szewczyk, E.; Rutkowska, E.; Kwiecień, I.; Korzeniowska, M.; Sołdacki, D.; Jahnz-Różyk, K. Patients with Common Variable Immunodeficiency Complicated by Autoimmune Phenomena Have Lymphopenia and Reduced Treg, Th17, and NK Cells. *J. Clin. Med.* **2021**, *10*, 3356. https://doi.org/10.3390/jcm10153356

Academic Editors: Rita Consolini and Giorgio Costagliola

Received: 26 May 2021
Accepted: 23 July 2021
Published: 29 July 2021

Publisher's Note: MDPI stays neutral with regard to jurisdictional claims in published maps and institutional affiliations.

1. Introduction

Common variable immunodeficiency (CVID) is the most common symptomatic primary antibody deficiency in adults. Most patients suffer from recurring respiratory tract infections; however, paradoxical autoimmunity, both systemic and organ related, is a secondary manifestation of CVID and affects 20–40% of cases [1]. Patients with a complicated CVID phenotype have the longest diagnostic delay, especially if autoimmune phenomena are the first manifestations of primary immunodeficiency [2]. Autoimmune phenomena might occur as the first symptom in 30% of patients with primary antibody deficiencies [3]. In a recent study focused on noninfectious CVID complications, autoimmune thrombocytopenic purpura (AITP) was most common (16.2%), followed by autoimmune hemolytic anemia (AIH 7.7%), amongst the 632 patients followed since 1974. Other associated autoimmune conditions include rheumatoid arthritis (2.7%) and uveitis (1%). Rarer autoimmune complications are psoriasis, psoriatic arthritis, vitiligo, alopecia, autoimmune thyroiditis, antiphospholipid syndrome, Sjogren syndrome, vasculitis, type 1 diabetes, myasthenia gravis, autoimmune pancreatitis, and severe oral ulcers [4]. Polyclonal lymphadenopathy occurred in 20–40% of patients [4]. Clinically, this often presents as generalized lymphadenopathy and splenomegaly. The challenge is to differentiate it from lymphoma [5]. In 20% of patients with a CVID-like phenotype, monogenic defects responsible for immune deregulation have been identified. Examples include CTLA4 and LRBA deficiency [6,7], nuclear factor kB (NFkB) mutations [8,9] and mutations of the catalytic subunit of phosphoinositide 3-kinase delta (PI3Kdeta) [10]. However, in the majority of patients with CVID, the pathogenesis of noninfectious phenomena is still unknown.

Similar symptoms were observed in systemic lupus erythematosus (SLE): cytopenia, generalized lymphadenopathy, hepatomegaly, splenomegaly, and interstitial lung disease. The clinical symptoms of CVID and systemic rheumatic diseases overlap. Evidence has accumulated that the coincidence of primary immunodeficiency (PID) and autoimmune diseases is high [11–13]. A secondary immunodeficiency-like state is present in a significant number of rheumatologic patients. Low serum levels of main immunoglobulins (Ig) and subclasses IgG3 and IgG4 are frequent, although in many cases laboratory abnormalities are not related to increased susceptibility to infections [14]. A recent study of patients with rheumatic diseases identified genetic variants that are responsible for PID in participants who developed hypogammaglobulinemia [11].

Immunophenotyping of the B cell compartment in the peripheral blood is a routine evaluation in patients with primary hypogammaglobulinemia. In CVID, detailed findings of B cell maturation can be classified according to several systems, of which two of the most popular are Freiburg [15] and EUROclass [16]. Characterization of CD19+ B cell subsets in CVID is classified according to low Ig switched memory (CD19+ CD27+ IgM− IgD−), B cell (smB) proportions, and abnormally high proportions of $CD21^{low}$ B cells. In addition to these two cell subsets, the EUROclass classification also uses an abnormal expansion of transitional B cells (CD19+ CD27− CD38+) for further subgrouping [16].

Low smB cell subsets are an abnormality present in 80% of patients with CVID; however, it is not specific for CVID. More detailed studies that assessed the correlation between B cell maturation and the phenotype of CVID have produced mixed results. In some studies, diminished smB cell count [17], reduced naïve B cells [18], and expansion of B cells with reduced CD21 expression ($CD21^{low}$ B cells) correlated with autoimmune phenomena or splenomegaly [16,19].

Although CVID is a disease of defective B cell maturation, various reports have associated CVID with T cell compartment abnormalities, such as CD4+ T cell lymphopenia with reduced subset counts of naive CD4+ T cells [20] and naive CD8+ T cells [21]. A reduced percentage of naïve CD4+ T cells was associated with complications and poor prognosis in CVID [22].

Regulatory T cells (Tregs), T helper 17 (Th17), and follicular T helper 17 (Tfh17) cells are reduced in patients with complicated CVID phenotypes [18]. T cells in patients with CVID have lower proliferative capacities [23] and abnormal cytokine production [24]. Recent studies have shown the involvement of follicular T cells in CVID pathogenesis [25] An increase in the circulating memory CD4+ T cells of CVID patients with noninfectious complications has been reported [26].

In contrast to PID, immunophenotyping of B and T cells in SLE and other autoimmune diseases is mainly used in scientific research and clinical trials [27]. Therefore, physicians are not familiar with the interpretation and utility of lymphocyte subset counts in clinical practice. The data showed that IgM memory B cells were significantly decreased in patients with SLE. In contrast, transitional B cells were significantly increased in SLE and other autoimmune disorders [28]. The population of plasmablasts also increased in active SLE [29].

Until now, studies comparing B and T cell subsets from patients with PID and patients with rheumatic diseases are limited. In one study, patients with primary and secondary hypogammaglobulinemia in the course of different rheumatic diseases were observed [30].

Another study involved the analysis of polymyalgia rheumatica patients treated with systemic glucocorticoids [31]. Both studies aimed to identify the distinction between primary and secondary hypogammaglobulinemia.

We analyzed the maturation of B and T lymphocytes in the peripheral blood of patients with CVID who were divided into two groups: patients with a phenotype limited to infections (CVID-OI) and patients with noninfectious, autoimmune complications (CVID-C). These results were compared with those of patients diagnosed with SLE and healthy controls (HCs). The aim of this study was to identify immunological markers of autoimmune phenomena associated with CVID.

2. Materials and Methods

2.1. Patients

The study population was selected from consecutive adult patients ($\geq$18 years old) who were under the care of the outpatient clinic of the Department of Internal Medicine, Pulmonology, Allergy, and Clinical Immunology, Central Clinical Hospital of the Ministry of National Defense, Military Institute of Medicine in Warsaw, Poland, between January 2016 and December 2019.

The study participants included confirmed CVID patients diagnosed according to the European Society for Immunodeficiencies' criteria [32] and were treated under the Polish Ministry of Health's drug programs, B.62 and B.78. CVID clinical phenotypes were defined according to the literature [32]. Group 1, or CVID-C (n = 24), included patients with CVID who suffered from increased susceptibility to infections and at least one other clinical event beyond increased susceptibility to infections attributable to PID [33]. Group 2, or CVID-OI (n = 9), included patients with only the infectious phenotype. Group 3 included patients with SLE (n = 17) who fulfilled the Systemic Lupus International Collaborating Clinics' (SLICCs') criteria [34] and had no clinical signs of immunodeficiency.

Healthy controls (HC) were selected from age-matched volunteers from hospital employees without any signs, symptoms, or history of immunodeficiency and/or autoimmunity.

2.2. Compliance with Research Ethics Standards

The study protocol was approved by the Bioethics Committee of the Military Institute of Medicine (approval no. 7/WIM/2020). All patients were informed in detail orally about the course, aims, and scope of this research. Blood sampling was limited to routine assessments. Separate written consent for blood sampling and review of records were not required by the IRB due to the retrospective nature of this study. All patient data were confidential, and the study procedures complied with the Declaration of Helsinki.

2.3. Flow Cytometry Analysis

All blood samples were drawn during routine visits. If the CVID patients were on immunoglobulin replacement therapy (IgRT), blood samples were drawn before the day of IgG infusion according to the national regulations for treatment reimbursement.

Lymphocyte subset percentages were determined according to literature [15,35] by flow cytometry using a panel of monoclonal antibodies using FACS Canto II BD flow cytometry (Becton Dickinson (BD) Biosciences, Franklin Lakes, NJ, USA). Subsequently, all eight-color surface staining panels for the basic subpopulation of lymphocytes were evaluated as follows: CD4 FITC, CD3 PerCP-Cy5-5, CD19 PE-Cy7, CD8 APC, CD16 APC-H7, and CD45 V500 (BD Biosciences).

B cell subpopulations were defined using the following antibodies: IgD PE, CD27 PerCP-Cy5-5, CD19 PE-Cy7, IgM APC, CD38 APC-H7, CD21 V450, and CD45 V500 (BD Biosciences).

CD4 T cell and CD8 maturation were defined using CD4 FITC, CD196 PE, CD197 PerCP-Cy5-5, CD45RO PE-Cy7, CD45RA APC, CD3 APC-H7, CD8 V450, and CD45 V500 (BD Biosciences) antibodies.

For Th17 cells, CD4 FITC, CD196 PE, CD45RO PE-Cy7, and CD45 V500 (BD Biosciences) antibodies were used.

Recent thymic emigrant (RTE) CD4+ or CD8+ cells were analyzed using CD4 FITC, CD62L PE, CD31 PerCP-Cy5-5, CD45RO PE-Cy7, CD45 RA APC, CD3 APC-H7, and CD8 V450 CD45 V500 (BD Biosciences) antibodies.

For Tregs, we used CD127 FITC, CD4 PerCP-Cy5-5, CD25 APC, CD3 APC-H7, CD45 V500 (BD Biosciences) antibodies, and for regulatory B cells (Bregs), CD1d PE, CD19 PE-Cy7, CD5 APC, and CD45 V500 (BD Biosciences) antibodies were used.

After surface staining for 15 min at 21 °C in the dark, erythrocytes were lysed with 2 mL of BD Pharm Lyse buffer (BD Biosciences) for 10 min. Following centrifugation and washing with Cell Wash buffer (BD Biosciences), the mixture was stored in the dark for analysis by flow cytometry within 2 h. Data were analyzed with DIVA Analysis software (version 8.0.1, BD Biosciences, San Jose, CA 95131 USA) and Infinicyt 1.8 Flow Cytometry (Cytognos, Salamanca, Spain).

Lymphocyte counts were obtained using a SYSMEX XN-1500 (Sysmex Corp., Kobe, Japan) hematological analyzer.

Internal quality control was performed daily by checking the optical detector, aligning lasers, and fluid systems using CS&T IVD Beads BD FACS Diva (BD Biosciences), San Jose, CA 95131 USA, respectively, according to the manufacturers' guidelines. Internal reference values of lymphocyte counts and proportions are presented in Supplementary Materials (Tables S1–S3).

We distinguished the following subpopulations in B cell maturation:

- transitional B cells: IgM++ IgD++ CD38++ CD27− CD19+ CD45+
- naïve B cells: IgM+ IgD++ CD38+ CD27− CD19+ CD45+
- nonswitched memory B cells (marginal zone-like B cells): IgM++ IgD+ CD38+ CD27+ CD19+ CD45+
- class-switched memory B cells: IgM− IgD− CD38+ CD27+ CD19+ CD45+
- plasmablasts: IgM−/+ IgD− CD38+++ CD27++ CD19+ CD45+
- CD21^{low} B cells: IgM+ IgD+ CD38+low CD27− CD21+low CD19+ CD45+

We distinguished the following subpopulations in T cell CD4+ or CD8+ maturation:

- RTE T cells: CD45RA+ CD62L+ CD31+ CD3+ CD45+
- naïve T cells: CD45RA+ CD197+ CD3+ CD45+
- effector T cells: CD45RA+ CD197− CD3+ CD45+
- central memory T cells: CD45RO+ CD197+ CD3+ CD45+
- effector memory T cells: CD45RO+ CD197− CD3+ CD45+
- RTE T cells: CD45RA+ CD62L+ CD31+

We distinguished the following other lymphocytes subpopulations:

- Bregs: CD19+ CD5+ CD1d^{high}
- Tregs: CD3+ CD4+ CD25^{high} FoxP3+ CD127−
- Th17: CD3+ CD4+ CD45RO+ CD196+

Representative B and T lymphocyte maturation gating strategies in patients are presented in Figures S1–S3 (Supplementary Materials).

2.4. Statistical Analysis

All statistical analyses were performed using Statistica® software (version 13.0; TIBCO Software, Palo Alto, CA, USA). Statistical significance was set at $p < 0.05$. The results are expressed as medians (Q1–Q3) of the lymphocyte populations. For group comparisons, the Kruskal–Wallis, analysis of variance (ANOVA), and post hoc analysis tests were used.

3. Results

3.1. The Clinical Characteristics of Patients

Among the 33 patients with CVID, 14 were women and 19 were men, and the overall mean age at the time of blood sampling was 37.75 years (min–max: 21–66 years). Among the 9 patients with CVID-OI, 4 (44%) were men, and among 24 patients with CVID-C, 16 (66%) were men. The mean age was 37.0 years (±17.2) and 38.4 years (±12.4) in patients with CVID-OI and CVID-C, respectively. By the time of blood sampling, two patients were receiving low-dose prednisolone (5 mg/day), and one was being treated with methotrexate and etanercept due to psoriatic arthritis.

All CVID patients had a positive history of increased susceptibility to infections; however, only 9 of 33 (27%) presented with a clinical phenotype limited to infections. The clinical characteristics of the patients are summarized in Table 1. All 17 SLE patients were female, mean age 43.05 years (min–max 18–60). The SLE patients had low or mild disease activity according to the Systemic Lupus Erythematosus Disease Activity Index 2000 SLEDAI2K (mean value 3.4). Sixteen patients with SLE were treated with antimalarials, and ten patients received prednisolone (5–15 mg/day).

Table 1. The clinical characteristics of patients with CVID and SLE.

Clinical Phenotypes and Organ Complications in CVID Patients (*n* = 33)	
No disease-related complications	9 (27%)
Bronchiectasis	4 (12%)
Splenomegaly	7 (21%)
Autoimmunity	20 (60%)
Thrombocytopenia	10 (33%)
Hemolytic anemia	6 (18%)
Addison–Biermer disease	2 (6%)
Vitiligo	1 (3%)
Chronic seronegative polyarthritis	2 (6%)
Alopecia areata	1 (3%)
Nonspecific inflammatory bowel disease	1 (3%)
Psoriasis	3 (6%)
Polyclonal lymphocytic infiltration	
Generalized lymphadenopathy	19 (57%)
Granulomatous lesions (histopathological confirmation)	9 (27%)
Immunoglobulin replacement therapy	29
Immunoglobulin naïve	3
Prednisolone	2; dose 5 mg/day
Methotrexate and etanercept	1
Rituximab in anamnesis	2
Clinical data of SLE patients (*n* = 17)	
SLEDAI2K	3.6 (min 0–max 9)
Treatment	16/17
Prednisolone	10 (58%); dose: 9 mg/day (min 5–max 15 mg)
Antimalarials	16 (94%)
Immunosuppressive medication	
Methotrexate	2 (11%)
Rituximab in anamnesis	1 (5%)

CVID: Common variable immunodeficiency, SLE: systemic lupus erythematosus, SLEDAI2K: Systemic Lupus Erythematosus Disease Activity Index 2000.

3.2. Peripheral Main Lymphocyte Subsets, Tregs, Bregs, and Th17 Cells

The proportion of total lymphocytes was low in the CVID-C and SLE groups. It differed significantly from that of the HC group. CVID-OI patients had all lymphocyte proportions similar to those of HC.

The proportion of T lymphocytes was the lowest in CVID-C and differed significantly from that in HCs. In SLE patients, T lymphocytes were also significantly lower than in the HC group. The proportion of T lymphocytes in CVID-OI was similar to that in HC. The results are summarized in Table 2.

Table 2. Differences in the median of lymphocyte subpopulations and absolute numbers between (a) the CVID with complicated phenotype (CVID-C) group, (b) CVID phenotype limited to only infections (CVID-OI group), (c) patients with systemic lupus erythematosus (SLE), and (d) healthy controls (HCs). Data expressed as median (Q1–Q3). * $p < 0.05$. ANOVA: analysis of variance, Post Hoc: post hoc analysis tests.

As Median (Q1–Q3)	CVID-C (a) $n = 24$	CVID-OI (b) $n = 9$	SLE (c) $n = 17$	HC (d) $n = 20$	$p < 0.05$ Group a-b-c-d ANOVA, Kruskal–Wallis	$p < 0.05$ between Groups Post Hoc Test
			% of all cells			
Lymphocytes	21.0 (16.6–29.8)	27.5 (22.7–36.3)	28.4 (12.1–36.6)	38.3 (33.2–46.3)	$p = 0.0002$	*a–d, *c–d
Lymphocytes T	17.6 (15.7–23.7)	22.4 (19.5–25.9)	19.3 (8.7–27.4)	29.5 (24.0–37.2)	$p = 0.0011$	*a–d, *c–d
CD4 cells	9.1 (4.3–11.4)	8.9 (6.8–9.9)	9.8 (4.9–12.0)	18.6 (13.6–22.0)	$p < 0.0001$	*a–d, *b–d, *c–d
CD8 cells	9.8 (7.0–12.0)	11.6 (8.3–11.9)	9.1 (3.3–11.5)	10.5 (7.8–13.2)	-	-
Lymphocytes B	0.8 (0.1–2.2)	3.3 (1.0–3.4)	2.4 (1.5–5.7)	3.9 (3.0–5.0)	$p < 0.0001$	*a–c, *a–d
NK cells	1.0 (0.5–1.9)	1.8 (1.4–2.3)	3.0 (2.0–3.7)	4.2 (2.8–7.0)	$p = 0.0001$	*a-d
Bregs	0.023 (0.008–0.113	0.050 (0.037–0.106)	0.041 (0.016–0.059)	0.065 (0.049–0.093)	-	-
Tregs	0.209 (0.093–0.366)	0.404 (0.356–0.613)	0.265 (0.167–0.478)	0.824 (0.711–1.109)	$p < 0.0001$	*a–d, *c–d
Th17	2.3 (1.0–4.3)	2.2 (1.8–3.7)	1.1 (0.4–1.9)	5.0 (3.8–7.1)	$p < 0.0001$	*a–d, *c–d
			(cells/µL)			
WBC	5575 (4605–7555)	6600 (5440–7370)	5690 (3630–8310)	6555 (4930–7535)	-	-
Lymphocytes	1201 (755–2145)	1986 (1119–2402)	1115 (1005–1576)	2037 (1838–2934)	$p = 0.0002$	*a–d, *c–d
Lymphocytes T	1071 (701–1614)	1457 (961–2093)	887 (570–1125)	1660 (1409–2292)	$p = 0.0004$	*a–d, *c–d
CD4 cells	458 (305–553)	574 (372–680)	418 (288–499)	978 (756–1559)	$p < 0.0001$	*a–d, *c–d
CD8 cells	580 (305–809)	814 (374–1089)	319 (162–582)	624 (457–791)	$p = 0.0269$	*c–d
Ratio CD4/CD8	0.8 (0.4–1.2)	0.8 (0.4–1.0)	1.3 (0.9–1.7)	1.8 (1.5–2.5)	$p = 0.0003$	*a–d, *b–d
Lymphocytes B	47 (12–127)	212 (117–332)	145 (69–222)	216 (190–284)	$p < 0.0001$	*a–d
NK cells	54 (32–100)	152 (96–488)	126 (87–234)	245 (204–447)	$p = 0.0001$	*a–d
Bregs	1 (0–4)	4 (2–9)	2 (1–3)	4 (3–7)	-	-
Tregs	15 (4–21)	26 (22–30)	20 (12–28)	55 (37–82)	$p < 0.0001$	*a–d, *c–d
Th17	130 (73–190)	131 (94–284)	77 (35–113)	256 (209–494)	$p < 0.0001$	*a–d, *c–d

Among the T lymphocytes, the proportion of CD4+ T cells was significantly reduced in patients with CVID-C, CVID-OI, and SLE compared to that in HC. There were no differences in the proportion of CD8+ T cells between the disease groups and HC. The CD4/CD8 ratio was reduced in CVID-C and CVID-OI compared with that in HCs.

The proportion of circulating B cells was reduced in CVID-C and differed significantly from that in HC and SLE.

The NK cell proportion was marked decreased in CVID-C patients compared to HC patients.

The proportion of Th17 cells was reduced in CVID-C and SLE and differed significantly from that in HCs. The Th17 cell counts were the lowest in SLE (Figure 1).

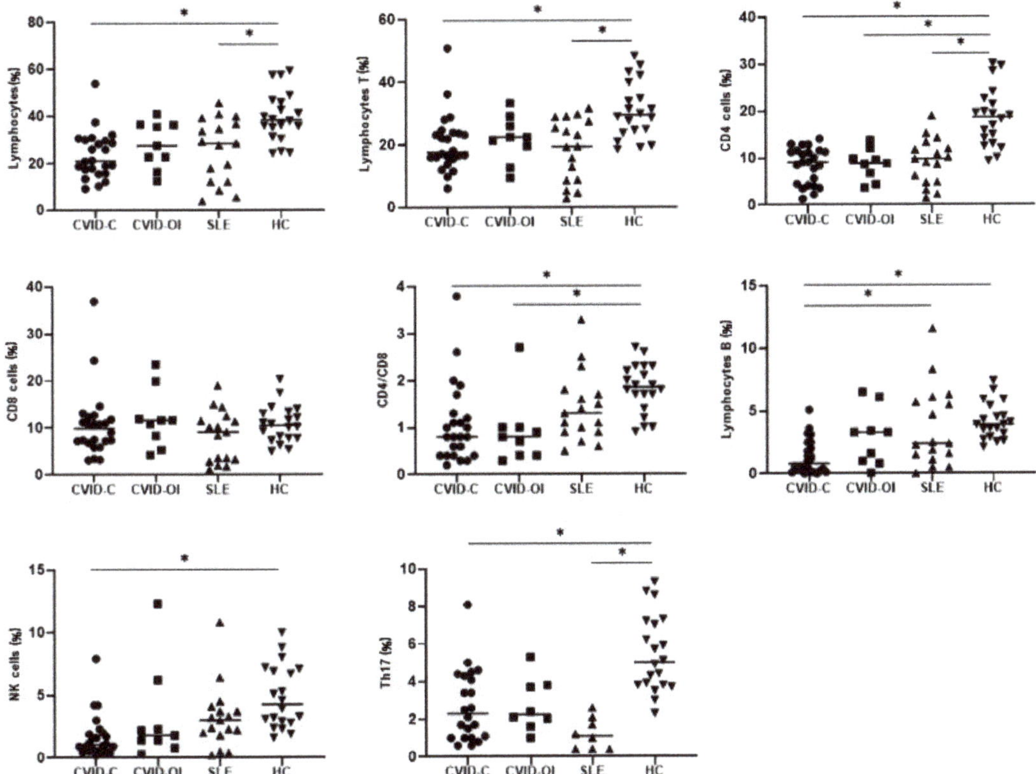

Figure 1. Differences in the median percentages of lymphocyte subpopulations between the CVID with complicated phenotype (CVID-C) group, the CVID phenotype limited to only infections (CVID-OI), patients with systemic lupus erythematosus (SLE), and healthy controls (HCs). Data expressed as median (Q1–Q3), * $p < 0.05$.

Treg counts were low in CVID-C and SLE patients and differed significantly from HCs. Breg counts were lowest in CVID-C patients, but no significant differences were noted between the disease groups and the HC group (Figure 2). Treg counts were low in CVID-C and SLE patients and differed significantly from HCs.

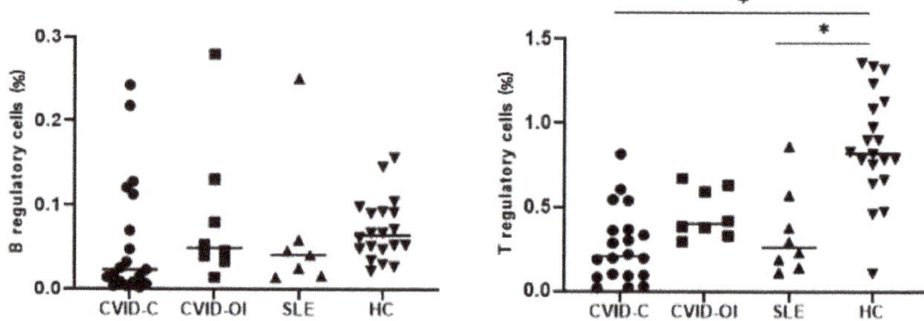

Figure 2. Differences in the median percentages of T regulatory cells (Tregs) and B regulatory cells (Bregs) between the CVID with complicated phenotype (CVID-C) group, CVID phenotype limited to only infections (CVID-OI group), patients with systemic lupus erythematosus (SLE), and healthy controls (HCs). Data are expressed as median (Q1–Q3) * $p < 0.05$.

Considering the absolute numbers of the above-mentioned lymphocyte populations, the same trends were observed. A difference was observed only in the CD8 lymphocytes of SLE patients, which were significantly less than in the HCs (Table 2).

3.3. B Lymphocyte Maturation

Analysis of the maturation of B-lymphocytes showed that the proportion of transitional B cells was highest in CVID-OI and differed significantly from that in HCs. The frequency of nonswitched memory B cells was the highest in patients with CVID-C. The difference in nonswitched memory B cell counts was significant between the CVID-C and SLE groups and between the CVID-C and HC groups. Class-switched memory B cell percentages were low in CVID-C and CVID-OI. The difference was significant in comparison with the percentage of smB cells in HCs and SLE. The proportion of CD21^{low} B cells was higher in CVID-C and CVID-OI than in HCs. Plasmablasts were significantly higher in SLE patients than in CVID-C and CVID-OI patients (Table 3, Figure 3). The proportions of B lymphocyte maturation for each patient with CVID-C, CVID-OI, SLE, and HC are presented on heat maps in Figure 4.

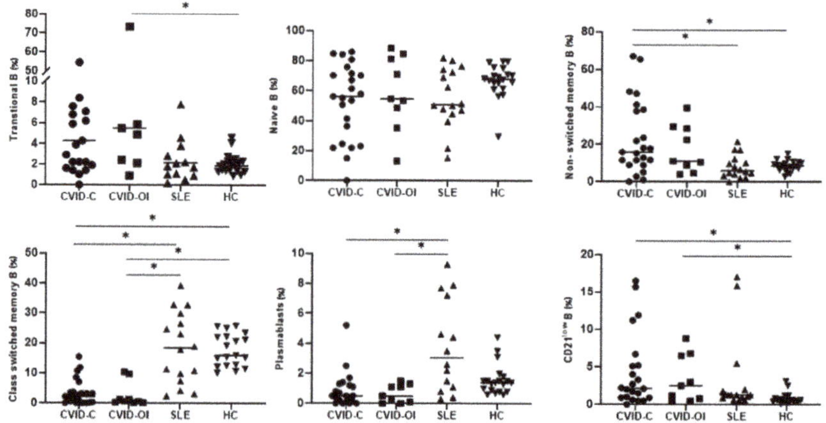

Figure 3. Differences in the proportions of B lymphocyte maturation between the CVID with complicated phenotype (CVID-C) group, CVID phenotype limited to only infections (CVID-OI group), patients with systemic lupus erythematosus (SLE), and healthy controls (HCs). Data expressed as median (Q1–Q3), * $p < 0.05$.

Table 3. Differences in the proportions of B lymphocyte maturation between (a) the CVID with complicated phenotype (CVID-C) group, (b) CVID phenotype limited to only infections (CVID-OI group), (c) patients with systemic lupus erythematosus (SLE), and (d) healthy controls (HCs). Data expressed as median (Q1–Q3). * $p < 0.05$. ANOVA: analysis of variance, Post Hoc: post hoc analysis tests.

As Median (Q1–Q3)	CVID-C (a) $n = 24$	CVID-OI (b) $n = 9$	SLE (c) $n = 17$	HC (d) $n = 20$	$p < 0.05$ Group a-b-c-d ANOVA, Kruskal–Wallis	$p < 0.05$ between Groups Post Hoc Test
			% of B cells			
Transitional B	4.3 (1.9–8.4)	5.5 (2.4–21.6)	2.1 (1.0–6.2)	1.8 (1.4–2.3)	$p = 0.0149$	*b–d
Naïve B	56.3 (24.6–71.5)	54.8 (48.8–81.3)	51.0 (46.0–73.4)	68.0 (63.5–73.1)	-	-
Nonswitched memory	15.9 (8.9–38.7)	11.0 (8.8–28.7)	6.0 (3.7–11.1)	8.6 (6.9–10.3)	$p = 0.0036$	*a–c, *a–d
Class-switched memory	2.2 (0.2–3.7)	0.6 (0.3–1.3)	18.4 (8.6–28.2)	17.6 (12.7–22.8)	$p < 0.0001$	*a–c, *a–d, *b–c, *b–d
Plasmablasts	0.5 (0.1–1.4)	0.5 (0.1–1.1)	3.9 (1.3–7.8)	1.4 (0.8–1.6)	$p = 0.0004$	*a–c, *b–c
CD21^{low} B cells	2.2 (0.9–6.7)	2.5 (0.8–6.5)	1.3 (0.9–1.9)	0.6 (0.4–0.9)	$p = 0.0005$	*a–d, *b–d
Transitional B	4.3 (1.9–8.4)	5.5 (2.4–21.6)	2.1 (1.0–6.2)	1.8 (1.4–2.3)	$p = 0.0149$	*b–d
Naïve B	56.3 (24.6–71.5)	54.8 (48.8–81.3)	51.0 (46.0–73.4)	68.0 (63.5–73.1)	-	-
Nonswitched memory	15.9 (8.9–38.7)	11.0 (8.8–28.7)	6.0 (3.7–11.1)	8.6 (6.9–10.3)	$p = 0.0036$	*a–c, *a–d

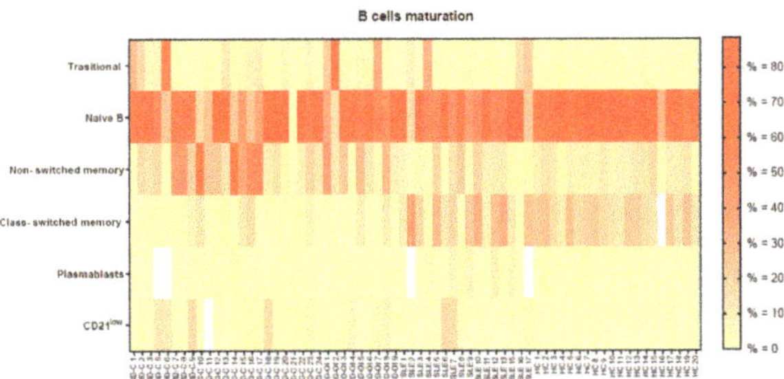

Figure 4. The proportions of B lymphocyte maturation for each patient with CVID with complicated phenotype (CVID-C) group, CVID phenotype limited to only infections (CVID-OI group), patients with SLE, and healthy controls (HCs). Data expressed as a median percentage (%).

3.4. T Lymphocyte Maturation

To examine the abnormalities in T cell maturation, we delineated CD4+ and CD8+ cells. RTE CD4+ and naïve CD4+ T cell percentages were significantly reduced in CVID-C patients compared to that in HCs. The proportion of effector memory CD45RO+ CD197-CD4+ T cells significantly increased in CVID-C compared to the HCs. The RTE CD8+ T cell proportion was low in CVID-C and differed significantly from SLE (the highest proportion) and HCs. Naïve CD8+ T cell counts were significantly reduced in patients with CVID-C compared to HCs. The proportion of effector memory CD8+ T cells was similar in the

CVID and HC groups. The proportion of effector memory CD8+ T cells was the lowest in the SLE group and significantly differed from CVID-C (Table 4, Figure 5). The proportions for the T lymphocyte maturation of each patient with CVID-C, CVID-OI, and SLE and HCs are presented on heat maps (Figure 6).

Table 4. Differences in the proportions of T lymphocytes maturation between (a) the CVID with complicated phenotype (CVID-C) group, (b) CVID phenotype limited to only infections (CVID-OI group), (c) patients with systemic lupus erythematosus (SLE), and (d) healthy controls (HCs). Data expressed as median (Q1–Q3), * $p < 0.05$. ANOVA: analysis of variance, Post Hoc: post hoc analysis tests.

As Median (Q1–Q3)	CVID-C (a) $n = 24$	CVID-OI (b) $n = 9$	SLE (c) $n = 17$	HC (d) $n = 20$	$p < 0.05$ Group a-b-c-d ANOVA, Kruskal–Wallis	$p < 0.05$ between Groups Post Hoc Test
% of CD4 cells						
Recent thymic emigrants (RTE) CD4	15.2 (5.2–27.3)	21.8 (6.3–29.7)	35.5 (18.5–48.7)	31.2 (26.3–37.6)	$p = 0.0031$	*a–d
Naïve CD4	28.7 (12.8–37.9)	31.1 (20.0–42.3)	40.6 (15.0–68.4)	50.0 (42.1–58.3)	$p = 0.0009$	*a–d
Effector CD4	2.4 (1.9–7.7)	2.8 (0.7–4.6)	4.5 (2.2–8.7)	1.8 (1.1–3.4)	-	-
Effector memory CD4	23.9 (12.1–34.8)	23.6 (15.2–29.7)	15.4 (6.3–34.9)	12.5 (9.2–15.0)	$p = 0.0126$	*a–d
Central memory CD4	40.2 (31.2–50.2)	44.9 (28.8–52.0)	25.3 (17.4–35.5)	33.2 (27.2–40.3)	-	-
CD21^{low} B cells	2.2 (0.9–6.7)	2.5 (0.8–6.5)	1.3 (0.9–1.9)	0.6 (0.4–0.9)	$p = 0.0005$	*a–d, *b–d
Transitional B	4.3 (1.9–8.4)	5.5 (2.4–21.6)	2.1 (1.0–6.2)	1.8 (1.4–2.3)	$p = 0.0149$	*b–d
Naïve B	56.3 (24.6–71.5)	54.8 (48.8–81.3)	51.0 (46.0–73.4)	68.0 (63.5–73.1)	-	-
Nonswitched memory	15.9 (8.9–38.7)	11.0 (8.8–28.7)	6.0 (3.7–11.1)	8.6 (6.9–10.3)	$p = 0.0036$	*a–c, *a–d
% of CD8 cells						
Recent thymic emigrants (RTE) CD8	27.6 (12.1–37.5)	30.4 (8.9–36.5)	45.6 (41.6–60.4)	39.5 (34.4–52.9)	$p = 0.0006$	*a–c, *a–d
Naïve CD8	25.4 (4.5–34.7)	29.3 (9.1–35.9)	41.6 (30.9–63.3)	41.3 (34.6–55.2)	$p = 0.0019$	*a–d
Effector CD8	35.3 (23.8–54.3)	41.1 (28.5–50.7)	24.8 (17.5–44.7)	25.5 (18.1–38.2)	-	-
Effector memory CD8	21.0 (17.2–35.7)	20.8 (18.2–31.5)	10.8 (7.4–19.8)	19.3 (16.2–22.9)	$p = 0.0490$	*a–c
Central memory CD8	7.3 (3.2–10.6)	7.9 (4.4–11.6)	6.8 (5.2–14.8)	7.8 (4.1–11.4)	-	-

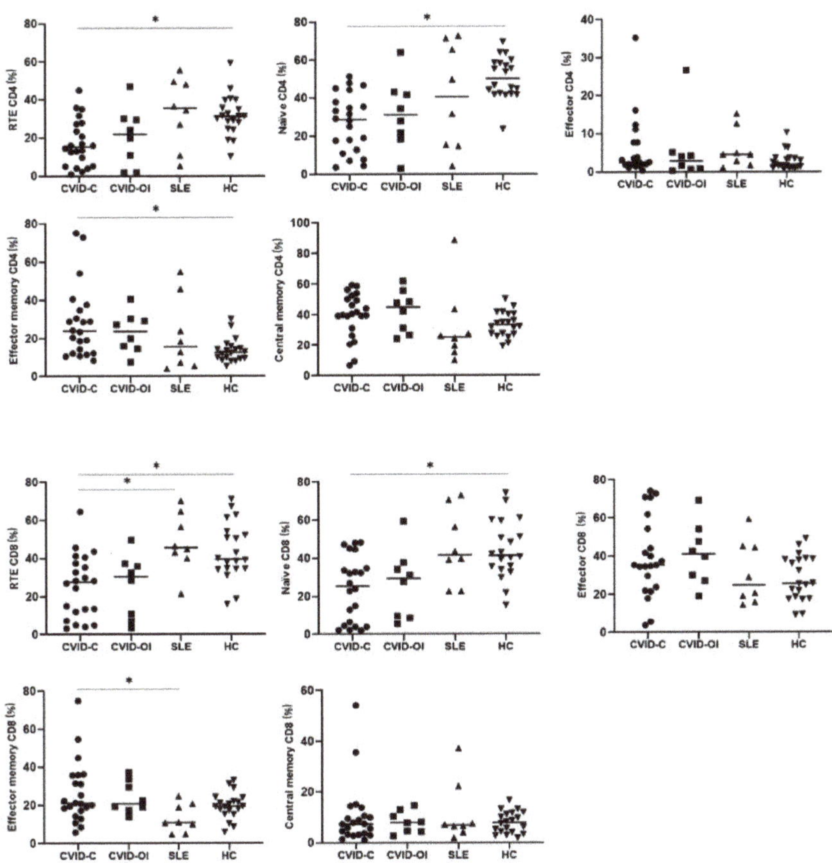

Figure 5. Differences in the proportions of T lymphocyte maturation between CVID with complicated phenotype (CVID-C) group, CVID phenotype limited to only infections (CVID-OI group), patients with SLE, and healthy controls (HCs). Data expressed as median (Q1–Q3), * $p < 0.05$.

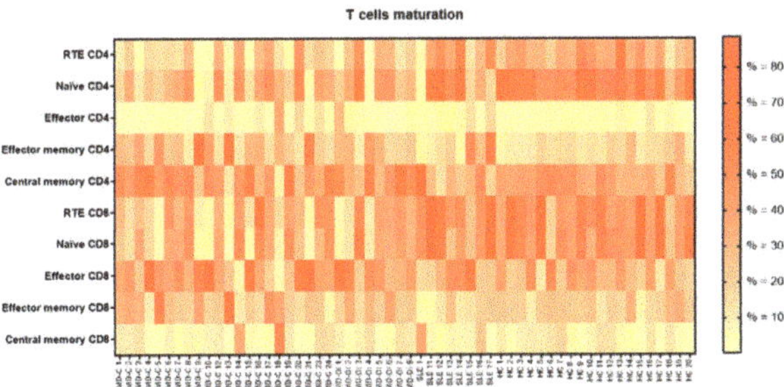

Figure 6. The proportions of T lymphocyte maturation for each patient with CVID with complicated phenotype (CVID-C) group, CVID phenotype limited to only infections (CVID-OI group), patients with systemic lupus erythematosus (SLE), and healthy controls (HCs). Data expressed as a median percentage (%).

4. Discussion

Using flow cytometry, we determined the lymphocyte profiles of four patient groups: CVID-C, CVID-OI, SLE, and HCs.

In our study, patients with a complicated CVID phenotype had low proportions of Tregs and NK cells. Decreased proportions of Tregs in CVID were first reported by Fevang et al. [36]. They included 26 patients diagnosed with CVID according to the WHO classification. In CVID patients, they found significantly reduced expression of the transcription factor FoxP3 in CD3+ cells and a decreased proportion of T CD4+CD25^{high}FoxP3+ cells in the CD4(+) population, as measured by flow cytometry. The lowest proportion of Tregs was found in patients with CVID and splenomegaly (spleen size > 13 cm on ultrasound examination). Treg proportions correlated negatively with neopterin levels as a marker of chronic inflammation [36]. Kofod-Olsen et al. in a study of 26 patients with CVID, demonstrated an association between decreased levels of Tregs and autoimmune phenomena [37]. Moreover, in patients with CVID, the functions of Tregs were disturbed [36,38]. One study that included 20 children with CVID (mean age 173 months) found no difference in Tregs between CVID cases and healthy controls [39]. This suggests different etiopathologies of childhood vs. adult-onset CVID.

We found low NK cells in the CVID-C group (median, 54 cells/μL). Similarly, patients with CVID from the French registry who had a severe reduction of NK cells (<50/μL at study inclusion) presented with a complicated phenotype [40]. Therefore, our results support the finding that low NK cell numbers can be a biomarker of complicated CVILD [41,42].

We found that patients with CVID-C had reduced Th17 proportions. Th17 cell involvement in CVID has not been well studied [18,43]. Barbosa et al. were the first to evaluate circulating Th17 cells in 30 patients with CVID [43]. They found a significant reduction in Th17 cells in CVID patients who had less than 2% smB cells and more than 10% of cd21low B cells. This suggested that there is a link between B cell maturation disturbances and maintenance of Th17 cells [43]. Edwards et al. found low Th17 numbers in patients with predominantly deficient antibodies and noninfectious complications [18].

Bregs play a critical role in immune homeostasis and tolerance. Despite extensive efforts to phenotypically characterize Bregs, we still lack a definitive set of phenotypic markers or a signature transcriptional regulator (equivalent to FoxP3 Tregs) that enables us to comprehensibly identify Bregs [44]. In our study, Bregs were identified as CD19+ CD5+ CD1d^{high}. CD1d is a major phenotypic marker highly expressed in many Breg cells, and it may play a crucial role in Breg-cell-mediated suppression [45]. The upregulation of CD1d on B cells is associated with B-cell-mediated protection against inflammation [44,46]. In our study, we observed a tendency for decreased Bregs in CVID-C patients; however, this was not statistically significant. Yesillic et al. found significantly lower proportions of Breg cells in 25 adult patients with CVID [47]. On the other hand, Kofod-Olsen et al. showed that an elevated frequency of pro-B10 cells in CVID correlated with autoimmunity and splenomegaly [37]. In another study that included 42 adult patients with CVID, the Breg number was low, and their function was disturbed [48]. Although 40% of patients in that study had autoimmune symptoms, there was no correlation between the reduction of Breg cells and autoimmunity. Discrepancies among the studies might be a result of different methodologies of Breg assessment and differences in the incidence and types of autoimmunity in published cohorts.

T cell subsets in our CVID-C patients were characterized by low naive CD4+T cells, low RTE CD4+ T cells, low naïve CD8+T cells, and low RTE CD8+T cells. These results illustrate that profound T cell pool abnormalities are part of the picture of complicated CVID. Signs of senescence in T cell maturation have been reported in CVID-C patients. Stuchlý et al. hypothesized that in CVID with autoimmune thrombocytopenia, naive CD4+ T cell pool depletion occurred and increased the likelihood of promoting autoreactive T cells to memory stages concurrently with B cell activation [49]. Activated B cells, which cannot progress through germinal center reactions, fail to produce isotype-switched antibodies [50].

To better understand autoimmunity, we compared the cytometry results between CVID and SLE patients, as a prototypic autoimmune disease. Patients with CVID and SLE both had lymphopenia. We found that only a more detailed analysis of peripheral lymphocyte subset counts showed differences; specifically, smB cells and plasmablasts were reduced in CVID, which is in agreement with other studies [30]. Therefore, low smB cells detected in patients with autoimmunity should increase the awareness of immunodeficiency. In a recently published study, retrospective evaluation of available peripheral lymphocyte subset counts revealed lower proportions of class-switched memory B cells [9] in patients with inflammatory rheumatic diseases. These patients also presented with mutations in PID genes and were subsequently revealed to have hypogammaglobulinemia.

The expansion of the CD21^{low} B cell compartment and increased plasmablasts are expected findings in autoimmune diseases. This was not the case in the SLE patients in our study. Moreover, they had the lowest Th17 percentages. This can be explained by the low disease activity in our patients [50]. However, Treg imbalance persists despite low SLE activity and might be a further attractive therapeutic goal [51].

Our study has some limitations. First, the sample size of each group was relatively small. Due to the limited sample size, we were not able to assess the relationships between specific clinical symptoms and lymphocyte subsets. However, the limited number of CVID patients is to be expected as it is a rare disease, and our sample size is similar to that of other single-center studies. On the other hand, we collected extensive data from each patient and only included patients with a confirmed diagnosis.

In conclusion, we confirmed that patients with CVID and autoimmune phenomena had reduced Treg, Th17, and NK cells.

Future studies are needed to confirm whether reductions of Treg, Th17, and NK cells might be a biomarker of more complicated CVID cases. Nevertheless, class-switched memory B cells can help distinguish patients with different causes of autoimmunity. Our results suggest that T and B cell maturation analyses should be performed routinely in clinical practice.

Supplementary Materials: The following are available online at https://www.mdpi.com/article/10.3390/jcm10153356/s1, Table S1: Internal laboratory reference values of lymphocyte subpopulations and absolute numbers, Table S2: Internal laboratory reference values if the proportions of B lymphocyte maturation, Table S3: Internal laboratory reference values if the proportions of T lymphocyte maturation, Figure S1: Representative gating strategy of main lymphocytes subpopulation and T regulatory (Tregs), B regulatory cells (B regs) and Th17 cells in study group, Figure S2: Representative B lymphocytes maturation gating strategy in study group. (A) B lymphocytes gating strategy: FSC-A vs. FSC-H plot: Gating the cells that have an equal area and height, thus removing clumps (greater FSC-A relative to FSC-H) and debris (very low FSC), CD45 vs. SSC-A plot: Selection of lymphocytes based on their SSC/CD45+ properties, CD19 vs. SSC-A plot: Selection of lymphocytes B based on their SSC/CD19+ properties. (B) Representative dot plots of each maturation B subsets cells: transitional B cells (orange), naïve B cells (blue), non-switched memory B cells (red), class switched memory B cells (yellow), CD21^{low} cells (purple) and plasmablasts (green) based on their IgD/IgM, IgD/CD38, CD21/CD38, IgM/CD38, CD27/CD21 properties (phenotypes of all cells described in section: material and method), Figure S3: Representative T lymphocytes maturation gating strategy in study group. (A) T lymphocytes gating strategy: FSC-A vs. FSC-H plot: Gating the cells that have an equal area and height, thus removing clumps (greater FSC-A relative to FSC-H) and debris (very low FSC), CD45 vs. SSC-A plot: Selection of lymphocytes based on their SSC/CD45+ properties, CD3 vs. SSC-A plot: Selection of lymphocytes T based on their SSC/CD3+ properties. CD4 vs. CD8 plot: Selection of lymphocytes T CD4+ (pink) and CD8+ (yellow) based on their CD4+ or CD8+ properties. (B) Representative dot plots of each maturation T CD4+ subsets cells: recent thymic emigrants T CD4+ cells, naïve CD4+ T cells, effector CD4+ T cells, central memory CD4+ T cells and effector memory CD4+ T cells based on CD197/CD45RO, CD62L/CD45RA, CD45RA/CD31 properties (phenotypes of all cells described in section: material and method). (C) Representative dot plots of each maturation T CD8+ subsets cells: recent thymic emigrants T CD8+ cells, naïve

CD8+ T cells, effector CD8+ T cells, central memory CD8+ T cells and effector memory CD8+ T cells based on their CD197/CD45RO, CD62L/CD45RA, CD45RA/CD31 properties (phenotypes of all cells described in section: material and method).

Author Contributions: Conceptualization, E.W.-S., E.R. and I.K.; methodology, E.W.-S., E.R. and I.K.; formal analysis, E.W.-S., E.R., I.K. and K.J.-R.; investigation, E.W.-S., E.R., I.K., M.K. and D.S.; resources, E.W.-S., E.R., I.K., M.K. and D.S.; data curation, E.W.-S. and E.R.; writing—original draft, E.W.-S.; writing—review and editing, E.W.-S., E.R., I.K. and K.J.-R.; visualization, E.W.-S., I.K. and E.R.; supervision, E.W.-S., E.R. and K.J.-R.; project administration, E.W.-S. and K.J.-R. All authors have read and agreed to the published version of the manuscript.

Funding: This research received no external funding.

Institutional Review Board Statement: The study protocol was approved by the Bioethics Committee of the Military Institute of Medicine (approval no. 7/WIM/2020). All patients were informed in detail orally about the course, aims, and scope of this research. Blood sampling was limited to routine assessments.

Informed Consent Statement: Separate written consent for blood sampling and review of records were not required by the IRB due to the retrospective nature of this study. All patient data were confidential, and the study procedures complied with the Declaration of Helsinki.

Data Availability Statement: Data supporting reported results can be made available by request from the corresponding author.

Conflicts of Interest: The authors declare no conflict of interest.

References

1. Więsik-Szewczyk, E.; Jahnz-Różyk, K. From Infections to Autoimmunity: Diagnostic Challenges in Common Variable Immunodeficiency. *World J. Clin. Cases* **2020**, *8*, 3942–3955. [CrossRef]
2. Ziętkiewicz, M.; Więsik-Szewczyk, E.; Matyja-Bednarczyk, A.; Napiórkowska-Baran, K.; Zdrojewski, Z.; Jahnz-Różyk, K. Shorter Diagnostic Delay in Polish Adult Patients With Common Variable Immunodeficiency and Symptom Onset After 1999. *Front. Immunol.* **2020**, *11*, 982. [CrossRef] [PubMed]
3. Bagheri, Y.; Vosughi, A.; Azizi, G.; Yazdani, R.; Kiaee, F.; Hafezi, N.; Alimorad, S.; Khoshmirsafa, M.; Seif, F.; Hassanpour, G.; et al. Comparison of Clinical and Immunological Features and Mortality in Common Variable Immunodeficiency and Agammaglobulinemia Patients. *Immunol. Lett.* **2019**, *210*, 55–62. [CrossRef]
4. Ho, H.E.; Cunningham-Rundles, C. Non-infectious Complications of Common Variable Immunodeficiency: Updated Clinical Spectrum, Sequelae, and Insights to Pathogenesis. *Front. Immunol.* **2020**, *11*, 149. [CrossRef]
5. Costagliola, G.; Consolini, R. Lymphadenopathy at the crossroad between immunodeficiency and autoinflammation: An intriguing challenge. *Clin. Exp. Immunol.* **2021**. [CrossRef] [PubMed]
6. Schubert, D.; Bode, C.; Kenefeck, R.; Hou, T.Z.; Wing, J.B.; Kennedy, A.; Bulashevska, A.; Petersen, B.S.; Schäffer, A.A.; Grüning, B.A.; et al. Autosomal dominant immune dysregulation syndrome in humans with CTLA4 mutations. *Nat. Med.* **2014**, *20*, 1410–1416. [CrossRef]
7. Lo, B.; Zhang, K.; Lu, W.; Zheng, L.; Zhang, Q.; Kanellopoulou, C.; Zhang, Y.; Liu, Z.; Fritz, J.M.; Marsh, R.; et al. Patients with LRBA deficiency show CTLA4 loss and immune dysregulation responsive to abatacept therapy. *Science* **2015**, *349*, 436–440. [CrossRef] [PubMed]
8. Klemann, C.; Camacho-Ordonez, N.; Yang, L.; Eskandarian, Z.; Rojas-Restrepo, J.L.; Frede, N.; Bulashevska, A.; Heeg, M.; Al-Ddafari, M.S.; Premm, J.; et al. Clinical and Immunological Phenotype of Patients With Primary Immunodeficiency Due to Damaging Mutations in NFKB2. *Front. Immunol.* **2019**, *10*, 297. [CrossRef] [PubMed]
9. Tuijnenburg, P.; Lango Allen, H.; Burns, S.O.; Greene, D.; Jansen, M.H.; Staples, E.; Stephens, J.; Carss, K.J.; Biasci, D.; Baxendale, H.; et al. Loss-of-function nuclear factor κB subunit 1 (NFKB1) variants are the most common monogenic cause of common variable immunodeficiency in Europeans. *J. Allergy Clin. Immunol.* **2018**, *142*, 1285–1296. [CrossRef] [PubMed]
10. Elgizouli, M.; Lowe, D.M.; Speckmann, C.; Schubert, D.; Hülsdünker, J.; Eskandarian, Z.; Dudek, A.; Schmitt-Graeff, A.; Wanders, J.; Jørgensen, S.F.; et al. Activating PI3Kδ mutations in a cohort of 669 patients with primary immunodeficiency. *Clin. Exp. Immunol.* **2016**, *183*, 221–229. [CrossRef] [PubMed]
11. Sogkas, G.; Dubrowinskaja, N.; Adriawan, I.R.; Anim, M.; Witte, T.; Schmidt, R.E.; Atschekzei, F. High Frequency of Variants in Genes Associated with Primary Immunodeficiencies in Patients with Rheumatic Diseases with Secondary Hypogammaglobulinaemia. *Ann. Rheum. Dis.* **2020**, *80*, 392–399, [CrossRef]
12. Perazzio, S.F.; Granados, Á.; Salomão, R.; Silva, N.P.; Carneiro-Sampaio, M.; Andrade, L.E.C. High Frequency of Immunodeficiency-like States in Systemic Lupus Erythematosus: A Cross-Sectional Study in 300 Consecutive Patients. *Rheumatology* **2016**, *55*, 1647–1655. [CrossRef]

13. Errante, P.R.; Perazzio, S.F.; Frazão, J.B.; da Silva, N.P.; Andrade, L.E.C. Primary Immunodeficiency Association with Systemic Lupus Erythematosus: Review of Literature and Lessons Learned by the Rheumatology Division of a Tertiary University Hospital at São Paulo, Brazil. *Rev. Bras. Reumatol.* **2016**, *56*, 58–68. [CrossRef]
14. Almaghlouth, I.; Su, J.; Johnson, S.R.; Pullenayegum, E.; Gladman, D.; Urowitz, M. Acquired Low Immunoglobulin Levels and Risk of Clinically Relevant Infection in Adult Patients with Systemic Lupus Erythematosus: A Cohort Study. *Rheumatol. Oxf. Engl.* **2020**, *60*, 1456–1464, [CrossRef] [PubMed]
15. Warnatz, K.; Denz, A.; Dräger, R.; Braun, M.; Groth, C.; Wolff-Vorbeck, G.; Eibel, H.; Schlesier, M.; Peter, H.H. Severe Deficiency of Switched Memory B Cells (CD27(+)IgM(-)IgD(−)) in Subgroups of Patients with Common Variable Immunodeficiency: A New Approach to Classify a Heterogeneous Disease. *Blood* **2002**, *99*, 1544–1551. [CrossRef] [PubMed]
16. Wehr, C.; Kivioja, T.; Schmitt, C.; Ferry, B.; Witte, T.; Eren, E.; Vlkova, M.; Hernandez, M.; Detkova, D.; Bos, P.R.; et al. The EUROclass Trial: Defining Subgroups in Common Variable Immunodeficiency. *Blood* **2008**, *111*, 77–85. [CrossRef] [PubMed]
17. Abolhassani, H.; Amirkashani, D.; Parvaneh, N.; Mohammadinejad, P.; Gharib, B.; Shahinpour, S.; Hirbod-Mobarakeh, A.; Mirghorbani, M.; Movahedi, M.; Gharagozlou, M.; et al. Autoimmune Phenotype in Patients with Common Variable Immunodeficiency. *J. Investig. Allergol. Clin. Immunol.* **2013**, *23*, 323–329. [PubMed]
18. Edwards, E.S.J.; Bosco, J.J.; Aui, P.M.; Stirling, R.G.; Cameron, P.U.; Chatelier, J.; Hore-Lacy, F.; O'Hehir, R.E.; van Zelm, M.C. Predominantly Antibody-Deficient Patients With Non-Infectious Complications Have Reduced Naive B, Treg, Th17, and Tfh17 Cells. *Front. Immunol.* **2019**, *10*, 2593. [CrossRef]
19. Warnatz, K.; Schlesier, M. Flowcytometric Phenotyping of Common Variable Immunodeficiency. *Cytom. B Clin. Cytom.* **2008**, *74*, 261–271. [CrossRef]
20. Giovannetti, A.; Pierdominici, M.; Mazzetta, F.; Marziali, M.; Renzi, C.; Mileo, A.M.; De Felice, M.; Mora, B.; Esposito, A.; Carello, R.; et al. Unravelling the Complexity of T Cell Abnormalities in Common Variable Immunodeficiency. *J. Immunol.* **2007**, *178*, 3932–3943. [CrossRef] [PubMed]
21. Bateman, E.A.L.; Ayers, L.; Sadler, R.; Lucas, M.; Roberts, C.; Woods, A.; Packwood, K.; Burden, J.; Harrison, D.; Kaenzig, N.; et al. T Cell Phenotypes in Patients with Common Variable Immunodeficiency Disorders: Associations with Clinical Phenotypes in Comparison with Other Groups with Recurrent Infections. *Clin. Exp. Immunol.* **2012**, *170*, 202–211. [CrossRef]
22. von Spee-Mayer, C.; Koemm, V.; Wehr, C.; Goldacker, S.; Kindle, G.; Bulashevska, A.; Proietti, M.; Grimbacher, B.; Ehl, S.; Warnatz, K. Evaluating Laboratory Criteria for Combined Immunodeficiency in Adult Patients Diagnosed with Common Variable Immunodeficiency. *Clin. Immunol.* **2019**, *203*, 59–62. [CrossRef]
23. Azizi, G.; Rezaei, N.; Kiaee, F.; Tavakolinia, N.; Yazdani, R.; Mirshafiey, A.; Aghamohammadi, A. T-Cell Abnormalities in Common Variable Immunodeficiency. *J. Investig. Allergol. Clin. Immunol.* **2016**, *26*, 233–243. [CrossRef]
24. Unger, S.; Seidl, M.; van Schouwenburg, P.; Rakhmanov, M.; Bulashevska, A.; Frede, N.; Grimbacher, B.; Pfeiffer, J.; Schrenk, K.; Munoz, L.; et al. The TH1 Phenotype of Follicular Helper T Cells Indicates an IFN-γ-Associated Immune Dysregulation in Patients with CD21low Common Variable Immunodeficiency. *J. Allergy Clin. Immunol.* **2018**, *141*, 730–740. [CrossRef]
25. Le Saos-Patrinos, C.; Loizon, S.; Blanco, P.; Viallard, J.-F.; Duluc, D. Functions of Tfh Cells in Common Variable Immunodeficiency. *Front. Immunol.* **2020**, *11*, 6. [CrossRef]
26. Turpin, D.; Furudoi, A.; Parrens, M.; Blanco, P.; Viallard, J.-F.; Duluc, D. Increase of Follicular Helper T Cells Skewed toward a Th1 Profile in CVID Patients with Non-Infectious Clinical Complications. *Clin. Immunol.* **2018**, *197*, 130–138. [CrossRef]
27. Robinson, G.A.; Peng, J.; Dönnes, P.; Coelewij, L.; Naja, M.; Radziszewska, A.; Wincup, C.; Peckham, H.; Isenberg, D.A.; Ioannou, Y.; et al. Disease-Associated and Patient-Specific Immune Cell Signatures in Juvenile-Onset Systemic Lupus Erythematosus: Patient Stratification Using a Machine-Learning Approach. *Lancet Rheumatol.* **2020**, *2*, e485–e496. [CrossRef]
28. Jin, W.; Luo, Z.; Yang, H. Peripheral B Cell Subsets in Autoimmune Diseases: Clinical Implications and Effects of B Cell-Targeted Therapies. *J. Immunol. Res.* **2020**, *2020*, 9518137. [CrossRef]
29. Wehr, C.; Eibel, H.; Masilamani, M.; Illges, H.; Schlesier, M.; Peter, H.-H.; Warnatz, K. A New CD21low B Cell Population in the Peripheral Blood of Patients with SLE. *Clin. Immunol.* **2004**, *113*, 161–171. [CrossRef] [PubMed]
30. Jablonka, A.; Etemadi, H.; Adriawan, I.R.; Ernst, D.; Jacobs, R.; Buyny, S.; Witte, T.; Schmidt, R.E.; Atschekzei, F.; Sogkas, G. Peripheral Blood Lymphocyte Phenotype Differentiates Secondary Antibody Deficiency in Rheumatic Disease from Primary Antibody Deficiency. *J. Clin. Med.* **2020**, *9*, 1049. [CrossRef] [PubMed]
31. Wirsum, C.; Glaser, C.; Gutenberger, S.; Keller, B.; Unger, S.; Voll, R.E.; Vach, W.; Ness, T.; Warnatz, K. Secondary Antibody Deficiency in Glucocorticoid Therapy Clearly Differs from Primary Antibody Deficiency. *J. Clin. Immunol.* **2016**, *36*, 406–412. [CrossRef]
32. ESID. Registry—Working Definitions for Clinical Diagnosis. Available online: https://Esid.Org/Education/Diagnostic-Criteria-Pid (accessed on 23 July 2021).
33. Chapel, H.; Lucas, M.; Lee, M.; Bjorkander, J.; Webster, D.; Grimbacher, B.; Fieschi, C.; Thon, V.; Abedi, M.R.; Hammarstrom, L. Common Variable Immunodeficiency Disorders: Division into Distinct Clinical Phenotypes. *Blood* **2008**, *112*, 277–286. [CrossRef] [PubMed]
34. Petri, M.; Orbai, A.-M.; Alarcón, G.S.; Gordon, C.; Merrill, J.T.; Fortin, P.R.; Bruce, I.N.; Isenberg, D.; Wallace, D.J.; Nived, O.; et al. Derivation and Validation of the Systemic Lupus International Collaborating Clinics Classification Criteria for Systemic Lupus Erythematosus. *Arthritis Rheum.* **2012**, *64*, 2677–2686. [CrossRef]

35. Boldt, A.; Borte, S.; Fricke, S.; Kentouche, K.; Emmrich, F.; Borte, M.; Kahlenberg, F.; Sack, U. Eight-color immunophenotyping of T-, B-, and NK-cell subpopulations for characterization of chronic immunodeficiencies. *Cytom. B Clin. Cytom.* **2014**, *86*, 191–206. [CrossRef]
36. Fevang, B.; Yndestad, A.; Sandberg, W.J.; Holm, A.M.; Müller, F.; Aukrust, P.; Frøland, S.S. Low Numbers of Regulatory T Cells in Common Variable Immunodeficiency: Association with Chronic Inflammation in Vivo. *Clin. Exp. Immunol.* **2007**, *147*, 521–525. [CrossRef]
37. Kofod-Olsen, E.; Jørgensen, S.E.; Nissen, S.K.; Westh, L.; Møller, B.K.; Østergaard, L.; Larsen, C.S.; Mogensen, T.H. Altered Fraction of Regulatory B and T Cells Is Correlated with Autoimmune Phenomena and Splenomegaly in Patients with CVID. *Clin. Immunol.* **2016**, *162*, 49–57. [CrossRef] [PubMed]
38. Arandi, N.; Mirshafiey, A.; Jeddi-Tehrani, M.; Abolhassani, H.; Sadeghi, B.; Mirminachi, B.; Shaghaghi, M.; Aghamohammadi, A. Evaluation of CD4+CD25+FOXP3+ Regulatory T Cells Function in Patients with Common Variable Immunodeficiency. *Cell. Immunol.* **2013**, *281*, 129–133. [CrossRef]
39. Kutukculer, N.; Azarsiz, E.; Aksu, G.; Karaca, N.E. CD4+CD25+Foxp3+ T Regulatory Cells, Th1 (CCR5, IL-2, IFN-γ) and Th2 (CCR4, IL-4, Il-13) Type Chemokine Receptors and Intracellular Cytokines in Children with Common Variable Immunodeficiency. *Int. J. Immunopathol. Pharmacol.* **2016**, *29*, 241–251. [CrossRef] [PubMed]
40. Ebbo, M.; Gérard, L.; Carpentier, S.; Vély, F.; Cypowyj, S.; Farnarier, C.; Vince, N.; Malphettes, M.; Fieschi, C.; Oksenhendler, E.; et al. Low Circulating Natural Killer Cell Counts Are Associated With Severe Disease in Patients With Common Variable Immunodeficiency. *EBioMedicine* **2016**, *6*, 222–230. [CrossRef] [PubMed]
41. Wehr, C. Trying to Understand NK Cell Function in Vivo Points towards a Severity Score for CVID Patients. *EBioMedicine* **2016**, *6*, 18–19. [CrossRef] [PubMed]
42. Haymore, B.R.; Mikita, C.P.; Tsokos, G.C. Common Variable Immune Deficiency (CVID) Presenting as an Autoimmune Disease: Role of Memory B Cells. *Autoimmun. Rev.* **2008**, *7*, 309–312. [CrossRef]
43. Barbosa, R.R.; Silva, S.P.; Silva, S.L.; Melo, A.C.; Pedro, E.; Barbosa, M.P.; Pereira-Santos, M.C.; Victorino, R.M.M.; Sousa, A.E. Primary B-Cell Deficiencies Reveal a Link between Human IL-17-Producing CD4 T-Cell Homeostasis and B-Cell Differentiation. *PLoS ONE* **2011**, *6*, e22848. [CrossRef] [PubMed]
44. Catalán, D.; Mansilla, M.A.; Ferrier, A.; Soto, L.; Oleinika, K.; Aguillón, J.C.; Aravena, O. Immunosuppressive Mechanisms of Regulatory B Cells. *Front. Immunol.* **2021**, *12*, 611795. [CrossRef]
45. Bosma, A.; Abdel-Gadir, A.; Isenberg, D.A.; Jury, E.C.; Mauri, C. Lipid-antigen presentation by CD1d(+) B cells is essential for the maintenance of invariant natural killer T cells. *Immunity* **2012**, *36*, 477–490. [CrossRef]
46. Chekol Abebe, E.; Asmamaw Dejenie, T.; Mengie Ayele, T.; Dagnew Baye, N.; Agegnehu Teshome, A.; Tilahun Muche, Z. The Role of Regulatory B Cells in Health and Diseases: A Systemic Review. *J. Inflamm. Res.* **2021**, *14*, 75–84. [CrossRef] [PubMed]
47. Yesillik, S.; Agrawal, S.; Gollapudi, S.V.; Gupta, S. Phenotypic Analysis of CD4+ Treg, CD8+ Treg, and Breg Cells in Adult Common Variable Immunodeficiency Patients. *Int. Arch. Allergy Immunol.* **2019**, *180*, 150–158. [CrossRef]
48. Barsotti, N.S.; Almeida, R.R.; Costa, P.R.; Barros, M.T.; Kalil, J.; Kokron, C.M. IL-10-Producing Regulatory B Cells Are Decreased in Patients with Common Variable Immunodeficiency. *PLoS ONE* **2016**, *11*, e0151761. [CrossRef] [PubMed]
49. Stuchlý, J.; Kanderová, V.; Vlková, M.; Heřmanová, I.; Slámová, L.; Pelák, O.; Taraldsrud, E.; Jílek, D.; Králíc Ková, P.; Fevang, B.; et al. Common Variable Immunodeficiency Patients with a Phenotypic Profile of Immunosenescence Present with Thrombocytopenia. *Sci. Rep.* **2017**, *7*, 39710. [CrossRef]
50. Koga, T.; Ichinose, K.; Kawakami, A.; Tsokos, G.C. Current Insights and Future Prospects for Targeting IL-17 to Treat Patients With Systemic Lupus Erythematosus. *Front. Immunol.* **2020**, *11*, 624971. [CrossRef]
51. Osnes, L.T.; Nakken, B.; Bodolay, E.; Szodoray, P. Assessment of Intracellular Cytokines and Regulatory Cells in Patients with Autoimmune Diseases and Primary Immunodeficiencies—Novel Tool for Diagnostics and Patient Follow-Up. *Autoimmun. Rev.* **2013**, *12*, 967–971. [CrossRef] [PubMed]

Journal of
Clinical Medicine

Article

The Evolutionary Scenario of Pediatric Unclassified Primary Antibody Deficiency to Adulthood

Mayla Sgrulletti [1,2], Giorgio Costagliola [3], Giuliana Giardino [4], Simona Graziani [1], Elisabetta Del Duca [1], Silvia Di Cesare [5], Gigliola Di Matteo [5], Rita Consolini [3], Claudio Pignata [4] and Viviana Moschese [1,*]

1   Pediatric Immunopathology and Allergology Unit, Policlinico Tor Vergata, University of Tor Vergata, 00133 Rome, Italy; maylasg@gmail.com (M.S.); simonagraziani@hotmail.com (S.G.); elisabetta.delduca@ptvonline.it (E.D.D.)
2   Ph.D. Program in Immunology, Molecular Medicine and Applied Biotechnology, University of Rome Tor Vergata, 00133 Rome, Italy
3   Section of Clinical and Laboratory Immunology, Division of Pediatrics, Department of Clinical and Experimental Medicine, University of Pisa, 56126 Pisa, Italy; giorgio.costagliola@hotmail.com (G.C.); rita.consolini@med.unipi.it (R.C.)
4   Pediatric Section, Department of Translational Medical Sciences, Federico II University, 80131 Naples, Italy; giuliana.giardino@unina.it (G.G.); pignata@unina.it (C.P.)
5   Department of Systems Medicine, University of Tor Vergata, 00133 Rome, Italy; di.cesare@med.uniroma2.it (S.D.C.); di.matteo@med.uniroma2.it (G.D.M.)
*   Correspondence: moschese@med.uniroma2.it

Citation: Sgrulletti, M.; Costagliola, G.; Giardino, G.; Graziani, S.; Del Duca, E.; Di Cesare, S.; Di Matteo, G.; Consolini, R.; Pignata, C.; Moschese, V. The Evolutionary Scenario of Pediatric Unclassified Primary Antibody Deficiency to Adulthood. *J. Clin. Med.* **2023**, *12*, 4206. https://doi.org/10.3390/jcm12134206

Academic Editor: David S. Fedson

Received: 17 May 2023
Revised: 17 June 2023
Accepted: 19 June 2023
Published: 22 June 2023

Abstract: Background: Unclassified primary antibody deficiency (unPAD) is a relatively novel inborn error of immunity (IEI) condition that can vary with time to more defined entities. Since long-term follow-up (FU) studies are scarce, we aimed to provide insight into the evolutionary clinical and immunological scenario of unPAD children to adulthood and identification of biomarkers of primary immune deficiency (PID) persistence. Methods: A total of 23 pediatric unPAD patients underwent clinical and immunological FU for a mean time of 14 years (range 3–32 years, median 16 years). Results: UnPAD diagnosis may change over time. At the last FU, 10/23 (44%) children matched the diagnosis of transient hypogammaglobulinemia of infancy and 13/23 (56%) suffered from a persistent PID. In detail, an unPAD condition was confirmed in 7/23 (30%) patients, whereas 3/23 (13%), 2/23 (9%), and 1/23 (4%) were reclassified as common variable immunodeficiency, selective IgA deficiency, and isolated IgM deficiency, respectively. Low IgA, low specific antibody response to pneumococcus, and lower respiratory tract infections at diagnosis were independently associated with IEI persistence. Conclusions: Long-term monitoring of unPAD patients is required to define their outcome and possible evolution towards a definitive IEI diagnosis.

Keywords: unclassified primary antibody deficiency; primary antibody deficiency; transient hypogammaglobulinemia of infancy; children; inborn errors of immunity; TNFRSF13B mutations; common variable immunodeficiency; selective IgA deficiency; isolated IgM deficiency

1. Introduction

Primary antibody deficiencies (PADs) consist of a varied group of conditions with different genetic etiologies characterized by an impairment of B cell development, differentiation or class switch recombination leading to hypogammaglobulinemia, and/or defective antibody production [1]. They represent the most common form of inborn errors of immunity (IEIs), counting for more than 60% of them [2]. PADs show a wide clinical spectrum, ranging from asymptomatic to severe forms. Common variable immunodeficiency disorder (CVID) represents the most frequent PAD, with a heterogeneous clinical phenotype, ranging from recurrent bacterial infections, mostly of the respiratory and gastrointestinal tracts to autoimmune disorders, allergy, lymphoproliferation, hyperinflammation, and/or malignancies [3]. Several other forms of milder PADs exist such as IgG deficiency, IgG subclass deficiency, selective IgA or IgM deficiency, and specific antibody defects. Some are in combination with each other [4–7]. In this context, a recent entity named "unclassified primary antibody deficiency" (unPAD) has been recognized and entered the European Society for Immunodeficiencies (ESID) definitions for clinical diagnosis. UnPAD patients match the following criteria: marked decrease in at least one of total IgG, IgG1, IgG2, IgG3, IgA, or IgM levels and/or failure of IgG antibody response(s) to vaccines, plus at least one of the following conditions: (I) recurrent or severe bacterial infections, (II) autoimmune phenomena (especially cytopenias), (III) polyclonal lymphoproliferation, (IV) affected family member. Secondary causes of hypogammaglobulinemia and clinical signs of T-cell-related diseases need to be excluded. UnPAD patients show a highly variable clinical spectrum as well [4–6]. Some children with unPAD may normalize their immunoglobulin levels within 4 years of age, framing the condition of transient hypogammaglobulinemia of infancy (THI); conversely, others may develop persistent or severe forms of PID. Although Ig levels might be less compromised in unPAD than in CVID, they may long remain unrecognized and undiagnosed. Moreover, they may suffer an underestimated risk of organ damage with a severe pulmonary involvement and a negative outcome. In fact, Janssen et al. reported that bronchial wall thickening, bronchiectasis, and atelectasis could be detected in 44%, 21%, and 19% of unPAD patients, respectively, similarly to CVID patients [7]. Immunoglobulin replacement therapy (IRT) is scarcely used in these patients, despite its efficacy being reported [1,8–11]. Recently, Karaman et al. found no significant difference in B-lymphocyte subset distribution of unPAD patients receiving Ig replacement therapy vs CVID patients [12]. In this study we provide a clinical, immunological, and genetic characterization of children with an early diagnosis of unPAD monitored for a mean time of 14 years (range 3–32 years, median 16 years) to outline their natural history and identify potential predictive and/or prognostic markers of final diagnosis.

2. Materials and Methods

Twenty-three pediatric patients (12–36 months) with an initial diagnosis of unPAD, attending the Pediatric Immunopathology and Allergology Unit/ Regional Referral Center for PIDs at Policlinico Tor Vergata in Rome, the Department of Pediatrics of University of Pisa, and the Pediatric Immunology Center of Federico II University in Naples were enrolled in the study. All patients matched ESID diagnostic criteria for unPAD (https://esid.org/Working-Parties/Registry-Working-Party/Diagnosis-criteria; accessed on 15 April 2023). A local ethical committee approved the study, and written informed consent was obtained from all participants or legal guardians.

According to general practice, immunological work-up of these patients included serum Ig levels by nephelometry, serum IgG subclass values by radial immunodiffusion, extended T and B cell immunophenotype by fluorescence-activated cell sorting (FACS), and specific IgG antibody response to tetanus and pneumococcal vaccines by ELISA (VaccZyme ^{TM} Tetanus toxoid IgG kit and VaccZyme ^{TM} Anti PCP- IgG, Binding Site, Birmingham, England). In patients with allergy, skin prick tests (SPT) and serum IgE (sIgE) were performed. Serum IgE was tested by ImmunoCAP FEIA, whereas airborne and/or food allergen extracts were used for SPT, according to clinical phenotype. In a limited number of patients, genetic analysis (next-generation sequencing—NGS) for the main PAD-associated genes was also performed. The NGS panel included the following genes: ICOS, TNFRSF13B, TNFRSF13C, TNFRSF12, CD19, CD81, CR2, CD20, CD27, IL21, IL21R, LRBA, CTLA4, PRKCD, PLCG2, NFKB2, NFKB1, PIK3CD, PIK3R1, PTEN, VAV1, RAC2, BLK, IKZF1, IRF2BP2, BTK, CD40L, SYK, LYK, FYK, MYD88, IRAK4, TNFSF13B, TNFSF17, TNFRSF17, RELB, REL, IKBE, IKBA, IKBB, IKK-alpha, IKK-beta, MAP3K14, RELA, STK4, AKT, LAT, IL12RB1, IL12B, IFNGR1, IFNGR2, ISG15, EVER1, EVER2, AICDA, CD40, UNG, CD79A, CD79B, PAX5, TCF3, BLNK. Sanger sequencing was used to confirm genetic variants detected by NGS as previously reported [13]. Clinical and immunological data were prospectively collected at enrollment and during follow-up (FU) at the following times: age 4 and then every 6–12 months according to common clinical practice. The last evaluation was performed on April 2023. The mean FU period was 14 years (range 3–32 years, median 16 years).

3. Statistical Analysis

Clinical and immunological data of patients were analyzed using Fisher's exact test and the chi-squared test. A p-value < 0.05 was considered statistically significant. Univariate statistical analysis was performed using Graphpad Prism software version 8.2.1, whereas multivariate logistic regression data analysis was performed by Epi Info™ CDC software version 7.2 (available at https://www.cdc.gov/epiinfo/index.html; accessed on 15 April 2023).

4. Results

Twenty-three children with an initial diagnosis of unPAD, (14 males (61%) and 9 females (39%) were included in the study. Positive family history for PIDs was reported in 6/23 (26%) of patients, of whom two had parental consanguinity. A positive family history for early deaths was present in one patient (4%).

4.1. Clinical and Immunological Findings at Initial Diagnosis, According to Age (<24 or >24 Months)

At diagnosis, 22/23 (96%) unPAD patients (10/10 < 24 months and 12/13 > 24 months) were symptomatic. As detailed in Table 1, recurrent infections were the major clinical manifestations (22/22, 100%), mainly involving the upper and lower respiratory tract (URTI, LRTI) and the gastrointestinal and urinary tracts. Seven patients (32%) suffered from allergic conditions. Skin prick tests were positive in 6/7 (86%) patients, and three of them showed high sIgE levels. Only one patient, suffering from atopic dermatitis, had both normal sIgE and negative SPTs (data not shown). Autoimmune neutropenia was observed in 2/22 (9%) unPAD subjects. Clinical manifestations at diagnosis did not significantly differ in the two age groups. Isolated or combined IgG, IgA, and IgM defects (compared with age appropriate values) were detected in 82%, 61%, and 39% patients, respectively. Combined or isolated IgM defects were more frequent in older patients than in children <24 months (8/13, 62% > 24 months vs 1/10, 10% < 24 months, p = 0.028). Seven out of nine unPAD patients presented IgG subclass deficiency, with no correlation with age at diagnosis. A poor specific antibody response to tetanus and pneumococcus was detected in 4/18 (22%) and 7/19 (37%) patients, respectively. Standard immunophenotypic analysis was normal in all 23 patients. Extended B cell immunophenotyping was performed in 18/23 patients, and low expression of switched memory B cells and IgM memory B cells was observed in 10/18 (55%) and 1/18 (5%), respectively. Percentages of activated CD21 low B cells and transitional B cells were within age-matched reference values in all patients.

Table 1. Clinical and immunological findings of 23 patients with an initial diagnosis of unPAD, according to age at diagnosis (<24 and >24 months).

	unPAD	<24 Months (10 pts)	>24 Months (13 pts)	p-Value
CLINICAL MANIFESTATIONS	22/23 (96%)	10/10 (100%)	12/13 (92%)	ns
Infections	22/22(100%)	9/10 (90%)	12/12 (100%)	ns
URTI	12/22 (55%)	6/10 (60%)	6/12 (50%)	ns
LRTI	10/22 (45%)	4/10 (40%)	6/12 (50%)	ns
GI	5/22 (23%)	1/10 (10%)	4/12 (33%)	ns
UTI	5/22(23%)	4/10 (40%)	1/12 (8%)	ns
FEVER	1/22 (5%)	1/10 (10%)	0/12 (0%)	ns
SKIN	1/22 (5%)	1/10 (10%)	0/12 (0%)	ns

Table 1. *Cont.*

	unPAD	<24 Months (10 pts)	>24 Months (13 pts)	*p*-Value
Allergy	7/22 (32%)	3/10 (30%)	4/12 (33%)	ns
Asthma	2/7 (29%)	0/3 (0%)	2/4 (50%)	ns
Atopic dermatitis	3/7 (43%)	2/3 (67%)	1/4 (25%)	ns
Food allergy	2/7 (29%)	1/3 (33%)	1/4 (25%)	ns
Conjiunctivitis	1/7 (14%)	1/3 (33%)	0/4 (0%)	ns
Rhinitis	4/7 (57%)	1/3 (33%)	3/4 (75%)	ns
Autoimmunity	2/22 (9%)	2/10 (20%)	0/12 (0%)	ns
Neutropenia	2/2 (100%)	2/2 (100%)	-	ns
IMMUNOLOGICAL ABNORMALITIES	23/23 (100%)	10/10 (100%)	13/13 (100%)	ns
Isolated or combined IgG defect	19/23 (82%)	9/10 (90%)	10/13 (77%)	ns
Isolated or combined IgA defect	14/23 (61%)	4/10 (40%)	8/13 (62%)	ns
Isolated or combined IgM defect	9/23 (39%)	1/10 (10%)	8/13 (62%)	0.028
Combined IgG defect	13/23 (56%)	6/10 (60%)	7/13 (54%)	ns
Combined IgA defect	13/23 (56%)	5/10 (50%)	8/13 (62%)	ns
Combined IgM defect	8/23 (35%)	0/10 (0%)	8/13 (62%)	0.0075
IgG subclasses defect	7/9 (77%)	1/2 (50%)	6/7 (86%)	ns
Low anti TT antibody response	4/18 (22%)	3/7 (43%)	1/11 (9%)	ns
Low anti PCP antibody response	7/19 (37%)	2/7 (29%)	5/12 (16%)	ns
Low switched memory B cells	10/18 (55%)	5/7 (71%)	5/11 (45%)	ns
Low IgM memory B cells	1/18 (5%)	1/7 (14%)	0/11 (0%)	ns

URTI upper respiratory tract infections; LRTI lower respiratory tract infections; GI gastrointestinal infections; UTI urinary tract infections; TT tetanus toxoid; PCP pneumococcal.

4.2. Clinical and Immunological Findings at 4 Years of Age and at Last FU (Mean 14 Years, Range 3–32 Years, Median 16 Years)

All 23 patients underwent clinical and immunological follow-up every 6 months up to 4 years of age and then yearly for a mean time of 14 years (range 3–32 years, median 16 years). As reported in Figure 1A, at 4 years of age, immunoglobulin analysis revealed age-appropriated values in 8/23 (35%) patients, matching the diagnosis of THI. In the remaining 15 patients (65%), a persistent PID condition was observed. Overall, a diagnosis of unPAD was confirmed in 10/23 (43%), while 3/23 (13%) and 2/23 (9%) patients developed a selective IgA deficiency (SIGAD) and a CVID, respectively.

At the last follow-up, two patients reached age-appropriate Ig values at the age of 16 years and 17 years, respectively, joining the category of THI for a total of 10/23 (44%) patients (Figure 1A,B). Among the remaining 13 patients, 7/23 (30%) confirmed the clinical and immunological features of unPAD, whereas 3/23 (13%), 2/23 (9%), and 1/23 (4%) shifted to CVID, SIGAD, and isolated IgM deficiency, respectively. Clinical and immunological findings of patients with persistent PIDs compared to THI patients at the last FU (mean 14 years, median 16 years) are reported in Table 2. In line with Ig normalization, half of THI patients only suffered from allergic manifestations. Conversely, recurrent infections were only observed in the persistent PID group vs THI (5/13, 38% vs 0/10, 0%; *p* value 0.04), mostly with URTI and LRTI (3/5, 60% vs 0/10, 0%; *p* value 0.02). Moreover, isolated or combined IgG, IgA, and IgM defects were detected in 46%, 61%, and 54% of persistent PID patients, respectively, whereas associated IgG subclass deficiency was observed in 7/13 (53%) of them. A poor specific antibody response to tetanus and pneumococcus was found in 1/9 (11%) and 2/10 (20%) of persistent PID patients, respectively. When the extended B cell immunophenotype was performed in the persistent PID cohort, low switched memory B cells and low IgM memory B cells were detected in 7/13 (54%) and 2/13 (15%) of patients, respectively. The percentage of activated CD21 low B cells and transitional B cells were within age-matched reference values in all patients (data not shown).

(A)

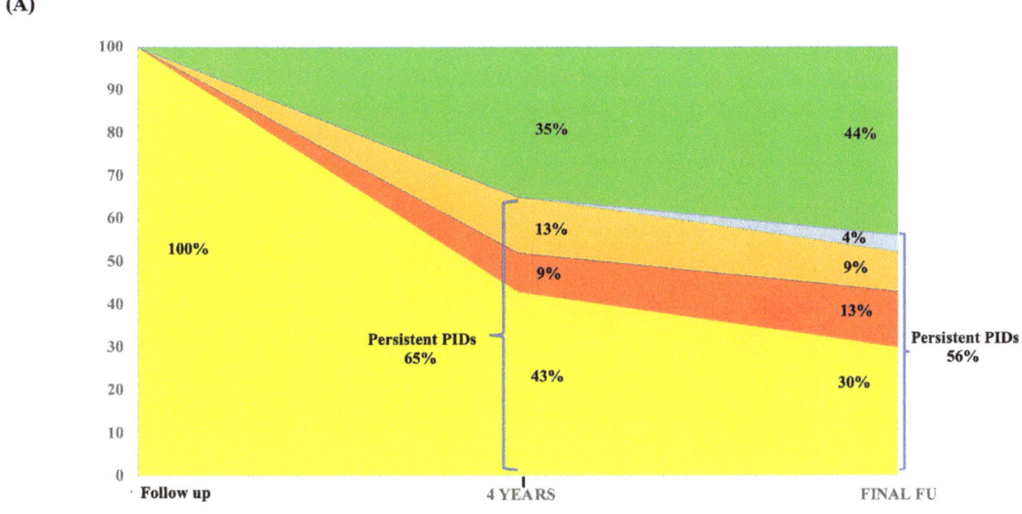

(B)

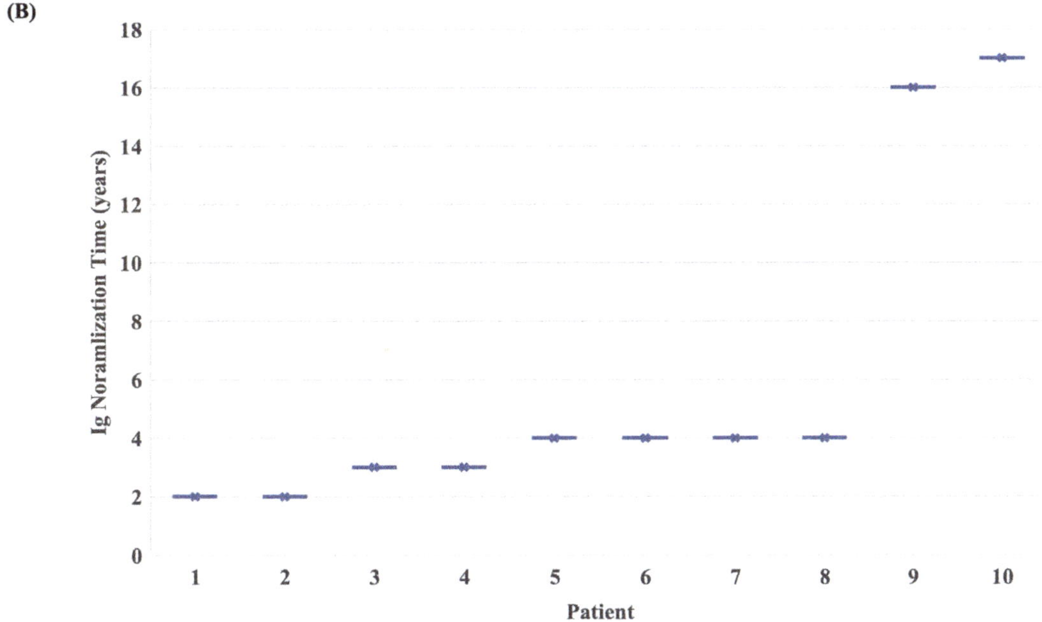

Figure 1. (**A**) Diagnostic reclassification of 23 unPAD patients at 4 years of age and at last FU (mean 14 years, median 16 years). Persistent PIDs: persistent primary immunodeficiencies; unPAD: unclassified primary antibody deficiency; CVID: common variable immunodeficiency disorder; SIGAD: selective IgA deficiency; IgMD: IgM deficiency; THI: transient hypogammaglobulinemia of infancy; Final FU: final follow-up. (**B**) Time of normalization of 10 THI patients.

Table 2. Clinical and immunological findings of 13 persistent PIDs patients vs 10 THI patients at last FU (mean 14 years, median 16 years).

	Persistent PIDs (13 pts)	THI (10 pts)	*p*-Value
CLINICAL MANIFESTATIONS			
Infections	5/13 (38%)	0/10 (0%)	0.04
URTI	3/5 (60%)	-	0.02
LRTI	3/5 (60%)	-	0.02
GI	1/5 (20%)	-	ns
Allergy	6/13 (46%)	5/10 (50%)	ns
Asthma	2/6 (33%)	0/5 (0%)	ns
Rhinitis	3/6 (50%)	4/5 (80%)	ns
Conjiunctivitis	1/6 (17%)	2/5 (40%)	ns
Autoimmunity	2/13 (15%)	0/10 (0%)	ns
Neutropenia	2/2 (100%)	-	ns
Vasculitis	1/2 (50%)	-	ns
IMMUNOLOGICAL ABNORMALITIES	13/13 (100%)	0/10 (0%)	0.0001
Isolated or combined IgG defect	6/13 (46%)	0/10 (0%)	0.02
Isolated or combined IgA defect	8/13 (61%)	0/10 (0%)	0.003
Isolated or combined IgM defect	7/13 (54%)	0/10 (0%)	0.007
IgG subclass defect	7/13 (54%)	0/10 (0%)	0.007
Low anti TT antibody response	1/9 (11%)	0/10 (0%)	ns
Low anti PCP antibody response	2/10 (20%)	0/10 (0%)	ns
Low switched memory B cells	7/13 (54%)	0/10 (0%)	ns
Low IgM memory B cells	2/13 (15%)	0/10 (0%)	ns
Genetic Characterization	8/13 (61%)	-	
TNFRSF13B mutations	4/8 (50%)		
TNFRSF13B c.301T>C plus TNFRSF13B c.204dupA	2/4 (50%)		
Heterozygous TNFRSF13B c.301T>C mutation	2/4 (50%)		

URTI upper respiratory tract infections; LRTI lower respiratory tract infections; GI gastrointestinal infections; UTI urinary tract infections; TT tetanus toxoid; PCP pneumococcal.

4.3. Clinical and Immunological Findings at Diagnosis of Patients with a Definitive Diagnosis of Persistent PID vs. THI

To identify potential predictive and/or prognostic markers of clinical outcome, a retrospective analysis of clinical and immunological findings at diagnosis of 13 patients with a final diagnosis of persistent PID vs 10 patients with a final diagnosis of THI was performed. As reported in Table 3, LRTI were more frequently observed in patients with persistent PID than with THI (8/12, 67% PID patients vs 2/10, 20% THI patients, *p* value 0.03). Isolated or combined IgA deficiency and low anti-PCP antibody response were also found to be associated with a final diagnosis of persistent PID (11/13, 85% PID patients vs 3/10, 30% THI patients, *p* value 0.0013, and 7/11, 64% PID patients vs 0/8, 0% THI patients, *p* value 0.0128). The variables that showed a significant ($p < 0.05$) association with PID persistence in univariate analysis were evaluated in a logistic regression model for multivariate analysis. Low IgA, low specific antibody response to pneumococcus, and LRTI at diagnosis were observed to be independently associated with a persistent PID diagnosis.

Table 3. Clinical and immunological findings at diagnosis of patients with a final diagnosis of persistent PID vs. THI.

	Persistent PIDs (13 pts)	THI (10 pts)	Univariate Analysis *p*-Value	Logistic Regression *p*-Value
Positive Family History for PID	5/13 (38%)	1/10 (10%)	ns	
CLINICAL MANIFESTATIONS				
Infections	12/13 (92%)	10/10 (100%)	ns	
LRTI	8/12 (67%)	2/10 (20%)	0.04	<0.05
Allergy	4/13 (31%)	3/10 (30%)	ns	
Autoimmunity	1/13 (8%)	1/10 (10%)	ns	
IMMUNOLOGICAL ABNORMALITIES				
Isolated or combined IgG defect	11/13 (85%)	8/10 (80%)	ns	
Isolated or combined IgA defect	11/13 (85%)	3/10 (30%)	0.0013	<0.05
Isolated or combined IgM defect	6/13 (46%)	3/10 (30%)	ns	
IgG subclass defect	5/5 (100%)	2/4 (50%)	ns	
Low anti PCP antibody response	7/11 (64%)	0/8 (0%)	0.0128	<0.05
Low switched memory B cells	7/13 (54%)	0/5 (0%)	ns	
Low IgM memory B cells	1/13 (8%)	0/5(0%)	ns	

LRTI lower respiratory tract infections; PCP pneumococcal.

4.4. Genetic Characterization

Next-generation sequencing analysis was performed in 8/13 (61%) of patients with persistent PID. Mutations in TNFRSF13B were identified in 4/8 patients (50%), belonging to two families. As reported in Figure S1, family A included three siblings, two sisters and one brother, in all of whom the final unPAD diagnosis was confirmed. A compound heterozygosis for two different TNF receptor superfamily member 13B (TNFRSF13B/TACI) gene mutations (TNFRSF13B c.301T>C plus TNFRSF13B c.204dupA) was identified in the two sisters and a heterozygous TNFRSF13B c.301T>C mutation in the brother. Their mother carried a TNFRSF13B c.301T>C mutation with the absence of clinical and immunological abnormalities; their father carried a TNFRSF13B c.204dupA and presented an asymptomatic isolated IgM defect. Regarding the clinical picture, both sisters suffered from both URTI and LRTI, whereas the brother had a milder phenotype. A heterozygous TNFRSF13B c.301T>C mutation was also detected in another patient with a final diagnosis of CVID. This patient showed a positive family history for early deaths of unknown-causes and a personal clinical history of URTI, LRTI, and gastrointestinal infections since the age of 9 months. Due to infectious recurrences, he initially received antibiotic prophylaxis and later immunoglobulin replacement therapy. His father carried the same mutation and was suffering from Hashimoto thyroiditis, psoriasis, and recurrent oral aphthosis.

5. Discussion

The onset and clinical spectrum of unPAD patients is widely heterogeneous, and knowledge of their natural history is still scarcely investigated. Some children might be identified at an early age and initially diagnosed as unPAD to enter the THI definition when normalization of their immunoglobulin levels occurs within 4 years of age. Some others may develop over time a defined antibody defect, reaching the definitive diagnosis of a classified IEI. However, at unPAD diagnosis, no markers are currently identified to allow the distinction between patients who will achieve a condition of transient hypogammaglobulinemia from those who will persist in the same condition or will develop other humoral defects, which would be crucial to timely initiate appropriate monitoring and treatment.

The aim of our study was to analyze the clinical, immunological, and genetic characterization of a cohort of children who received an early diagnosis of unPAD and their long-term clinical and immunological monitoring and outcome. To our knowledge, this is the first prospective study describing the evolution of unPAD children towards adulthood, with the clinical and immunological evaluation of their long-term follow-up.

Except for one asymptomatic child, all symptomatic children suffered from recurrent infections, mostly of the respiratory tract, and to a lesser extent from allergy and autoimmunity (32 and 9%, respectively). Indeed, as shown for other PAD patients [6,14–21], in unPAD children, recurrent respiratory infections represent the clinical hallmark, particularly in the first decade of life [18]. Conversely, immune dysregulation may complicate the clinical course mostly in early and late adulthood [14,18]. In the overall cohort, we observed that at diagnosis, an isolated or combined IgG defect was more frequently observed than an isolated or combined IgA defect. Instead, an isolated or combined IgM defect was more frequently observed in children >24 months of age than at a younger age.

In a previous paper, we reported that in children with an initial suspicion of THI, nowadays pertaining into the category of unPAD, a milder clinical and immunological profile and a benign course over time was observed in comparison to those children who did not normalize their IgG levels [22]. In the present study, at 4 years of age, 8/23 (35%) patients matched the diagnosis of THI, and, at last follow-up, two more patients (44% of the total cohort) enriched the THI category at 16 and 17 years of age, respectively. In accordance with previous studies by us and other groups [22,23], IgG normalization occurs more frequently, despite not doing so exclusively, within the first 4 years of age. Indeed, there is no univocal evidence in literature on the time to recovery and Ig normalization that might occur until the third or fourth decade of life [23]. In line with IgG normalization, THI patients maintained a milder clinical picture, mostly allergic manifestations, whereas patients with persistent quantitative and/or qualitative antibody defects featured recurrent/severe infections, mostly of the lower respiratory tract. At final follow-up, 13 patients (56%) suffered from a persistent PID. In detail, an unPAD condition remained in 7/23 (30%) patients, whereas some moved to different primary antibody defects, comprising CVID (3/23; 13%), SIGAD (2/23; 9%), and isolated IgM deficiency (1/23; 4%). Interestingly, one patient who, at enrollment, matched the clinical diagnostic criteria for unPAD, moved to a SIGAD condition at 4 years of age to later develop a CVID at 5 years of age. This patient, harboring a C104R (TNFRSF13B c.301T>C) heterozygous TACI variant, who continued to suffer from recurrent and severe infections, firstly required antibiotic prophylaxis and, after receiving a CVID diagnosis, benefited of additional immunoglobulin replacement therapy. The multi-step diagnostic evolution of this patient clearly outlines the wide spectrum of antibody disorders associated with TACI mutations, as previously reported [24]. Mutations in TNFRSF13B have been identified in a family comprising three siblings, two sisters and one brother, who maintained the unPAD diagnosis. Particularly, the two sisters carried a compound heterozygosis for two different TNFRSF13B gene mutations (TNFRSF13B c.301T>C plus TNFRSF13B c.204dupA), whereas a heterozygous TNFRSF13B c.301T>C mutation was found in the brother. Their mother carried the TNFRSF13B c.301T>C mutation, and their father the TNFRSF13B c.204dupA. The same compound heterozygosis found in the two sisters has been previously reported by Salzer et al. [25] and associated with normal expression of switched and IgM memory B cells, as in our female patients.

Moreover, the asymptomatic unPAD patient who was diagnosed at 15 months, whose brother was suffering from CVID, over time developed recurrent respiratory infections, receiving the final diagnosis of a selective IgM deficiency. It is still unclear why some patients may be asymptomatic. Functional immunological factors as well as epigenetic or environmental factors might be compensative for a certain time.

When we investigated at diagnosis potential markers of dissection between PID persistence vs THI, we observed that in addition to LRTI, low IgA and low specific antibody response were independently associated with a persistent condition. Conclusively, we demonstrated that a subgroup of children with unPAD shares common B cell abnormalities with patients suffering from a range of antibody defects, mostly CVID, according to previous reports [12,26,27]. We point out that a critical long-time clinical and immunological follow-up of unPAD patients is recommended to monitor their evolution (towards a definitive antibody defect in most cases), which may be variable over time, and also among family members with the same genetic mutation, as herein described. As the reservoir for CVID lacks molecular definition, the identification of unPAD patients potentially evolving to a CVID diagnosis during the routine monitoring process is of relevance. UnPad patients are usually underestimated and often suffer from a long diagnostic lag and delayed optimal management. However, a significant proportion of unPad patients may emerge with pulmonary and extrapulmonary complications that might affect several aspects of life. Similarly to CVID patients, it has been reported that bronchial wall thickening, bronchiectasis, and atelectasis could be detected in 44%, 21%, and 19% of unPAD patients, respectively [7]. Their autoimmune and inflammatory complications and whether they might be addressed with immunoglobulin replacement therapy are largely unknown, although it has recently been reported that B-lymphocyte subset distribution did not significantly differ between unPAD patients who received Ig replacement therapy from those with CVID [8]. The lack of extensive and comparable studies describing unPAD cohorts hampers the draft of international networks for the assessment of their diagnostic approach and monitoring. Although the progressive advances in genetic and functional technology will allow for the identification of IEI disorders currently assembled in the unPAD reservoir, increasing awareness for their early diagnosis, the molecular pathways and epigenetic changes underlying the complex heterogeneity of unPAD disorders deserve further investigation.

6. Conclusions

Our study, despite the limited cohort, is the longest FU study analyzing a pediatric cohort of unPAD patients. A continuous long-term clinical and immunological monitoring of unPAD children is necessary to define their outcome and possible evolution towards a definitive IEI defect. We focused on distinct clinical-immunological markers, suitable for identifying patients at higher risk of PID persistence. Furthermore, as the genetic and functional characterization of patients with persistent hypogammaglobulinemia might provide us with valuable information on the pathogenic role of distinct molecules, potentially indicating specific treatment, we emphasized the need to structure national and international networks for the diagnostic approach and monitoring of the heterogeneous condition of unPAD patients.

Supplementary Materials: The following supporting information can be downloaded at: https://www.mdpi.com/article/10.3390/jcm12134206/s1, Figure S1: Pedigree analysis of patients with TACI mutations.

Author Contributions: Conceptualization V.M.; methodology, M.S., S.G., E.D.D., G.C., G.G., S.D.C. and G.D.M.; software, M.S.; validation, V.M.; formal analysis, V.M.; investigation, V.M. and M.S.; resources, V.M. and M.S.; data curation M.S., S.G. and E.D.D.; writing—original draft preparation, M.S.; writing—review and editing, V.M., R.C. and C.P.; visualization, V.M., R.C. and C.P.; supervision, V.M., R.C. and C.P.; project administration, V.M. All authors have read and agreed to the published version of the manuscript.

Funding: This research received no external funding.

Institutional Review Board Statement: The study was conducted according to the guidelines of the Declaration of Helsinki and approved by Policlinico Tor Vergata Ethics Committee, Viale Oxford 81, 00133 Rome, Italy (protocol code 247.20; date of approval: 10 February 2021).

Informed Consent Statement: Informed consent was obtained from all subjects involved in the study.

Data Availability Statement: The data are available upon request from the corresponding author.

Acknowledgments: This work was carried out in the tutorial framework of a Master in Advanced Pediatric Allergy and Immunology at University of Rome Tor Vergata as well as of the Ph.D. program in Immunology, Molecular Medicine, and Applied Biotechnology at the University of Rome Tor Vergata.

Conflicts of Interest: The authors declare no conflict of interest.

Abbreviations

CVID	common variable immunodeficiency
ELISA	enzyme-linked immunosorbent assay
ESID	European Society for Immunodeficiencies
FACS	fluorescence-activated cell sorting
FU	follow-up
GI	gastrointestinal infections
IEI	inborn error of immunity
IgMD	IgM deficiency
IRT	immunoglobulin replacement therapy
LRTI	lower respiratory tract infection
NGS	next-generation sequencing
PAD	primary antibody deficiency
PID	primary immunodeficiency
PCP	pneumococcal
SIGAD	selective IgA deficiency
sIgE	serum IgE
SPT	skin prick tests
THI	transient hypogammaglobulinemia of infancy
TNFRSF13B/TACI	TNF receptor superfamily member 13B
TT	tetanus toxoid
UnPAD	unclassified primary antibody deficiency
UTI	urinary tract infections
URTI	upper respiratory tract infections

References

1. Vivarelli, E.; Matucci, A.; Bormioli, S.; Parronchi, P.; Liotta, F.; Cosmi, L.; Almerigogna, F.; Vultaggio, A. Effectiveness of low-dose intravenous immunoglobulin therapy in minor primary antibody deficiencies: A 2-year real-life experience. *Clin. Exp. Immunol.* **2021**, *205*, 346–353. [CrossRef] [PubMed]
2. Lougaris, V.; Pession, A.; Baronio, M.; Soresina, A.; Rondelli, R.; Gazzurelli, L.; Benvenuto, A.; Martino, S.; Gattorno, M.; Biondi, A.; et al. The Italian Registry for Primary Immunodeficiencies (Italian Primary Immunodeficiency Network; IPINet): Twenty Years of Experience (1999–2019). *J. Clin. Immunol.* **2020**, *40*, 1026–1037. [CrossRef] [PubMed]
3. Yazdani, R.; Habibi, S.; Sharifi, L.; Azizi, G.; Abolhassani, H.; Olbrich, P.; Aghamohammadi, A. Common Variable Immunodeficiency: Epidemiology, Pathogenesis, Clinical Manifestations, Diagnosis, Classification, and Management. *J. Investig. Allergol. Clin. Immunol.* **2020**, *30*, 14–34. [CrossRef] [PubMed]
4. Filion, C.A.; Taylor-Black, S.; Sweater, P.J.; Radigan, L.; Cunningham-Rundles, C. Differentiation of Common Variable Immunodeficiency from IgG Deficiency. *J. Allergy Clin. Immunol. Pract.* **2018**, *7*, 1277–1284. [CrossRef]
5. Keles, S.; Artac, H.; Kara, R.; Gokturk, B.; Ozen, A.; Reisli, I. Transient hypogammaglobulinemia and unclassified hypogammaglobulinemia: 'Similarities and differences'. *Pediatr. Allergy Immunol.* **2010**, *21*, 843–851. [CrossRef]
6. Kutukculer, N.; Gulez, N. The outcome of patients with unclassified hypogammaglobulinemia in early childhood. *Pediatr. Allergy Immunol.* **2009**, *20*, 693–698. [CrossRef] [PubMed]
7. Janssen, L.M.A.; Bassett, P.; Macken, T.; Van Esch, J.; Pruijt, H.; Knoops, A.; Sköld, M.; Parker, A.; DeVries, J.; de Vries, E. Mild Hypogammaglobulinemia Can Be a Serious Condition. *Front. Immunol.* **2018**, *9*, 2384. [CrossRef]
8. Shehata, N.; Palda, V.; Bowen, T.; Haddad, E.; Issekutz, T.B.; Mazer, B.; Schellenberg, R.; Warrington, R.; Easton, D.; Anderson, D.; et al. The Use of Immunoglobulin Therapy for Patients With Primary Immune Deficiency: An Evidence-Based Practice Guideline. *Transfus. Med. Rev.* **2010**, *24* (Suppl. 1), S28–S50. [CrossRef]
9. Modell, V.; Gee, B.; Lewis, D.B.; Orange, J.S.; Roifman, C.M.; Routes, J.M.; Sorensen, R.U.; Notarangelo, L.D.; Modell, F. Global study of primary immunodeficiency diseases (PI)–diagnosis, treatment, and economic impact: An updated report from the Jeffrey Modell Foundation. *Immunol. Res.* **2011**, *51*, 61–70. [CrossRef]
10. Wasserman, R. L Personalized Therapy: Immunoglobulin Replacement for Antibody Deficiency. *Immunol. Allergy Clinic. North Am.* **2019**, *39*, 95–111. [CrossRef]
11. Khokar, A.; Gupta, S. Clinical and Immunological Features of 78 Adult Patients with Primary Selective IgG Subclass Deficiencies. *Arch. Immunol. Et Ther. Exp.* **2019**, *67*, 325–334. [CrossRef] [PubMed]
12. Karaman, S.; Erdem, S.B.; Gülez, N.; Genel, F. The Significance of B-cell Subsets in Patients with Unclassified Hypogammaglobulinemia and Association with Intravenous Immunoglobulin Replacement Requirement. *Iran. J. Immunol.* **2018**, *15*, 1–13. [PubMed]

13. Cifaldi, C.; Bridget, I.; Barzaghi, F.; Zoccolillo, M.; Ferradini, V.; Petricone, D.; Cicalese, M.P.; Lazarevic, D.; Cittaro, D.; Omrani, M.; et al. Targeted NGS Platforms for Genetic Screening and Gene Discovery in Primary Immunodeficiencies. *Front. Immunol.* **2019**, *10*, 316. [CrossRef]
14. Thalhammer, J.; Kindle, G.; Nieters, A.; Rusch, S.; Seppänen, M.R.; Fischer, A.; Grimbacher, B.; Edgar, D.; Buckland, M.; Mahlaoui, N.; et al. Initial presenting manifestations in 16,486 patients with inborn errors of immunity include infections and noninfectious manifestations. *J. Allergy Clin. Immunol.* **2021**, *148*, 1332–1341.e5. [CrossRef]
15. Raje, N.; Dinakar, C. Overview of Immunodeficiency Disorders. *Immunol. Allergy Clinic. North Am.* **2015**, *35*, 599–623. [CrossRef]
16. Modell, V.; Quinn, J.; Ginsberg, G.; Gladue, R.; Orange, J.; Modell, F. Modeling strategy to identify with primary immunodeficiency patients utilizing risk management and outcome measurement. *Immunol. Res.* **2017**, *65*, 713–720. [CrossRef] [PubMed]
17. Bahrami, A.; Sayyahfar, S.; Soltani, Z.; Khodadost, M.; Moazzami, B.; Rezaei, N. Evaluation of the frequency and diagnostic delay of primary immunodeficiency disorders among suspected patients based on the 10 warning sign criteria: A cross-sectional study in Iran. *Allergol. Immunopathol.* **2020**, *48*, 711–719. [CrossRef]
18. Slade, C.A.; Bosco, J.J.; Giang, T.B.; Kruse, E.; Stirling, R.G.; Cameron, P.U.; Hore-Lacy, F.; Sutherland, M.F.; Barnes, S.L.; Holdsworth, S.; et al. Delayed Diagnosis and Complications of Predominantly Antibody Deficiencies in a Cohort of Australian Adults. *Front. Immunol.* **2018**, *9*, 694. [CrossRef]
19. Janssen, L.M.A.; Reijnen, I.C.G.M.; Milito, C.; Edgar, D.; Chapel, H.; de Vries, E. The unPAD consortium Protocol for the unclassified primary antibody deficiency (unPAD) study: Characterization and classification of patients using the ESID online Registry. *PLoS ONE* **2022**, *17*, e0266083. [CrossRef]
20. Pickett, G.; Motazedi, T.; Kutac, C.; Cahill, G.; Cunnigham-Rundles, C.; Fuleihan, R.L.; Sullivan, K.E.; Rider, N.L. Infection Phenotypes Among Patients with Primary Antibody Deficiency Mined from a US Patient Registry. *J. Clin. Immunol.* **2021**, *41*, 374–381. [CrossRef]
21. Demirdag, Y.Y.; Gupta, S. Update on Infections in Primary Antibody Deficiencies. *Front. Immunol.* **2021**, *12*, 634181. [CrossRef] [PubMed]
22. Moschese, V.; Graziani, S.; Avanzini, M.; Carsetti, R.; Marconi, M.; LaRocca, M.; Chini, L.; Pignata, C.; Soresina, A.; Consolini, R.; et al. A Prospective Study on Children with Initial Diagnosis of Transient Hypogammaglobulinemia of Infancy: Results from the Italian Primary Immunodeficiency Network. *Int. J. Immunopathol. Pharmacol.* **2008**, *21*, 343–352. [CrossRef] [PubMed]
23. Ameratunga, R.; Ahn, Y.; Steele, R.; Woon, S. Transient hypogammaglobulinaemia of infancy: Many patients recover in adolescence and adulthood. *Clin. Exp. Immunol.* **2019**, *198*, 224–232. [CrossRef] [PubMed]
24. Seijas, A.B.; Graziani, S.; Cancrini, C.; Finocchi, A.; Ferrari, S.; Miniero, R.; Conti, F.; Zuntini, R.; Chini, L.; Chiarello, P.; et al. The Impact of TACI Mutations: From Hypogammaglobulinemia in Infancy to Autoimmunity in Adulthood. *Int. J. Immunopathol. Pharmacol.* **2012**, *25*, 407–414. [CrossRef] [PubMed]
25. Salzer, U.; Bacchelli, C.; Buckridge, S.; Pan-Hammarström, Q.; Jennings, S.; Lougaris, V.; Bergbreiter, A.; Hagena, T.; Birmelin, J.; Plebani, A.; et al. Relevance of biallelic versus monoallelic TNFRSF13B mutations in distinguishing disease-causing from risk-increasing TNFRSF13B variants in antibody deficiency syndromes. *Blood* **2009**, *113*, 1967–1976. [CrossRef] [PubMed]
26. Eroglu, F.K.; Kaya, F.A.; Cagdas, D.; Özgur, T.T.; Yilmaz, T.; Tezcan, I.; Sanal, O. B lymphocyte subsets and outcomes in patients with an initial diagnosis of transient hypogammaglobulinemia of infancy. *SCAN J. Immunol.* **2018**, *88*, e12709. [CrossRef]
27. Moschese, V.; Carsetti, R.; Graziani, S.; Chini, L.; Soresina, A.R.; LaRocca, M.; Bossi, G.; DiCesare, S.; Plebani, A.; Italian Primary Immunodeficiency Network. Memory B-cell subsets as a predictive marker of outcome in hypogammaglobulinemia during infancy. *J. Allergy Clin. Immunol.* **2007**, *120*, 474–476. [CrossRef]

MDPI

St. Alban-Anlage 66

4052 Basel

Switzerland

Tel. +41 61 683 77 34

Fax +41 61 302 89 18

www.mdpi.com

Journal of Clinical Medicine Editorial Office

E-mail: jcm@mdpi.com

www.mdpi.com/journal/jcm